MATRIX ANALYSIS OF DISCONTINUOUS CONTROL SYSTEMS

MATRIX ANALYSIS OF DISCONTINUOUS CONTROL SYSTEMS

P. V. Bromberg

Edited by J. O. Flower, B.Sc.(Eng.)
Queen Mary College, University of London

MACDONALD TECHNICAL & SCIENTIFIC
LONDON

SBN 356 02684 1

Translation, editing and cold composition by
Transcripta Service, 138 New Bond Street, London W.1

First published under the title
Matrichnye metody v teorii releinogo i
impul'snogo regulirovaniya
by Nauka, Moscow, in 1967

English edition published in 1969 by
Macdonald & Co. (Publishers) Ltd., 49 Poland Street, London W.1

Printed in England
at The Gresham Press
UNWIN BROTHERS LIMITED
Old Woking, Surrey

CONTENTS

PREFACE

It is preferable when solving scientific problems to be able to consider them from a number of different points of view and to apply different methods of investigation. In this way one can make use of the whole gamut of mathematical procedures and devices characteristic of the chosen method, and bring to bear on the analysis results already available in related fields. Furthermore, this approach often suggests analogies which enable the investigator to penetrate more deeply into the problem he is studying.

In the study of the dynamics of automatic control systems, a very fruitful approach has been one involving the application of the methods of matrix calculus. These methods permit the investigator to use not only the classical theory of differential and finite difference equations but also the methods of operational calculus. Conversion formulae are available which allow the final results to be represented either in their classical form or in a form characteristic of the operational methods, depending on whichever happens to be more convenient in actual applications. The compression of the basic expressions and calculations coupled with the use of the special procedures of the matrix calculus lead to a considerable economy of space and to a simplification of the intermediate steps in the calculations. Such methods also make it easier to anticipate the kind of results the chosen line of investigation will yield. These quantitative gains can add what is essentially a new quality to the investigations, which can sometimes be very useful in obtaining new results.

This book is concerned with the application of matrix methods to the solution of problems pertaining to relay and other discontinuous systems. Relay systems are widely used nowadays because of their simplicity and because of the possibility of obtaining what can be virtually an optimum control process. Pulse-type systems arose out of the requirements of practical engineering. Built up of very simple components operating in the relay mode, they provide a proportional control and at the same time resolve the conflicting requirements of precision and signal power which arise during the time that information is being picked off from the delicate transducers of automatic control systems.

Because of their use in control systems such as radar, digital computers and telemetering devices, certain pulse-type control systems have in recent times come into their own again, for in many such new systems the information generally only exists in discrete, cyclically repeating instants of time.

The special characteristics of the pulsed control principle are such that the designer is able to find operating conditions that make the best possible use of the equipment from both a technological and economic viewpoint; he is able to build systems operating with precision, and can at the same time improve the dynamics of the system. This is particularly true of multi-channel control systems incorporating a single digital computer, systems with a single communication channel for transmitting information about several independent control systems, and also in systems containing a delay.

Such control systems have an external repetition cycle with a constant period which is associated with the beginning or end of the period during which new information is being introduced into the system. In relay systems operating under certain conditions this cycle is provided by the internal properties of the system itself. In both types of system, the solution of many dynamical problems reduces to an analysis of the solutions of equations in finite differences with a discrete or 'floating' argument.

This work is devoted to the application of the methods of matrix calculus to the solution of problems pertaining to the theory of relay and other discontinuous control systems. The layout of the book is as follows.

In the discussion of both relay and pulse-type control system an actual example is chosen from among the various schemes for stabilising the course of an aircraft. In these examples those special technical peculiarities in the functioning of individual control elements which reflect most on the chosen principle of control are emphasised. In all the examples the expressions 'control function' and 'argument of the control' are used.

The control function determines the form of the signal (quantised with respect to amplitude and time) supplied to the servomotor; the argument of the control indicates the dependence of the control signal on a linear combination of the coordinates of the control system. Each automatic pilot scheme leads to a particular choice of mathematical model which in this book is investigated by the methods of matrix calculus.

A review gives, in condensed form, the basic rules and methods of the matrix calculus. Considerable attention is given to the representation of matrices in the canonical Jordan form because such a representation has shown itself to be a powerful means of carrying out the intermediate calculations and provides a useful framework within which to introduce the various steps in the argument.

A two-way approach to the problem of inverting the characteristic matrix is given by establishing identities connecting the adjoint matrix and its first derivatives with the matrices of the direct and inverse transformation to the canonical form. These identities permit one to represent the final results in a form convenient for practical application. Different forms of matrix solutions of linear differential and finite difference equations with constant coefficients are given.

The theorems of the direct Lyapunov method are extended to include the case of discrete dynamical systems whose perturbed motion is determined by equations in finite differences which do not depend explicitly on time. First, the basic theorems of Lyapunov are generalised; then the problem of the stability of the solutions of linear finite difference equations with constant coefficients is solved.

Stability and the processes of control in a linear system operating according to the pulse amplitude modulation principle are investigated. The analysis here is made for control functions with finite pulse height and pause width and also for impulse and stepped control functions. Characteristic equations and rational functions are constructed which reduce the investigation of stability to the application of the Hurwitz criterion and the argument principle.

The natural motions and reactions of linear pulsed systems to the more important types of external actions are found not only at the boundary points of the recurrence cycles but also at all intermediate points. From the Lyapunov function the process for the natural oscillations of the system with respect to any coordinate is determined. An analysis is given of the stability in the small for nonlinear systems with pulse-width modulation.

The natural and forced oscillations of relay systems are discussed, and the self-oscillation and sliding modes are investigated. The stability of the periodic modes is determined by constructing and analysing the point transformation of the switching plane in the neighbourhood of the invariant point, or of the equations of the perturbed motion in linear approximation.

A symmetric form of the equations describing the sliding motions is suggested permitting the solution to be represented in a simple form; by means of a direct comparison the sliding modes are shown to be identical to the motions of the corresponding linear system with an infinitely large gain.

The final results are presented in closed form and expressed in terms of the parameters of the control systems independently of whether the mathematical model is given as a system of differential equations in normal form or as second order equations which are insoluble with respect to the highest derivatives.

The author first applied the methods of matrix calculus to the study of the stability and self-oscillations of pulse-type control systems in the years 1953-1954 [30, 31].

The bibliography given in this book does not pretend to be exhaustive. Apart from references to work mentioned directly by the author, most of the works listed are monographs devoted to certain aspects of the theory which either are not covered in this book, or which present a different approach to the solution of a problem similar to one treated here.

In conclusion, the author takes pleasure in thanking Ya. Z. Tsipkin for help and encouragement during the preparation of this book, and E. N. Rozeivasser for reading through the manuscript and for valuable comment.

P. Bromberg

Chapter 1

AUTOMATIC CONTROL SYSTEMS

1.1 FUNCTIONAL SCHEME

Control theory is concerned with the investigation of mathematical models of actual control systems, but in a manner that takes no account of the dependence of these models on the physical nature of the components making up the system. Such mathematical models take the form of ordinary or partial differential equations, difference-differential equations and equations in finite differences having discrete or continuous arguments which, with some idealisation, describe the behaviour about the neighbourhood of the position being controlled. There are also mathematical models in the form of logical schemes that determine the strategy of behaviour of the control system as a function of various external factors.

We shall be considering systems whose motion is described by ordinary differential equations; these equations represent control systems with lumped parameters.

The usual basic problem of an automatic control system is to keep one or more of the variables of a physical system at some constant level, or to restrict the variations of these variables so that they remain within certain limits when the system is acted upon by disturbances. The physical system is usually called the 'plant' while the automatic device connected to it for the purpose of solving a given problem is called the 'regulator'. The plant and the regulator together constitute the control system.

Present-day control systems can be very complex, and a variety of physical principles are used in their design. However, provided one ignores the dependence on the physical nature of the individual elements, one can always divide the system up to a greater or lesser extent into a number of individual blocks which fulfil the same functions in different control systems.

As a concrete example of our problem, consider a very popular form of functional scheme for a system providing indirect control of a controlling element. The system is illustrated in Fig 1.1. The sensing element block consists of one or more devices which measure deviations of one or more plant variables from the specified values. It may, in addition, measure the deviations of the first derivatives. The signals from the sensing element block are fed to the input of a summing device. Feedback signals, which are proportional to the output of a servo-motor (and sometimes also to the derivative), are fed into the summer in such a way that they neutralise the action of the signals arriving from the sensing element block.

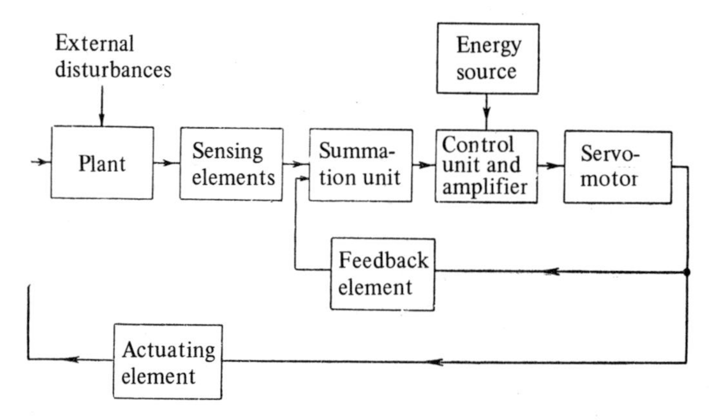

Fig 1.1 Functional diagram of a control system.

In the summer all these signals are added in definite proportions, and the summed signal is fed to the input of the controlling element and its amplifier. Generally, the sensing elements and the summer do not produce sufficient power to energise the actuator of the plant directly, without further amplification. The required action is taken by a special servo-motor, powered from an external source.

Depending on the magnitude and sign of the input signal, the controlling element and its amplifier regulate the amount of energy passing from the external source to the servo-motor in accordance with some specified law. The servo-motor sets the actuator in motion; it then operates in the required manner on the plant. We thus have a closed loop automatic control system.

Later we shall be dealing with open loop control systems – that is, systems in which the connection between the controlling element and the servo-motor is broken.

In practice, the summer and the controlling element may be incorporated in the same equipment. However, it is convenient to consider them as two different functional elements, and we shall refer to their output signals as the argument of the control and the controlling function, respectively. These will be seen to play a determining role in the classification of control systems.

Automatic control systems have access to a virtually unlimited supply of external energy. In closed systems, therefore – despite the inevitable dissipation of energy – periodic motions or oscillations that build up in amplitude, and which demand a continual inflow of external energy, can occur. This possibility of instability of the system is one of the most important problems in control theory.

1.2 RELAY SYSTEMS FOR STABILISING A NEUTRAL AIRCRAFT

The problem of the automatic stabilisation of the course of a neutral aircraft has played a prominent role in the development of certain methods of analysis of present-day control theory. This problem has been studied by many authors [2, 5, 12, 14, 32 and 34].

We shall use it as a concrete example to demonstrate the type of mathematical model and to answer the kind of questions that lie at the basis of the study of control theory. Moreover, such concepts as the argument of the control and the controlling function, which play an important role in the classification of models, can be clearly formulated in relation to the problem of the automatic control of an aircraft.

Figure 1.2 shows a possible mechanical model of a neutral aircraft. The model consists of a massive platform A which can be rotated about a vertical axis, and a light beam B which is free to rotate about an axis parallel to the axis of the platform. The

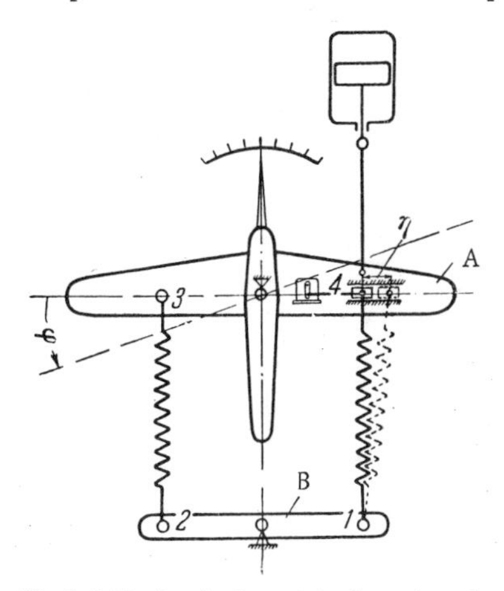

Fig 1.2 Mechanical model of an aircraft.

platform is provided with a damper and is connected to the beam by two identical springs. The springs are adjusted to some initial tension F_0 acting on the points *1*, *2*, *3* and *4*. Points *1*, *2* and *3* are fixed relative to both the platform and the beam, and point *4* is fastened to a rod which can be moved to either side by the servomotor of the automatic pilot. In the normal state, points *1*, *2*, *3* and *4* (at which the springs are anchored to the platform and beam) form the vertices of a parallelogram so that, for any rotation of the system, the total couple formed by the tension of the springs acting on the platform is equal to zero. The system is neutral with respect to any angle of rotation of the platform about the vertical axis.

Any displacement of point *4* produces a couple that acts on the platform. This couple is the output of the actuator. The deviation of point *4* is usually restricted by end-stops (not shown in the drawing) which place a limit on the controlling force available for stabilising the aircraft.

Let φ and η represent the turning angle (yaw) of the platform and the displacement of point *4* (the diagram indicates the positive sense of direction of these two quantities). The equation of motion of the platform can, with sufficient accuracy, be represented in the form

$$\ddot{\varphi} + M\dot{\varphi} = -N\eta + g_1' \tag{1.1}$$

where the dot notation is used for differentiation with respect to time.

Let M be the natural damping coefficient and N the efficiency factor of the rudder control. The angle φ is the controlled variable and η is the demanded variable. The quantity g_1', which has been reduced to the dimensions of an angular acceleration, characterises the effect of the external couple disturbing the motion of the aircraft.

A similar model could, for example, be made to simulate the action of a constant external couple by giving the springs a different preliminary tension. For an actual aircraft, a couple of this kind might arise if it is subject to an eccentric pull. For example, an effect which is particularly strong is that of an engine failure in a multi-engine aircraft. One could mention several other factors that can give rise to disturbances which depend on the aerodynamics of the aircraft, the variability of meteorological factors etc.

An automatic pilot is a controller connected to the plant (in this case, an aircraft). The control problem consists of providing the system with an additional characteristic, namely the capacity to recognise a previously specified direction in space and turn the aircraft towards the direction should it stray from it.

Figure 1.3 shows a schematic diagram of an automatic pilot

capable of solving this problem.

The sensing element which registers the deviation of the aircraft from the specified direction is a course gyroscope. This is an astatic gyroscope mounted within a gymbols. The azimuthal position of the gyroscope axis is corrected by a magnetic compass (this correction is not shown in Fig 1. 3), and the position in the vertical plane is stabilised by a correction taken from the casing.

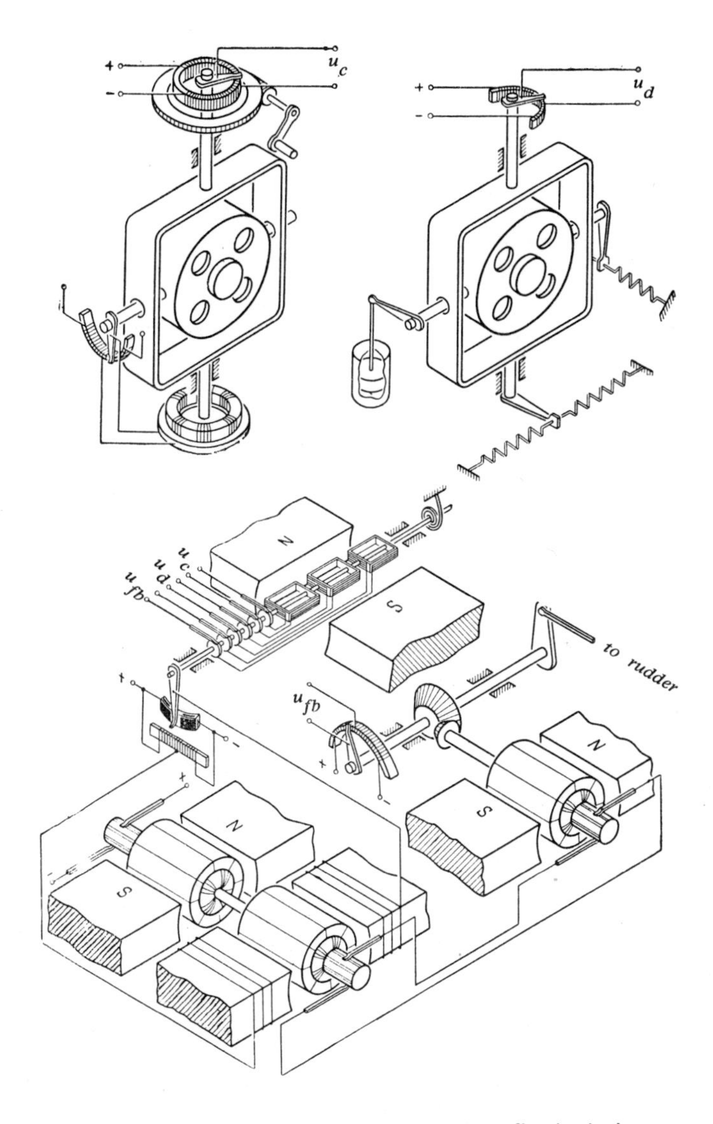

Fig 1.3 Arrangement for an automatic pilot with a fixed-velocity servo-motor.

An electric signal in the form of voltage is taken from a potentiometer whose wiper is fastened to the outer casing of the gyroscope. A screw drive allows the potentiometer to rotate relative to the aircraft, and it is by means of this rotation that the specified flight direction is fed into the aircraft. The processes involved in correcting the position of the gyroscope take place slowly enough to neglect their effect upon the control process. Thus the voltage picked off the course gyroscope potentiometer will be proportional to the angle φ. We shall always measure the angle φ from the base line representing the specified direction of flight which is the subject of the control.

The so-called 'damping gyroscope' is a second sensing element. It is shown in the diagram in the form of an astatic three-stage gyroscope whose freedom of motion about both axes of its gymbols is restricted by elastic constraints. From gyrodynamic theory, the motion of a damping gyroscope relative to the outer casing may be represented by a second order, inhomogeneous differential equation whose right-hand side is a linear combination of the first two derivatives of φ, the controlled variables, i.e. $\dot{\varphi}$ and $\ddot{\varphi}$.

The output potentiometer delivers a voltage proportional to the angle of turn of the outer casing of the gyroscope. Within limits, this voltage will be proportional to a linear combination of $\dot{\varphi}$ and $\ddot{\varphi}$.

The summer is shown in the form of an electromagnetic galvanometer with several independent frames. The ends of the windings on the frames are brought out to sliding contacts. Also connected to these contacts are the outputs of the potentiometer devices of the course and damping gyroscopes, and the negative feedback signal. The torque of the galvanometer frames, which is neutralised by the elastic hair springs, is proportional to a linear combination of the voltages taken from the potentiometers. The output signal is taken from a contact device consisting of a brush mounted on the galvanometer by two slip rings. In control terminology, the galvanometer may be considered as a second order dynamic element.

The system includes a rotating machine amplifier in the form of a motor generator which amplifies the low-power signal from the galvanometer contact device and applies it to the winding of a separately excited generator. The voltage from the generator armature brushes is applied to the collector brushes of the servo-motor armature, which is shown in the diagram as a constant current motor with independent excitation. Through a reduction gear, the servo-motor drives the actuator of the system – i.e. the rudder of the aircraft. In working out a functional representation of the generator and servo-motor, account is taken of the effect of the inductance of the exciter winding and the inertia of the armature.

The amplifier and servo-motor can then be considered as stable, aperiodic elements.

The equations defining the functional representation of the automatic pilot can be put in the following form:

the equation for the damping gyroscope

$$T_1^2\ddot{\alpha} + 2d_1T\dot{\alpha} + \alpha = k_{11}\dot{\varphi} + k_{12}\ddot{\varphi} \tag{1.2}$$

the equation for the summer

$$T_2^2\ddot{\sigma} + 2d_2T_2\dot{\sigma} + \sigma = k_{21}\varphi + k_{22}\alpha - k_{23}\eta \tag{1.3}$$

the equation for the amplifier

$$\Theta_1\mu + \mu = k_{31}\psi(\sigma) \tag{1.4}$$

the equation for the servo-motor

$$\Theta\ddot{\eta} + \dot{\eta} = k_{41}\mu \tag{1.5}$$

Here, α and σ are the turning angles of the outer casing of the damping gyroscope and the framework of the galvanometer, respectively; μ is the current in the generator exciter winding; Θ_1 is the time constant of the generator exciter winding and Θ is the time constant of the servo-motor – which is proportional to the moment of inertia of its armature, referred to the actuator displacement.

If the frequency characteristics of the damping gyroscope and galvanometer are significantly higher than those of the system being controlled, one can neglect the dynamics of these devices and assume that they ideally reproduce the input quantities. One may also neglect the time constant of the amplifier.

The equation of a closed system for stabilising the course of a aircraft may then be written in the form

$$\left.\begin{aligned} \ddot{\varphi} + M\dot{\varphi} &= -N\eta + g_1' \\ \Theta\ddot{\eta} + \dot{\eta} &= h_2'\psi(\sigma) \\ \sigma &= \varphi + \alpha_1\dot{\varphi} + \alpha_2\ddot{\varphi} - \frac{1}{a}\eta \end{aligned}\right\} \tag{1.6}$$

where the positive constants α_1, α_2, a and h_2' will be called the coefficients of velocity control, acceleration control, negative feedback and gain, respectively. We note that, for ideal damping gyroscopes and an ideal summer, the time constants Θ and Θ_1 have identical effects upon the dynamics of the control.

The third part of (1.6) defines the argument of the control. The controlling function $\psi(\sigma)$ can assume three fixed values. It has a relay characteristic – either maximum voltage of either sign or zero voltage is applied to the servo-motor. In our case, we have a control system with a constant velocity servo-motor. By suitably

Table 1.1

No	Name of characteristic	Schematic diagram	Form of characteristic
1	Ideal relay	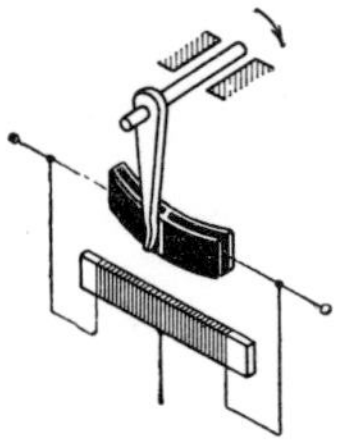	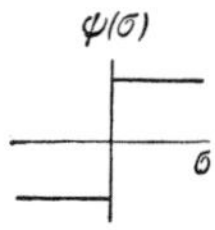 $\psi(\sigma)$, σ
2	With hysteresis (backlash in transmission)	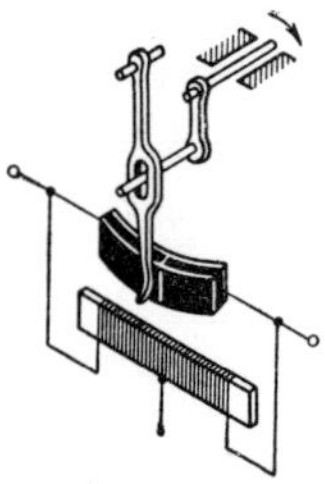	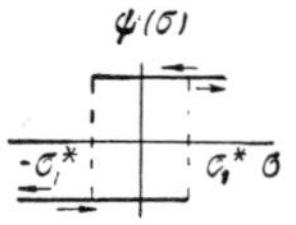 $\psi(\sigma)$, $-\sigma_1^*$, σ_1^*, σ
3	Hysteresis (friction on the axis)	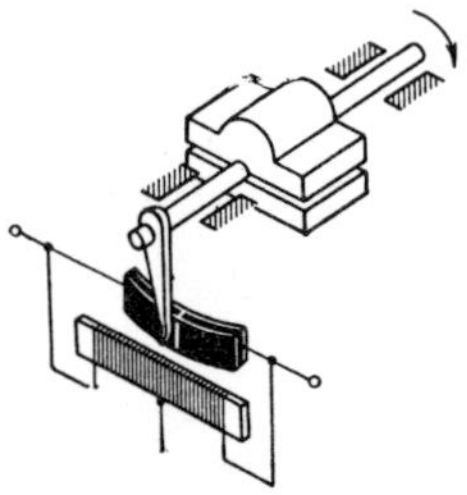	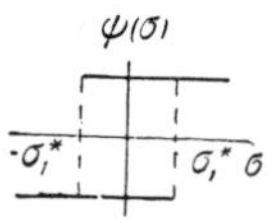 $\psi(\sigma)$, $-\sigma_1^*$, σ_1^*, σ

No	Name of characteristic	Schematic diagram	Form of characteristic
4	With a dead zone		$\varphi(\sigma)$ $-\sigma_2^*$ σ_2^* σ
5	Hysteresis with dead zone		$\varphi(\sigma)$ σ
6	Asymmetric, hysteresis with dead zone		$\varphi(\sigma)$ σ

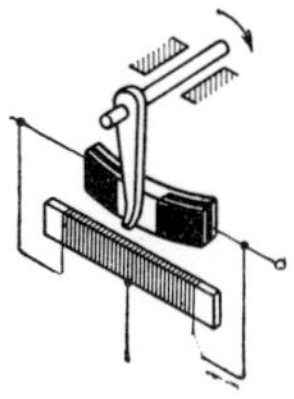

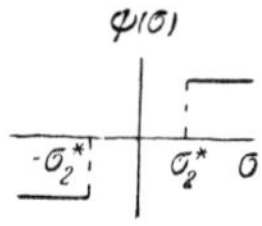

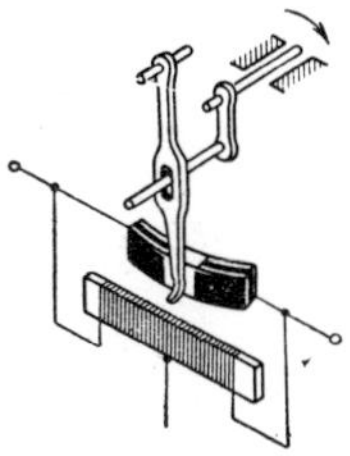

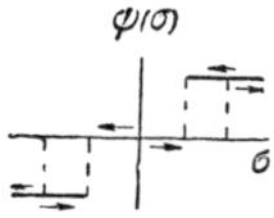

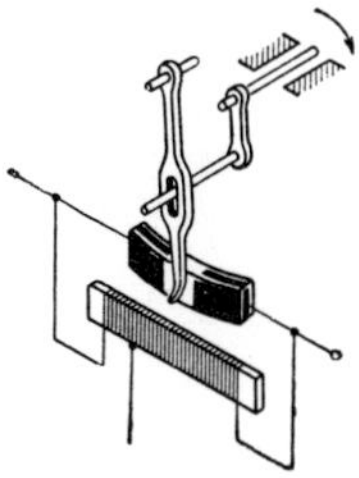

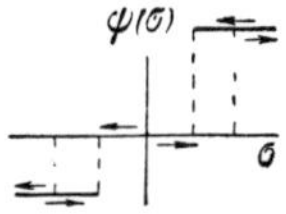

choosing the gain h_2' we can represent the ideal control function in the form

$$\psi(\sigma)=\begin{cases} 1 & \text{for } \sigma>0 \\ 0 & \text{for } \sigma=0 \\ -1 & \text{for } \sigma<0 \end{cases} \qquad (1.7)$$

In practice, the control function deviates from the ideal relay characteristic. Table 1.1 presents a number of characteristics of the control function and indicates the reasons for these specific forms. Naturally, these reasons have been chosen to apply to the output contact device of the galvanometer shown in Fig 1.3.

For hysteresis-type characteristics, the magnitude of the controlling function $\psi(\sigma)$ depends not only on the magnitude and sign of the argument of the control σ but also on the previous history of its variation. This fact is indicated on the characteristic diagrams by arrows.

In the automatic pilot system we have been considering, the servo-motor operates at a finite speed of rotation. In this system it is assumed that relay control by the speed of the servo-motor has been adopted.

However, for better utilisation of the limited controlling forces available, use has been made in some cases of control from the position of the actuator [32]. In this kind of system, the servo-motor (which takes the form of a pulling mechanism) adjusts the actuating element from one extreme to the other, virtually instantaneously.

Figure 1.4 illustrates one possible scheme for an automatic pilot using this principle. In the diagram, the servo-motor is shown in the form of two solenoids with a steel plunger mechanically connected to the actuator. When the pick-up makes contact with the left- or right-hand contact, the corresponding solenoid throws the plunger of the servo-motor from one stop to the other. When the pick-up is in contact with the insulated gap, the centreing springs return the plunger to its middle position.

With the restrictions indicated above, the equations of motion of the control system and relay servo-motor may be put in the form

$$\left.\begin{aligned} \ddot{\varphi}+M\dot{\varphi} &= -h_1'\psi(\sigma)+g_1' \\ \sigma &= \varphi+\alpha_1\dot{\varphi}+\alpha_2\ddot{\varphi} \end{aligned}\right\} \qquad (1.8)$$

There is now no term in the argument of the equation defining the negative feedback with respect to the coordinate of the actuator.

The automatic pilots considered so far are nonlinear controllers. Figure 1.5 shows an automatic pilot arrangement which, in spite of the presence of contact and relay devices, operates for practical purposes under linear conditions. In this arrangement,

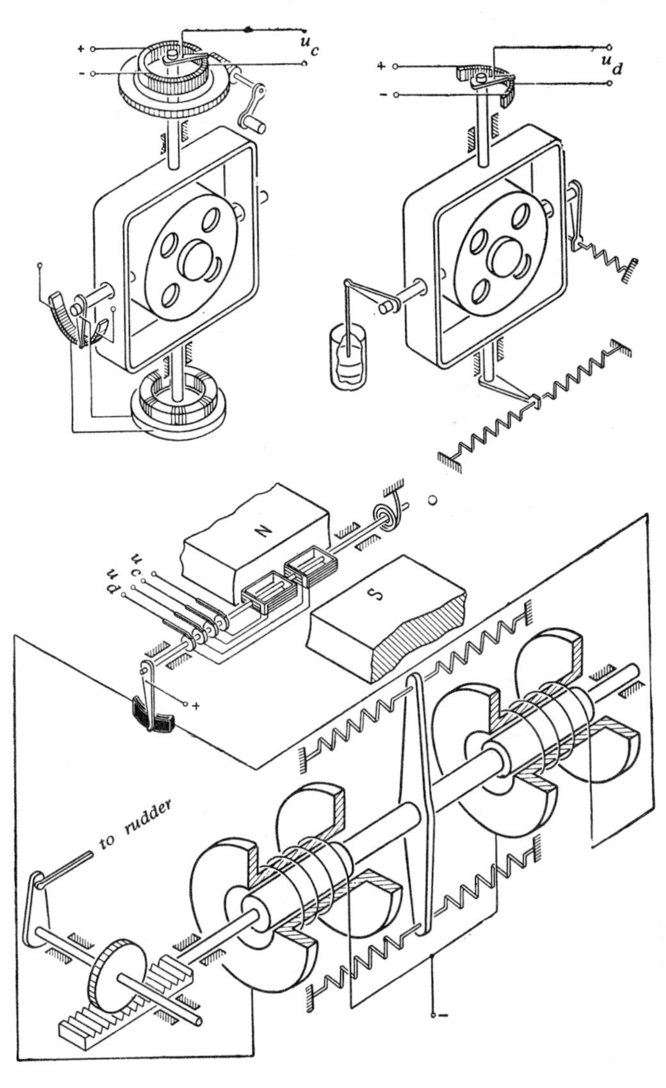

Fig 1.4 Arrangement for an automatic pilot with a relay-operated servo-motor.

the automatic pilot functions in the following way: a motor operating in the stationary regime rotates two drive pinions of the servo-device at constant speed and in opposite directions. Solenoids, operating through friction clutches, couple one or other of the drive pinions to a shaft rigidly coupled to the actuator (rudder), which begins to turn at constant velocity in the required direction.

The control by the solenoids is brought into action by the output signal from a summer, via a relay which plays the part of

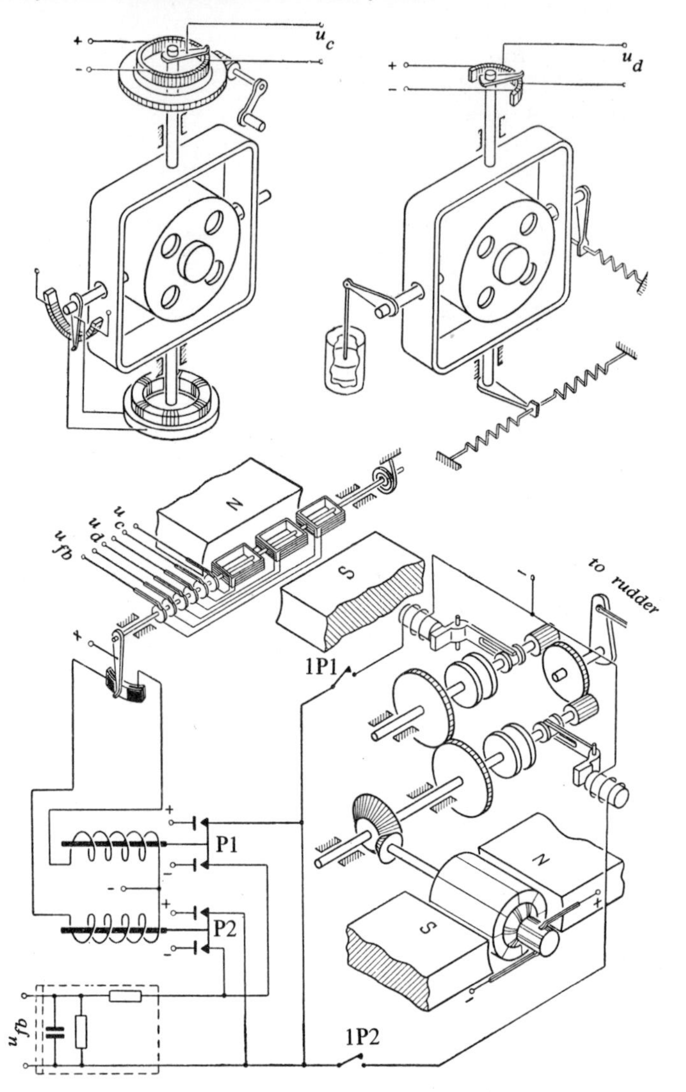

Fig 1.5 Arrangement for an automatic pilot with supplementary, relay-operated negative feedback.

an intermediate amplifier. The control element is encompassed by a relay-operated negative feedback loop.

The maximum signal in the feedback circuit is always greater than the input signal to the galvanometer (received from the sensing elements and servo-motor), and it causes the output brush of the galvanometer to turn towards the neutral position. When the brush reaches the insulated gap of the contact block, the solenoid coil is de-energised and the corresponding clutch is pushed against a braking shoe by a spring (not shown in Fig 1.5). The rudder is thereby locked in its current position. At the same time, the feedback circuit is de-energised. The galvanometer input signal then returns the galvanometer frame to the previous position, or to a slightly different position, and the whole cycle of operation is repeated. A resistor/capacitor network controls the repetition frequency of the cycle. The ratio of the time interval during which the rudder is being moved at constant velocity to the overall period of the cycle is proportional to the input signal of the galvanometer, i.e. to the expression

$$\varphi + \alpha_1\dot{\varphi} + \alpha_2\ddot{\varphi} - \frac{1}{a}\eta$$

During a period of one cycle, the speed of the servo-motor output shaft is – on average – proportional to this signal. The time constant of the network and the maximum voltage in the relay-operated feedback circuit are chosen so that the frequency is considerably (10-20 times) greater than the fundamental frequency of the process being controlled. Consequently, for all practical purposes, the speed of the output shaft of the servo-motor can be regarded as a continuous, linear function of the input signal of the summing galvanometer.

With these simplifications, the equations of motion of the system can be written in the form (in Fig 1.5 the feedback potentiometer attached to the relay device is not shown)

$$\left.\begin{aligned} &\ddot{\varphi} + M\dot{\varphi} = -N\eta + g_1' \\ &\eta = h_2'\psi(\sigma) \\ &T_c\dot{\mu} + \mu = h_3'\psi(\sigma) \\ &\sigma = \varphi + \alpha_1\dot{\varphi} + \alpha_2\ddot{\varphi} - \frac{1}{a}\eta - b\mu \end{aligned}\right\} \qquad (1.9)$$

Here, μ denotes the voltage on the capacitor plate, T_c and h_3' are the electrical time constant and maximum voltage in the relay-operated negative feedback circuit, respectively.

In the general theory of relay systems, the conditions for linearisation just considered are referred to as the 'sliding mode. Sliding modes are dependent upon the structure of the relay circuit. If the conditions for the existence of sliding modes are fulfilled, they can arise only in a certain region of variation of the states of the system – including the position being controlled. The role played by sliding modes in the general functional representation of relay control systems depends on the size of this region. Where such regions are sufficiently large, the sliding modes will be the operating conditions of the system. This, for example, is the case in an automatic pilot system in which the controlling element is implemented by a relay-operated negative feedback circuit. In the presence of sliding modes, the relay system can be replaced by an equivalent linear control system.

1.3 DISCONTINUOUS CONTROL SYSTEMS

As mentioned in the Introduction, in the case of pulsed control systems one can resolve the conflict between the precision and the power of the controlling signal by applying simple engineering methods. Figure 1.6 illustrates one scheme of this kind, the so-called discontinuous control system with forced rhythm of pulse alternation. The output signal of the summing galvanometer is produced by a potentiometer device incorporating a spring brush which, in the free state, does not come into contact with the potentiometer. The galvanometer brush is brought into contact with the potentiometer by a special bow-shaped contact which, after equal intervals of time, is thrown from the extreme lower position to the extreme upper position and back. This operation is carried out by a servo-motor provided with a small cam operating against a restoring spring.

The time the brush spends in contact with the potentiometer and the pressure on the brush are determined by the profile of the cam and the spring cam-follower. The over-all repetition period of a cycle is controlled by the rotation speed of the synchronous motor. By making a proper choice of the pressure and area of brush contact, it is possible to ensure that the output signal power delivered to the servo-motor is sufficient to accomplish the required control. Thus, during the 'signal-on' periods, the galvanometer is locked in position and a signal taken from it which is constant in magnitude and of sufficient power. During the 'signal-off' periods, the galvanometer is isolated from the output device and is free to assume with great precision a new position corresponding to the new input signal. Consequently, there is no rotating machine amplifier in Fig 1.6 like the one shown in Fig 1.3; in the earlier example it was not possible to increase the power of the signal taken from the galvanometer without at the same time making the readings of the summing galvanometer more coarse. Thus, in discontinuous control systems one must find a compromise solution between the two contradictory requirements laid down at the beginning of this section.

Figure 1.7 gives a graphical representation of the output signal from the galvanometer as a function of time. The output signal takes the form of rectangular pulses having a constant width T_1 which follow one another after equal time intervals T (in Fig 1.7, T_1 = constant, T_2 = constant, $T = T_1 + T_2$). The heights of the pulses are proportional to the values of the input signal at the beginning of the 'signal-on' periods. The area of the pulses is proportional to

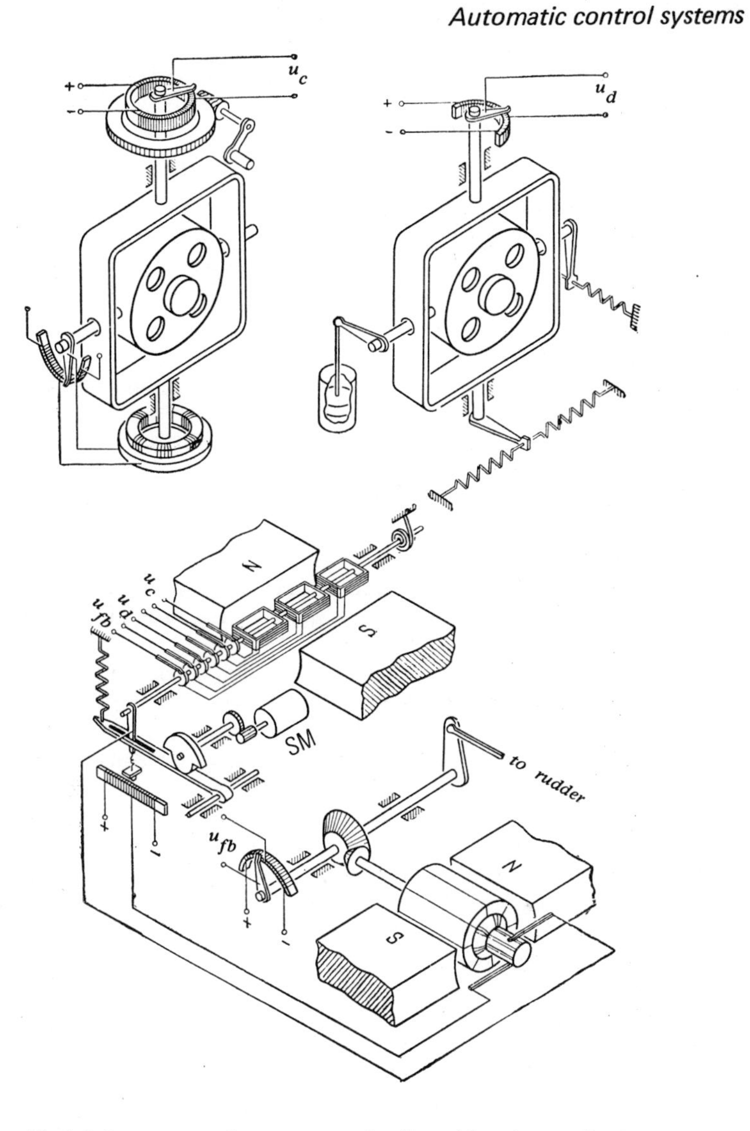

Fig 1.6 Arrangement for an automatic pilot with pulse-amplitude modulation of the control signal.

the values of the input quantities indicated above. The area is a more adequate and meaningful characteristic of the pulses. If the galvanometer is assumed to be an ideal instrument and its input is denoted by σ (the argument of the control), the output signal can be denoted by $\psi(\sigma)$ – the controlling function. The graph in Fig 1.7 illustrates the case corresponding to this assumption. Analytically, the controlling function $\psi(\sigma)$ can be written in the form

$$\psi(\sigma)=\begin{cases} \sigma(mT) & \text{for} \quad mT \leqslant t < mT+T_1 \\ & \quad T_1=\text{const},\ T=\text{const} \\ 0 & \text{for} \quad mT+T_1 \leqslant t < (m+1)T \\ & \quad T-T_1=T_2=\text{const} \end{cases} \tag{1.10}$$

It is assumed here that, at the beginning of the 'signal-on' period, the controlling function equals the corresponding discrete value of the argument of the control $\sigma(mT)$; any different proportionality factor can be accounted for in the equations of the system by inserting a suitable factor in front of $\psi(\sigma)$.

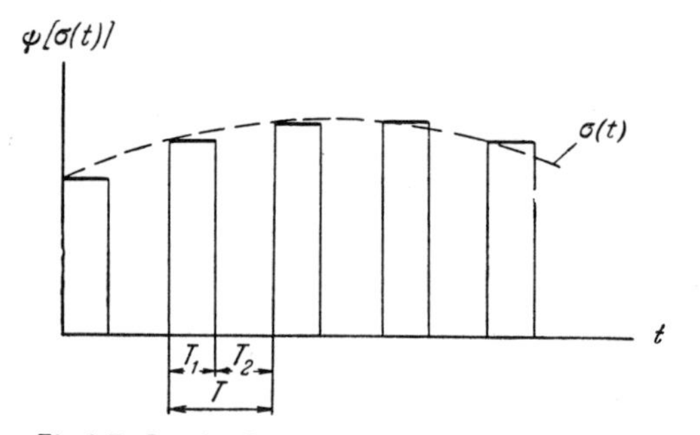

Fig 1.7. Graph of a control function with pulse-amplitude modulation.

To apply to an automatic pilot the servo-motor-operated pulsed control system we have been considering, the equations of motion of the system for stabilising a neutral aircraft can be formally written in the form (1.6). But it is now necessary to define $\psi(\sigma)$ in these equations by (1.10), or by the graph in Fig 1.7.

The output device of a galvanometer equipped with a chopper bar is a pulse element which introduces pulse amplitude modulation of the control signal. Control systems with this type of pulse element belong to the class of linear pulsed systems, since their state at discrete, equally spaced instants of time is determined by linear finite difference equations.

Figure 1.8 illustrates another type of pulse element which brings about a pulse-width modulation of the control signal.

In this arrangement, the potentiometer is replaced by two contact bars which are electrically insulated from each other and whose profile on the operating side is such that their widths increase in proportion to the distance from the centre towards the edges. The absolute magnitude of the output signal is now constant, but the time for which contact is made is proportional to the angle of rotation of the galvanometer frame, i.e. to the absolute magnitude of its input signal. In this case the width of the pulses is modulated but the pulse repetion rate is constant.

Let $T_{1,m}$ denote the variable width of the pulses, and $T_{2,m}$ the duration of the break, with $T_{1,m}+T_{2,m}=T=\text{constant}$. The width

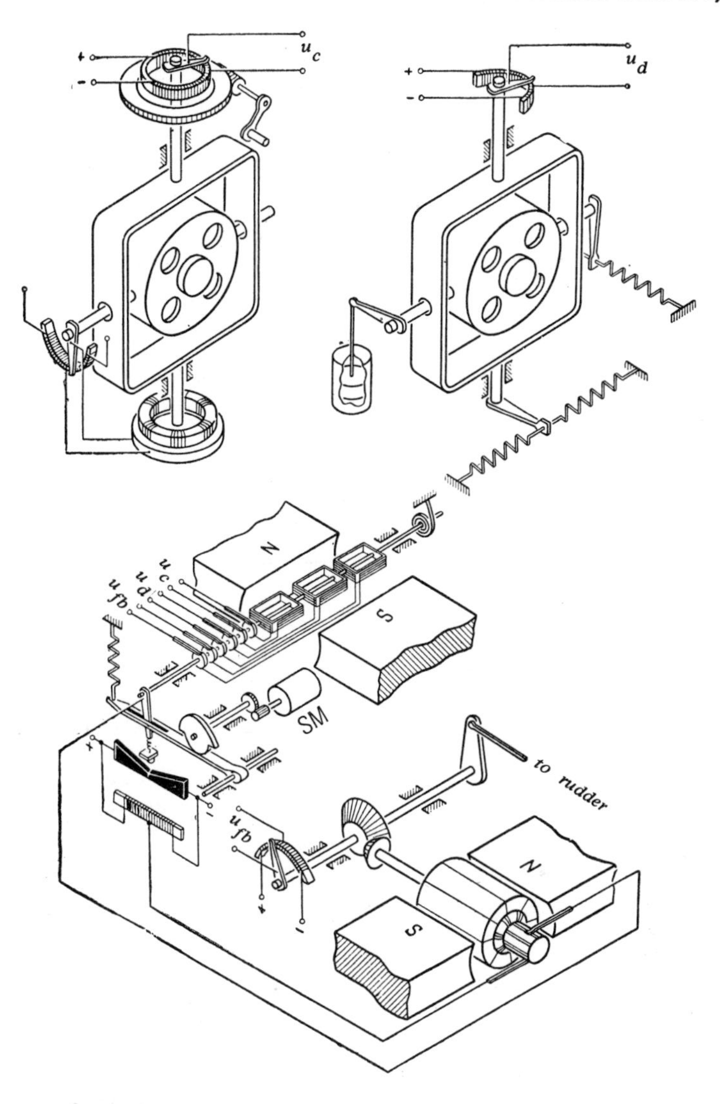

Fig 1.8 Arrangement for an automatic pilot with pulse-width modulation of the control signal.

of a pulse $T_{1,m}$ will be proportional to the value of the input signal corresponding to the beginning of the contact period at the m-th repetition period. Figure 1.9 illustrates the control function $\psi(\sigma)$ and the argument of the control when pulse-width modulation is used.

Analytically, $\psi(\sigma)$ can be put in the form

$$\psi(\sigma)=\begin{cases}\chi_m & \text{for} \quad mT \leqslant t < mT + T_{1,m} \\ 0 & \text{for} \quad mT + T_{1,m} \leqslant t < (m+1)T\end{cases} \tag{1.11}$$

where

$$T_{1,m} = T_1 \chi_m \sigma(mt),$$

$$\chi_m = \begin{cases} 1 & \text{for} \quad \sigma > 0 \\ 0 & \text{for} \quad \sigma = 0 \\ -1 & \text{for} \quad \sigma < 0 \end{cases}$$

Here, T_1 is a constant having the dimensions of time, so that the argument of the control σ is a dimensionless quantity. It is convenient to suppose that $T_1 < T$.

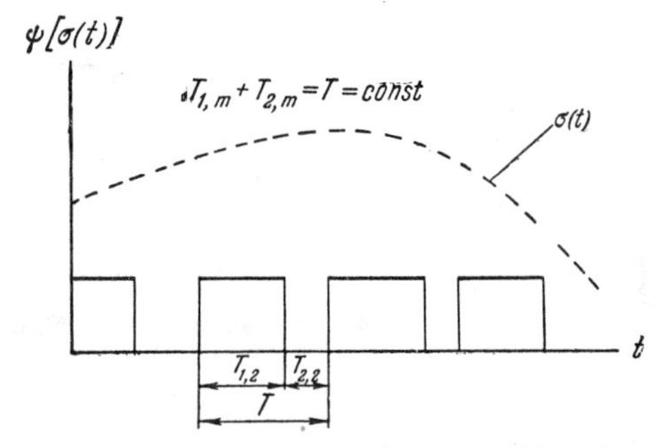

Fig 1.9 Graph of a control function with pulse-width modulation.

Control systems with pulse-width modulation of the control signal are nonlinear. It is interesting to note the special feature of pulse-width modulation which enables it to be used in digital computers. Quantities whose amplitudes are constant but whose durations differ can be measured comparatively simply with the aid of elementary pulses of standard width, i. e. it is relatively simple to convert them to a form which is very much more convenient for discrete counting.

1.4 GENERAL EQUATIONS OF MOTION AND THE CLASSIFICATION OF CONTROL SYSTEMS

The pulsed and relay control systems discussed in the preceding sections can be described by a single type of differential equation. It is therefore advantageous to use as our mathematical model a system of second order differential equations represented in a general form so that all the particular cases can be obtained by a suitable choice of coefficients and form for the control function.

Let us write these equations in the form

$$\left.\begin{aligned} \sum_{j=1}^{m} (a_{ij}\ddot{y}_j + b_{ij}\dot{y}_j + c_{ij}y_j) = h'_i\psi(\sigma) + g'_i \\ \sigma = \sum_{j=1}^{m} (\gamma'_j y_j + \gamma''_j \dot{y}_j) \\ i = 1, 2, \ldots, m \end{aligned}\right\} \quad (1.12)$$

Here, the y_j denote the coordinates of the system expressed as the distance sway from the position being controlled; a_{ij}, b_{ij}, c_{ij}, h'_i, γ'_j and γ''_j are real constant coefficients; the magnitudes of g'_i determine the external action. Since in all cases $\psi(0) = 0$, the sequence of equalities

$$y_1(t) = y_2(t) = \ldots = y_m(t) \equiv 0 \quad (1.13)$$

is a trivial solution to the system of equations (1.12) in the absence of an external action. The equalities (1.13) determine the stationary state of the system to be controlled. Consequently, (1.12) may be considered as equations for the perturbed motion of the control system about the stationary state.

Let us suppose that the determinant of the coefficients in front of the higher derivatives of system (1.12) is non-zero. Then it can be solved with respect to the higher derivatives and, with the help of the new variables

$$\left.\begin{aligned} x_1 = y_1, \quad x_2 = y_2, \ \ldots, \ x_m = y_m \\ x_{m+1} = \dot{y}_1, \quad x_{m+2} = \dot{y}_2, \ \ldots, \ x_n = \dot{y}_m \\ n = 2m \end{aligned}\right\} \quad (1.14)$$

it can be put in the normal Cauchy form

$$\left.\begin{aligned} \dot{x}_1 &= p_{11}x_1 + p_{12}x_2 + \ldots + p_{1n}x_n + h_1\psi(\sigma) + g_1 \\ \dot{x}_2 &= p_{21}x_1 + p_{22}x_2 + \ldots + p_{2n}x_n + h_2\psi(\sigma) + g_2 \\ &\ldots\ldots\ldots\ldots\ldots\ldots\ldots\ldots \\ \dot{x}_n &= p_{n1}x_1 + p_{n2}x_2 + \ldots + p_{nn}x_n + h_n\psi(\sigma) + g_n \end{aligned}\right\} \quad (1.15)$$

$$\sigma = \gamma_1 x_1 + \gamma_2 x_2 + \ldots + \gamma_n x_n$$

where p_{ij}, h_i and γ_i are constant coefficients which are determined as the calculations proceed; g_i are linear combinations of g'_i. In the new coordinates the stationary state is defined by the equalities

$$x_1(t) = x_2(t) = \ldots = x_n(t) \equiv 0 \quad (1.16)$$

The classification of the control systems considered in this book will be made in accordance with the form of the control function $\psi(\sigma)$ and of the argument of the control σ. We shall distinguish between relay and pulse systems according to the form of the control function, the latter being sub-divided into linear and nonlinear systems. In this cases the control function is represented by the graphs given in Table 1.1 and Figs 1.7 and 1.9.

We shall use the form of the argument of the control σ to distinguish systems operating according to the principle of control by coordinate and first derivative from systems with feedback loops of various kinds.

Equations (1.12) and (1.15) embrace all cases lying within the scope of the systems being considered. In particular, if the instrument forming the argument of the control cannot be regarded as inertialess, than its output will play the part of σ. This case can be obtained from the general case, when all but one of the coefficients γ_i – see (1.15) – are equal to zero. However, the undoubted merit of the concept of an argument of the control in a general form is that it permits one to find the limiting potentialities of a given principle of control.

In (1.12) and (1.15), all the coordinates enter symmetrically. Symmetric equations permit one to find general regularities in the behaviour of the solutions and to formulate the final results in a simple and convenient form.

1.5 ARGUMENTS OF THE CONTROL FOR A DESIRED RESPONSE

The automatic pilot schemes we have discussed provide a solution to the classical problem of control: that is, the problem of maintaining the control variable at some constant level, or of restricting the variations of that variable arising from external disturbances.

These control systems make use of the principle of negative feedback. The sensing element of the system measures the deviations of the variables and their derivatives to be controlled from a specified level while the actuating element, in accordance with the signals received from the controller, corrects such deviations, regardless of their cause. An automatic system which solves this type of problem is sometimes called a regulator problem.

However, any control system can easily be converted to another mode of operation, called the tracking mode. In the tracking mode the operator introduces a desired response into the system and, by means of a controller, forces the plant to change its position continuously in the desired direction. The system can be transferred to the tracking mode by adjusting the sensing elements to a new level which can either be fixed or continuously variable, depending on the information supplied to the operator. The control system will then compel the controlled coordinate to follow the continuously changing level and will simultaneously stabilise its variation about the new position. The desired responses are described by arbitrary functions of time since they depend on changes in the external environment and on the demands of the operator.

As a rule, however, these functions are subject to certain restrictions in regard to the magnitude and rate of change of the signal etc, due to the special nature of the complex tasks which a system operating in the tracking mode is called upon to solve. A control system working in the tracking mode is often called a servo-system.

Figure 1.10 shows one possible schematic arrangement for introducing a desired response or control action into an automatic

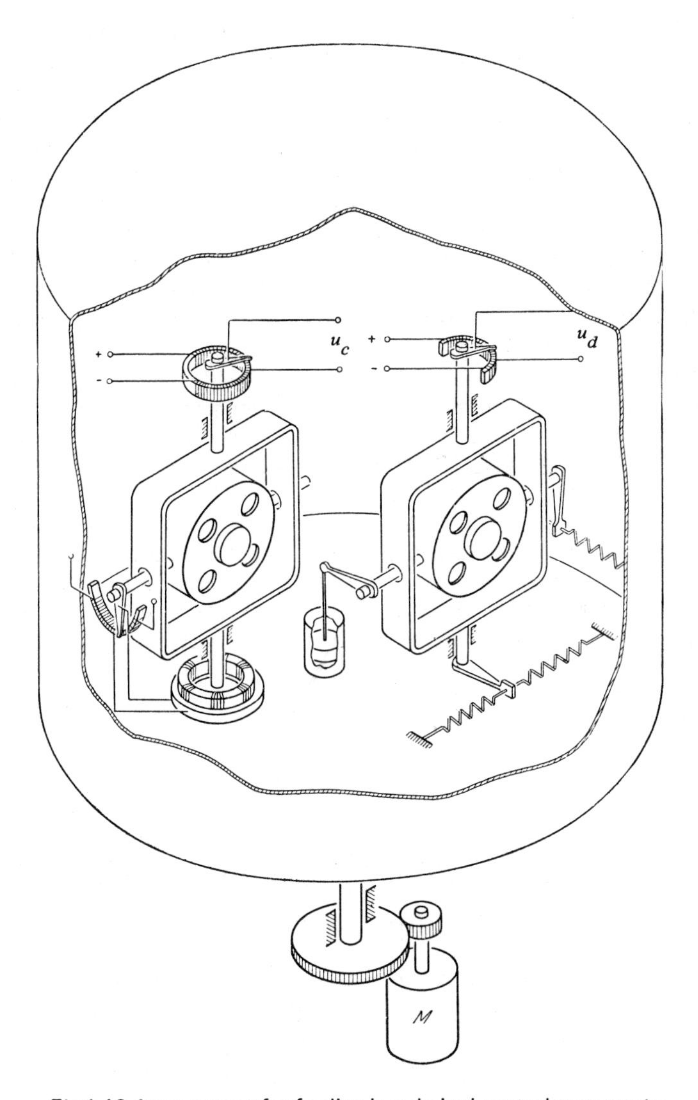

Fig 1.10 Arrangement for feeding in a desired control response to an automatic pilot.

pilot. This scheme is very convenient for explaining the fundamental principles of the problem, but for practical application other solutions may be more acceptable.

In Fig 1.10 a damping gyroscope and a course gyroscope are set up on a special disc, which can be rotated relative to the aircraft about a vertical axis by an electric motor and reduction gear. The motor is set into operation by a desired response voltage u_3' which varies in accordance with the decision taken by the operator.

We shall denote the angular displacement of the disc relative to the aircraft by φ_3 and assume that it is positive when it changes in a direction opposite to that of the positive direction of variation of the yaw angle of the aircraft, φ. The disc takes part in two motions: a rotational motion as it moves with the aircraft, and a relative motion with respect to the aircraft. These motions are defined by the corresponding angles $\varphi(t)$ and $\varphi_3(t)$, which are functions of time. Having regard to the chosen directions for reading off the angles φ and φ_3, the position of the disc with respect to absolute motion is given by the angle $(\varphi - \varphi_3)$.

The axis of the pattern of the course gyroscope preserves this direction in inertial space so that the voltage u_c picked off the potentiometer is proportional to the difference $(\varphi - \varphi_3)$. A damping gyroscope determines the angular velocity and angular acceleration relative to inertial space. In the arrangement shown in Fig 1.10, the damping gyroscope is set up on a disc so that the voltage u_d taken off the potentiometer is proportional to a linear combination of the difference $(\dot{\varphi} - \dot{\varphi}_3)$ and $(\ddot{\varphi} - \ddot{\varphi}_3)$. When a desired response signal is fed into the arrangement of Fig 1.10, the argument of the control has the normal structure but, in the expression that determines it, it is obviously necessary to substitute $(\varphi - \varphi_3)$, $(\dot{\varphi} - \dot{\varphi}_3)$ and $(\ddot{\varphi} - \ddot{\varphi}_3)$, respectively, in place of φ, $\dot{\varphi}$ and $\ddot{\varphi}$. Thus, when a desired response signal is fed into the automatic pilot, the argument of the response is given by the formula

$$\sigma = (\varphi - \varphi_3) + \alpha_1 (\dot{\varphi} - \dot{\varphi}_3) + \alpha_2 (\ddot{\varphi} - \ddot{\varphi}_3) - \frac{1}{a} \eta \tag{1.17}$$

For general equations of motion of the control system having the form (1.12) and (1.15), the argument of the control σ – allowing for the desired response – can be written in the form

$$\sigma = \sum_{j=1} \gamma_j' (y_j - y_{3j}) + \gamma_j'' (\dot{y}_j - \dot{y}_{3j}) \tag{1.18}$$

or

$$\sigma = \gamma_1 (x_1 - x_{31}) + \gamma_2 (x_2 - x_{32}) + \ldots + \gamma_n (x_n - x_{3n}) \tag{1.19}$$

for (1.12) and (1.15), respectively.

Chapter 2

ELEMENTS OF MATRIX ANALYSIS

2.1 BASIC DEFINITIONS AND OPERATIONS

2.1.1 General remarks

There are a number of books on the subject of matrix theory and its application to the solution of practical problems. We would mention in particular the monographs by B. V. Bylgakov [5], F. R. Gantmaker [6], R. A. Frazer, W. J. Duncan and A. R. Collar [24]. However, it seems desirable to bring together the fundamental formulae of matrix algebra and to give an account of methods of the matrix analysis which will form the basis of the material to be discussed. This will enable us to present concise solutions to control problems, and this approach should make the book easier to use.

2.1.2 Basic definitions

An ordered array of complex or real numbers forming a rectangular table of m rows and n columns

$$A = \begin{Vmatrix} a_{11} & a_{12} & \dots & a_{1n} \\ a_{21} & a_{22} & \dots & a_{2n} \\ \cdot & \cdot & \cdot & \cdot \\ a_{m1} & a_{m2} & \dots & a_{mn} \end{Vmatrix} \tag{2.1}$$

is called a rectangular matrix of dimensions $m \times n$. The numbers a_{ij} are called the matrix elements; the first index indicates the number of the row and the second the number of the column at whose intersection the element is found in the table. A matrix is said to be square of order n if $m=n$. When $n=1$, we have a column matrix

$$h=\begin{Vmatrix} h_1 \\ h_2 \\ \cdot \\ \cdot \\ \cdot \\ h_m \end{Vmatrix} \tag{2.2}$$

and when $m=1$, we have a row matrix

$$\gamma=\|\gamma_1\ \gamma_2\ \dots\ \gamma_n\| \tag{2.3}$$

A table consisting of r ($<m$ rows) and s ($<n$) columns of the matrix (2.1) is called a sub-matrix. If a sub-matrix is a square matrix of order r, then its determinant is called a minor of order r. We say that a matrix has rank r if the highest order of a non-zero minor is equal to r. A square matrix is said to be non-singular if its rank is the same as the order of the matrix; i.e. the determinant of a non-singular matrix is not zero.

Two matrices A and B are said to be equal if they have the same dimensions and if corresponding elements are all equal to each other, i.e. if $a_{ij}=b_{ij}$. The matrix $A=0$ if all its elements $a_{ij}=0$. A matrix A^T with elements a^T_{ij} is said to be transposed with respect to A with elements a_{ij}, if $a^T_{ji}=a_{ij}$. The rows of a transposed matrix are equal to the columns of the original matrix, and vice versa.

A square matrix A is said to be symmetric if it is identical to its transposed matrix, i.e. if $A=A^T$. Elements in such a matrix which are symmetrically located with respect to the main diagonal are equal to each other, i.e. $a_{ij}=a_{ji}$.

A square matrix A is said to be diagonal if all its elements except those lying on the main diagonal are equal to zero, i.e. $a_{ij}=0$ if $i\neq j$.

A diagonal matrix of form

$$E=\begin{Vmatrix} 1 & 0 & \dots & 0 \\ 0 & 1 & \dots & 0 \\ \cdot & \cdot & \cdot & \cdot \\ 0 & 0 & \dots & 1 \end{Vmatrix} \tag{2.4}$$

is called a unit matrix. One sometimes writes E_n if one wishes to emphasise the order of a unit matrix.

2.1.3 Addition of matrices

Matrices A and B having the same dimensions $m\times n$ can be added.

The matrix $C = A + B$, having the same dimensions as A and B, and whose elements c_{ij} are given by

$$c_{ij} = a_{ij} + b_{ij} \tag{2.5}$$

is called the sum of A and B. From (2.5) it follows that the addition of matrices is subject to the same laws as those for the addition of scalar quantities.

2.1.4 Multiplication of matrices

If the number of columns of the first matrix A of dimensions $m \times n$ is equal to the number of rows of the second matrix B of dimensions $k \times l$, i.e. if $n = k$, then it is possible to multiply the matrices in the direction indicated.

A matrix C having dimensions $m \times l$ is the product of the above two matrices

$$C = AB \tag{2.6}$$

if its elements c_{ij} are the sum of the product of the elements of the ith row of matrix A and the corresponding elements of the jth column of matrix B, i.e. if

$$c_{ij} = \sum_{r=1}^{n=k} a_{ir} b_{rj} \tag{2.7}$$

A matrix with dimensions $m \times n$ can be multiplied from the right by an $n \times 1$ column matrix or from the left by a $1 \times m$ row matrix; the results obtained are a column matrix and a row matrix with dimensions $m \times 1$ and $1 \times n$, respectively. Matrices with dimensions $m \times n$ and $n \times m$ and, in particular, square matrices of the same order, can be multiplied in either direction.

Matrix multiplication satisfies the associative and distributive laws, i.e.

$$\left.\begin{aligned} (AB)C &= A(BC) = ABC \\ (A + B)C &= AC + BC \end{aligned}\right\} \tag{2.8}$$

but the commutative law is, in general, not satisfied, so that

$$AB \neq BA \tag{2.9}$$

Two matrices A and B are said to commute if

$$AB = BA \tag{2.10}$$

When a matrix is multiplied by a scalar quantity, all its elements are multiplied by this quantity. Diagonal matrices of the same

order commute.

If two matrices A and B are square and one of them is non-singular, then the equality $AB=0$ can hold if – and only if – the other matrix is zero. The determinant of the product of two square matrices is equal to the product of the determinants of the matrices being multiplied.

$$\det AB = \det A \cdot \det B$$

Matrix transposition obeys the following law:

$$(AB)^T = B^T A^T$$

In particular, it follows that the product of two symmetric and commuting matrices is also a symmetric matrix

$$(AB)^T = B^T A^T = BA = AB$$

The rank of the product of two rectangular matrices cannot be greater than the rank of any of its cofactors. If $C=AB$ and r_A, r_B and r_C are the ranks of matrices A, B and C, then $r_C \leqslant r_A, r_B$. In particular, the product of the column matrix h with dimensions $n\times 1$ and the row matrix γ with dimensions $1\times n$

$$h\gamma = \begin{Vmatrix} h_1\gamma_1 & h_1\gamma_2 & \dots & h_1\gamma_n \\ h_2\gamma_1 & h_2\gamma_2 & \dots & h_2\gamma_n \\ \cdot & \cdot & \cdot & \cdot \\ h_n\gamma_1 & h_n\gamma_2 & \dots & h_n\gamma_n \end{Vmatrix} \tag{2.11}$$

has a rank equal to unity – i.e. all the minors of this matrix of order above the first are equal to zero.

Adjoint and inverse matrices. Let A_{ij} be the cofactor of the elements a_{ij} of a square matrix A. If in matrix A the elements a_{ij} are replaced by the cofactors A_{ij}, and the matrix is then transposed, we obtain the adjoint matrix A'. The elements a'_{ij} of matrix A' satisfy the conditions

$$a'_{ij} = A_{ji}$$

From the well-known property of determinants

$$\sum_{j=1}^{n} a_{ij}a'_{jk} = \sum_{j=1}^{n} a'_{ij}a_{jk} = \begin{cases} \det A & \text{for } i=k \\ 0 & \text{for } i\neq k \end{cases} \tag{2.12}$$

Whence, remembering the rule for multiplying matrices by a scalar quantity, there follows the equality

$$A\cdot A' = A'A = (\det A)E \tag{2.13}$$

If a matrix A is non-singular, i.e. if $\det A\neq 0$, then $\det A' = (\det A)^{n-1}$.

A matrix A^{-1} with elements α_{ij} satisfying the equality

$$A \cdot A^{-1} = A^{-1} A = E \tag{2.14}$$

is called an inverse matrix. A non-singular matrix A always has an inverse matrix A^{-1} given by the formula

$$A^{-1} = \frac{A'}{\det A}$$

so that the elements α_{ij} of matrix A^{-1} are equal to the transposed cofactors A_{ji} divided by the determinant of matrix A. From (2.12) and the properties of the elements α_{ij} indicated above we find the equality

$$\sum_{j=1}^{n} a_{ij}\alpha_{jk} = \sum_{j=1}^{n} \alpha_{ij} a_{jk} = \begin{cases} 1 & \text{for } i = k \\ 0 & \text{for } i \neq k \end{cases} \tag{2.15}$$

The matrix inverse to the product AB is equal to the product of the inverse matrices A^{-1} and B^{-1} taken in the reverse order, i. e.

$$(AB)^{-1} = B^{-1} A^{-1} \tag{2.16}$$

A square matrix A is said to be orthogonal if $A^{-1} = A^{T}$. The determinant of an orthogonal matrix is equal to ± 1. In an orthogonal matrix the elements are identical to their cofactors. From this fact, and from the equalities (2.12) and (2.15), it follows that the elements of an orthogonal matrix satisfy the conditions

$$\sum_{j=1}^{n} a_{ij} a_{kj} = \sum_{j=1}^{n} a_{ji} a_{jk} = \begin{cases} 1 & \text{for } i = k \\ 0 & \text{for } i \neq k \end{cases} \tag{2.17}$$

2.1.5 Power of a matrix

The product of k square matrices A is called the kth power of matrix A and is denoted A^k. For any integral non-negative k and l the associative law yields

$$A^k \cdot A^l = A^{k+l} \qquad (A^k)^l = A^{kl} \tag{2.18}$$

In the case of non-singular matrices, this equality is valid for any whole number since one can write

$$A^{-k} = (A^{-1})^k \tag{2.19}$$

Consider a square matrix E_{ij} whose element in the ith row and jth column is equal to unity and whose remaining elements are equal to zero. Matrices of this type are subject to the following rule of multiplication:

$$E_{ij}E_{kl}=\begin{cases}E_{il} & \text{for } j=k\\ 0 & \text{for } j\neq k\end{cases} \tag{2.20}$$

Any square matrix A with elements a_{ij} can be expanded as a series of these matrices and represented in the form

$$A=\sum_{i,\,j=1}^{n} a_{ij}E_{ij} \tag{2.21}$$

Consider two square matrices K and K^{-1} with elements k_{ij} and $\varkappa_{ij}$, and introduce the notation

$$k_i=\left\|\begin{matrix}k_{1i}\\ k_{2i}\\ \vdots\\ k_{ni}\end{matrix}\right\|,\quad \varkappa_j=\|\varkappa_{j1}\varkappa_{j2}\dots\varkappa_{jn}\| \tag{2.22}$$

for the ith column and jth row of K and K^{-1}, respectively. Then it is easy to demonstrate that

$$KE_{ij}K^{-1}=k_i\varkappa_j \tag{2.23}$$

and, in general, for the three square matrices A, K and K^{-1} the expansion

$$KAK^{-1}=\sum_{ij=1}^{n} a_{ij}k_i\varkappa_j \tag{2.24}$$

is valid. [(2.23) and (2.24) are true also in the case when, instead of K and K^{-1}, two arbitrary square matrices are taken].

In particular, if A is a unit matrix we have

$$KEK^{-1}=KK^{-1}=\sum_{i=1}^{n} k_i\varkappa_i \tag{2.25}$$

2.1.6 Oblique rows

A square matrix of the n-th order

$$H^i = \left\| \begin{matrix} 0 & 0 & \dots & 0 \\ \cdot & \cdot & \cdot & \cdot \\ 0 & 0 & \dots & 0 \\ 1 & 0 & \dots & 0 \\ 0 & 1 & \dots & 0 \\ \cdot & \cdot & \cdot & \cdot \\ 0 & 0 & \dots & 0 \end{matrix} \right\| \begin{matrix} \left.\vphantom{\begin{matrix}0\\0\\0\end{matrix}}\right\} i \text{ rows} \\ \\ \\ \\ \end{matrix} \tag{2.26}$$

in which the elements situated on the straight line parallel to the main diagonal but shifted i places to the left are equal to unity, and whose remaining elements are all equal to zero, is called a unit (left) oblique row matrix. The index i can run from 1 to $n-1$ and the corresponding matrices are called the first, second etc. unit oblique rows.

By direct calculation, one can verify that the following rule for multiplying unit oblique rows is valid:

$$H^i \cdot H^j = H^j \cdot H^i = \begin{cases} H^{i+j} & \text{for } i+j < n \\ 0 & \text{for } i+j \geqslant n \end{cases} \tag{2.27}$$

In multiplying unit rows, the indices behave formally like powers when their sum is less than the order of the matrix.

One can, with the help of unitary oblique rows, express a left triangular matrix of order n

$$A_\Delta = \left\| \begin{matrix} \alpha_0 & 0 & 0 & \dots & 0 \\ \alpha_1 & \alpha_0 & 0 & \dots & 0 \\ \alpha_2 & \alpha_1 & \alpha_0 & \dots & 0 \\ \cdot & \cdot & \cdot & \cdot & \cdot \\ \alpha_{n-1} & \alpha_{n-2} & \alpha_{n-3} & \dots & \alpha_0 \end{matrix} \right\| \tag{2.28}$$

with equal elements in the oblique rows in the form

$$A_\Delta = \alpha_0 E + \alpha_1 H^1 + \alpha_2 H^2 + \dots + \alpha_{n-1} H^{n-1} \tag{2.29}$$

By virtue of (2.27), the expansion obtained is formally subject to the rules applying to the algebra of scalar quantities. Using (2.29) it is easy to form the power of matrix (2.28). Consider the special case when $\alpha_2 = \alpha_3 = \dots = \alpha_{n-1} = 0$, i. e.

$$A_\Delta = \alpha_0 E + \alpha_1 H^1 \tag{2.30}$$

Taking into account the fact that $H^i = 0$ when $i \geqslant n$, and using the normal binomial formula, we obtain for the mth power of the matrix (2.30):

when $m < n$

$$A_\Delta^m = (\alpha_0 E + \alpha_1 H^1)^m = \alpha_0^m E + m\alpha_0^{m-1}\alpha_1 H^1$$
$$\dots + \frac{m!}{(m-k)!\,k!}\alpha_0^{m-k}\alpha_1^k H^k + \dots + \alpha_1^m H^m \tag{2.31}$$

and when $m \geqslant n$

$$A_{\Delta}^{m} = (\alpha_0 E + \alpha_1 H^1)^m = \alpha_0^m E + m\alpha_0^{m-1}\alpha_1 H^1$$
$$\dots + \frac{m!}{(m-k)!\,k!}\alpha_0^{m-k}\alpha_1^k H^k$$
$$\dots + \frac{m!}{(m-n+1)!\,(n-1)!}\alpha_0^{m-n+1}\alpha_1^{n-1}H^{n-1} \tag{2.32}$$

The expansion (2.29) permits us to systematise the process of forming the inverse matrix A_{Δ}^{-1} [5]. Formally applying the binomial expansion to (2.30), we obtain

$$A_{\Delta}^{-1} = (\alpha_0 E + \alpha_1 H^1)^{-1}$$

$$= \alpha_0^{-1}E - \alpha_0^{-2}\alpha_1 H^1 + \alpha_0^{-3}\alpha_1^2 H^2 - \dots + (-1)^{n-1}\alpha_0^{-n}\alpha_1^{n-1}H^{n-1} \tag{2.33}$$

Multiplying the right- and left-hand sides of this expansion by (2.30), we obtain a unit matrix. This proves that (2.33) does indeed define the inverse matrix A_{Δ}^{-1}.

2.1.7 Characteristic matrix

Let A be a square matrix of order n with elements a_{ij}. Then the matrix

$$\lambda E - A = \begin{Vmatrix} \lambda - a_{11} & -a_{12} & \dots & -a_{1n} \\ -a_{21} & \lambda - a_{22} & \dots & -a_{2n} \\ \dots & \dots & \dots & \dots \\ -a_{n1} & -a_{n2} & \dots & \lambda - a_{nn} \end{Vmatrix} \tag{2.34}$$

is called the characteristic matrix and its determinant $\det(\lambda E - A)$ is an nth degree polynomial in λ; it is the characteristic polynomial of matrix A. The roots $\lambda_1, \lambda_2, \dots, \lambda_n$ of the characteristic polynomial are called the characteristic values or 'eigenvalues' of matrix A

Two square matrices A and B of the same order are said to be similar if they are connected by the relation $A = CBC^{-1}$. Similar matrices have identical characteristic equations. This follows from the sequence of equalities

$$\det(\lambda E - A) = \det(\lambda E - CBC^{-1})$$

$$= \det C(\lambda E - B)C^{-1} = \det(\lambda E - B) = 0$$

It follows that similar matrices have identical characteristic values $\lambda_1, \lambda_2, \dots, \lambda_n$.

2.1.8 Composite matrices

By appropriate partitioning, a matrix can be divided into blocks. As an example, let us take a matrix of the fifth order

$$A = \left\| \begin{array}{ccc|cc} a_{11} & a_{12} & a_{13} & a_{14} & a_{15} \\ a_{21} & a_{22} & a_{23} & a_{24} & a_{25} \\ a_{31} & a_{32} & a_{33} & a_{34} & a_{35} \\ \hline a_{41} & a_{42} & a_{43} & a_{44} & a_{45} \\ a_{51} & a_{52} & a_{53} & a_{54} & a_{55} \end{array} \right\| = \left\| \begin{array}{cc} A_{11} & A_{12} \\ A_{21} & A_{22} \end{array} \right\| \tag{2.35}$$

The sub-matrices A_{ij} will be called blocks, and the matrix represented by the block will be called the composite matrix. Composite matrices may be added and multiplied if they satisfy the conditions of dimension stipulated above (pp. 24–25). These conditions must be satisfied not only by the composite matrix containing the block elements but also by the corresponding sub-matrix blocks. Operations are then carried out in cascade, first on the blocks of the composite matrices and then on the elements of the blocks.

The composite matrix

$$J = \left\| \begin{array}{cccc} J_1 & & & \\ & J_2 & & \\ & & \ddots & \\ & & & J_s \end{array} \right\| \tag{2.36}$$

all of whose block elements not lying on the main diagonal are equal to zero, is called a quasidiagonal matrix. It is not difficult to see that two quasidiagonal matrices J and J' of suitable dimensions can be multiplied together in accordance with the rule

$$JJ' = \left\| \begin{array}{cccc} J_1 J'_1 & & & \\ & J_2 J'_2 & & \\ & & \ddots & \\ & & & J_s J'_s \end{array} \right\| \tag{2.37}$$

Matrices J and J' are commutative, i.e. $JJ' = J'J$ if all the blocks J_i and J'_i are commutative matrices when taken in pairs. In particular, for square blocks J_i, we have

$$J^m = \begin{Vmatrix} J_1^m & & & & \\ & J_2^m & & & \\ & & \cdot & & \\ & & & \cdot & \\ & & & & \cdot \\ & & & & J_s^m \end{Vmatrix} \tag{2.38}$$

If the blocks J_l are non-singular square matrices, then the inverse matrix $J^{-'}$ has the form

$$J^{-'} = \begin{Vmatrix} J_1^{-'} & & & & \\ & J_s^{-'} & & & \\ & & \cdot & & \\ & & & \cdot & \\ & & & & \cdot \\ & & & & J_s^{-'} \end{Vmatrix} \tag{2.39}$$

It is not difficult to see that (2.37)-(2.39) are natural generalisations of the corresponding operations on diagonal matrices.

2.2 CANONICAL REPRESENTATION OF MATRICES

Consider a square matrix B of order n with elements b_{ij}. One can always find a non-singular matrix

$$K = \begin{Vmatrix} k_{11} & k_{12} & \dots & k_{1n} \\ k_{21} & k_{22} & \dots & k_{2n} \\ \cdot & \cdot & \cdot & \cdot \\ k_{n1} & k_{n2} & \dots & k_{nn} \end{Vmatrix} \quad (\det K \neq 0) \tag{2.40}$$

and its inverse

$$K^{-1} = \begin{Vmatrix} \varkappa_{11} & \varkappa_{12} & \dots & \varkappa_{1n} \\ \varkappa_{21} & \varkappa_{22} & \dots & \varkappa_{2n} \\ \cdot & \cdot & \cdot & \cdot \\ \varkappa_{n1} & \varkappa_{n2} & \dots & \varkappa_{nn} \end{Vmatrix} \tag{2.41}$$

with the help of which matrix B can be represented in the form

$$B = KJK^{-1} \tag{2.42}$$

where J is, in general, a quasidiagonal matrix of the type (2.36) which depends on the properties and type of the eigenvalues of matrix B. A matrix J is called a canonical form of matrix B and K is called the transformation matrix. (There exists a non-denumerable set of quasidiagonal matrices J. We shall use the so-called Jordan canonical form, as that indicated below.)

It is sometimes convenient to use the canonical form of a

matrix to form the power of a matrix B and its inverse B^{-1}. Raising the left- and right-hand sides of (2.42) to the m th power and using the property of associativity, we obtain

$$B^m = KJ^mK^{-1} \tag{2.43}$$

When the eigenvalues $\lambda_1, \lambda_2, \ldots, \lambda_n$ of matrix B are distinct, the canonical form is a diagonal matrix of the eigenvalues (for the case of distinct eigenvalues the canonical matrix will be denoted by Λ rather than J).

$$\Lambda = \left\| \begin{matrix} \lambda_1 & 0 & \ldots & 0 \\ 0 & \lambda_2 & \ldots & 0 \\ \cdot & \cdot & \cdot & \cdot \\ 0 & 0 & \ldots & \lambda_n \end{matrix} \right\| \tag{2.44}$$

Therefore,

$$\Lambda^m = \left\| \begin{matrix} \lambda_1^m & 0 & \ldots & 0 \\ 0 & \lambda_2^m & \ldots & 0 \\ \cdot & \cdot & \cdot & \cdot \\ 0 & 0 & \ldots & \lambda_n^m \end{matrix} \right\| \tag{2.45}$$

Making use of (2.24) and noting the forms of (2.43) and (2.45), we can represent B^m in the form

$$B^m = \sum_{i=1}^{n} \lambda_i^m k_i \varkappa_i \tag{2.46}$$

where k_i and $\varkappa_i$ are the i th column and row of matrices K and K^{-1}, respectively – (see (2.22)). If the matrix B is non-singular, its canonical matrix J will also be non-singular. In that case the inverse matrix B^{-1} will be obtained if, in (2.43) and (2.46), we put $m = -1$. Thus, from (2.46), we obtain

$$B^{-1} = \sum_{i=1}^{n} \lambda_i^{-1} k_i \varkappa_i \tag{2.47}$$

If any of the eigenvalues are repeated, the expansion of B^m and B^{-1} assumes a more complex form. In this case it is convenient to take for the canonical matrix J the quasidiagonal matrix (2.36) with Jordan cells (blocks) J_σ. The number and form of the Jordan cells depend on the number and degree of the so-called elementary divisors corresponding to the repeated eigenvalues. Let λ_σ be a characteristic value of matrix B, with multiplicity q_σ. We will count the characteristic value λ_σ as many times as there are different elementary divisory $(\lambda - \lambda_\sigma)^{e_\sigma}$ corresponding to it.

The number e_σ is called the power of the elementary divisor. The Jordan cell J corresponding to the elementary divisor $(\lambda - \lambda_\sigma)^{e_\sigma}$ may be found in the form

$$J_\sigma = \begin{Vmatrix} \lambda_\sigma & 0 & \dots & 0 & 0 \\ 1 & \lambda_\sigma & \dots & 0 & 0 \\ \cdot & \cdot & \cdot & \cdot & \cdot \\ 0 & 0 & \dots & 1 & \lambda_\sigma \end{Vmatrix} \tag{2.48}$$

The Jordan cell is a left triangular square matrix of order e_σ; it can be represented as a unit first oblique row in the form

$$J_\sigma = \lambda_\sigma E_{e_\sigma} + H^1_{e_\sigma} \tag{2.49}$$

The lower index e_σ here indicates the order of the corresponding matrix. If $e_\sigma = q_\sigma$, one elementary divisor and one cell – whose order in this case is the same as the multiplicity q_σ – corresponds to the characteristic value λ_σ. In particular, for the case when the characteristic values $\lambda_1, \lambda_2, \dots, \lambda_{n-2}$ are simple, and $\lambda_{n-1} = \lambda_n$ is a binary root with a single elementary divisor of the second degree, the canonical matrix J will assume the form

$$J = \begin{Vmatrix} \lambda_1 & \dots & 0 & \dots & 0 & 0 \\ \cdot & \cdot & \cdot & \cdot & \cdot & \cdot \\ 0 & \dots & \lambda_{n-2} & \dots & 0 & 0 \\ 0 & \dots & 0 & \dots & \lambda_{n-1} & 0 \\ 0 & \dots & 0 & \dots & 1 & \lambda_{n-1} \end{Vmatrix} \tag{2.49a}$$

The degree of the Jordan cell J_σ^m for $m < e_\sigma$ and $m \geqslant e_\sigma$ can be found from (2.31) and (2.32) if, in accordance with (2.30) and (2.49), one inserts $\alpha_0 = \lambda_\sigma$ and $n = e_\sigma$.

Let us write J_σ^m in the form of a left triangular matrix of order e_σ with equal elements $b_\rho^{(\sigma)}$ in oblique rows:

$$J_\sigma^m = \begin{Vmatrix} b_0 & 0 & \dots & 0 & 0 \\ b_1 & b_0 & \dots & 0 & 0 \\ \cdot & \cdot & \cdot & \cdot & \cdot \\ b_\rho & b_{\rho-1} & \dots & 0 & 0 \\ \cdot & \cdot & \cdot & \cdot & \cdot \\ b_{e_\sigma-1} & b_{e_\sigma-2} & \dots & b_1 & b_0 \end{Vmatrix} \tag{2.50}$$

where

$$\left.\begin{aligned} & b_\rho^{(\sigma)} = \frac{m!\,\lambda_\sigma^{m-\rho}}{(m-\rho)!\,\rho!}, \\ & \rho = 0, 1, 2, \dots, m \qquad \text{for} \quad m < e_\sigma \\ & \rho = 0, 1, 2, \dots, e_\sigma - 1 \qquad \text{for} \quad m \geqslant e_\sigma \end{aligned}\right\} \tag{2.51}$$

Here, for simplicity, the upper index σ of the coefficients $b_\rho^{(\sigma)}$ has been omitted from (2.50). In accordance with (2.38), J^m is a quasi-diagonal matrix with blocks J_σ^m, given by (2.50) and (2.51). Let us expand $B^m = KJ^mK^{-1}$ (2.43) with respect to the elements of matrix J^m For this purpose we denote the sum of the orders of the Jordan cells preceding J_σ through

$$e_1 + e_2 + \dots + e_{\sigma-1} = s^{(\sigma)} \tag{2.52}$$

Then the numbers of the ith row and jth column of the matrix J^m at whose intersection lie the elements of the ρth oblique row of the Jordan cell J_σ^m are given by the formula $i - j = \rho$, where j assumes successive values from $s^{(\sigma)} + 1$ to $s^{(\sigma)} + e_\sigma - \rho$. Taking this and also (2.23) into account, we shall represent B^m in the form

$$\begin{aligned} B^m = \sum_\sigma{}'' b_0^{(\sigma)} &(k_{s+1}\varkappa_{s+1} + k_{s+2}\varkappa_{s+2} + \dots + k_{s+e_\sigma}\varkappa_{s+e_\sigma}) \\ &+ b_1^\sigma (k_{s+2}\varkappa_{s+1} + k_{s+3}\varkappa_{s+2} + \dots + k_{s+e_\sigma}\varkappa_{s+e_\sigma-1}) \\ &\dots + b_{e_\sigma-1}^{(\sigma)} k_{s+e_\sigma}\varkappa_1 \end{aligned} \tag{2.53}$$

where, for simplicity, s has been written instead of $s^{(\sigma)}$. In this Equation, $\sum''$ means that the sum is taken over roots having different elementary divisors. The coefficients $b_\rho^{(\sigma)}$ are found from (2.51), and s is found from (2.52). In particular, for the case when $\lambda_1, \lambda_2, \dots, \lambda_{n-2}$ are simple roots and $\lambda_{n-1} = \lambda_n$ is a binary root with a single elementary divisor of second degree, we have

$$e_\sigma = 1 \quad \text{for} \quad \sigma = 1, 2, \dots, n-2$$
$$e_\sigma = 2 \quad \text{for} \quad \sigma = n-1$$

and

$$s = \sigma - 1 \quad \text{for} \quad \sigma \leqslant n-1$$

Therefore,

$$\begin{aligned} B^m &= \sum_{i=1}^{n-2} \lambda_i^m k_i\varkappa_i + \lambda_{n-1}^m (k_{n-1}\varkappa_{n-1} + k_n\varkappa_n) \\ &\quad + m\lambda_{n-1}^{m-1} k_n\varkappa_{n-1} \end{aligned} \tag{2.53a}$$

If the matrix B is non-singular, the inverse matrix B^{-1} is obtained from (2.51) and (2.53) by setting $m = -1$. We will therefore have

$$\begin{aligned} B^{-1} = \sum_\sigma{}'' \lambda_\sigma^{-1} &(k_{s+1}\varkappa_{s+1} + k_{s+2}\varkappa_{s+2} + \dots + k_{s+e_\sigma}\varkappa_{s+e_\sigma}) \\ &- \lambda_\sigma^{-2} (k_{s+2}\varkappa_{s+1} + k_{s+3}\varkappa_{s+2} + \dots + k_{s+e_\sigma}\varkappa_{s+e_\sigma-1}) \\ &\dots + (-1)^{e_\sigma-1} \lambda_\sigma^{-e_\sigma} k_{s+e_\sigma}\varkappa_{s+1} \end{aligned} \tag{2.54}$$

We must emphasise once more that among the roots λ_σ there may also be some equal roots having different elementary divisors.

2.3 EXPANSION OF INVERSE AND CHARACTERISTIC MATRICES INTO ELEMENTARY FRACTIONS

Consider the characteristic matrix $f(\lambda) = \lambda E - B$ for a square matrix of order n. We will denote the matrix adjoint to $F(\lambda)$ by $f(\lambda)$. Then, in accordance with (2.14), we have

$$f(\lambda)^{-1} = (\lambda E - B)^{-1} = \frac{F(\lambda)}{\Delta(\lambda)} \tag{2.55}$$

where $\Delta(\lambda) = \det(\lambda E - B)$ is a characteristic polynomial of the nth degree. The elements F_{ji} of the adjoint matrix $F(\lambda)$ are polynomials of degree not higher than $(n-1)$. This follows from the fact that the F_{ji} are the algebraic components of the elements of matrix $f(\lambda)$. Consequently, the expressions $F_{ji}(\lambda)/\Delta(\lambda)$ are proper fractions; they can be expanded in terms of elementary fractions with the help of the usual formulae and represented in matrix form. If the roots $\lambda_1, \lambda_2, \ldots, \lambda_n$ of the characteristic polynomial $\Delta(\lambda)$ are simple, the expansion can be represented in the form

$$\frac{F(\lambda)}{\Delta(\lambda)} = \sum_{i=1}^{n} \frac{F(\lambda_i)}{\Delta'(\lambda_i)} \cdot \frac{1}{\lambda - \lambda_i} \tag{2.56}$$

where

$$\Delta'(\lambda_i) = \left[\frac{d\Delta(\lambda)}{d\lambda}\right]_{\lambda=\lambda_i}$$

If λ_σ is a root of multiplicity q_σ, this expansion can be represented in the form

$$\frac{F(\lambda)}{\Delta(\lambda)} = \sum_{\sigma}{}' \sum_{\rho=1}^{q_\sigma} \frac{1}{(q_\sigma - \rho)!} \left[\frac{F(\lambda_\sigma)}{\Delta\sigma(\lambda_\sigma)}\right]^{(q_\sigma - \rho)} \cdot \frac{1}{(\lambda - \lambda_\sigma)^\rho} \tag{2.57}$$

where

$$\left.\begin{aligned} \Delta_\sigma(\lambda) &= \frac{\Delta(\lambda)}{(\lambda - \lambda_\sigma)^{q_\sigma}} \\ \left[\frac{F(\lambda_\sigma)}{\Delta_\sigma(\lambda_\sigma)}\right]^{(\rho)} &= \left[\frac{d^\rho}{d\lambda^\rho} \frac{F(\lambda)}{\Delta_\sigma(\lambda)}\right]_{\lambda=\lambda_\sigma} \end{aligned}\right\} \tag{2.58}$$

and the prime indicates that the sum should be taken over all the different roots λ_σ. The derivative of a matrix of any order is the matrix formed from the corresponding derivatives of its elements. Formally, (2.56) and (2.57) are identical to their scalar analogues.

We shall now apply the procedure for constructing the inverse matrix $f(\lambda)^{-1}$ discussed in Section 2.2. Assuming that

$$KEK^{-1} = KK^{-1} = E$$

we write

$$f(\lambda) = \lambda E - B = \lambda E - KJK^{-1} = K[\lambda E - J]K^{-1} \tag{2.59}$$

Whence, according to (2.15), we have

$$f(\lambda)^{-1} = (\lambda E - B)^{-1} = K[\lambda E - J]^{-1}K^{-1} \tag{2.60}$$

The characteristic matrix $J = \lambda E - J$ has the same structure as the canonical matrix J. When all the eigenvalues λ_i are simple, it is a diagonal matrix with elements $\lambda - \lambda_i$ and, by analogy with (2.42)-(2.47), one can write

$$f(\lambda)^{-1} = (\lambda E - B)^{-1} = \sum_{i=1}^{n} k_i \varkappa_i \frac{1}{\lambda - \lambda_i} \tag{2.61}$$

If the λ_σ are characteristic values with elementary divisors of degree e_σ, then by analogy with (2.49), we can write

$$J'_\sigma = (\lambda - \lambda_\sigma)E_{e_\sigma} - H^1_{e_\sigma} \tag{2.62}$$

where J'_σ is the corresponding Jordan cell in the quasidiagonal matrix $J' = \lambda E - J$. It is then easy to see that $(\lambda E - B)^{-1}$ can be obtained from the right-hand side of (2.54) if we replace λ_σ by $\lambda - \lambda_\sigma$ and instead of the minus sign we put a plus sign everywhere. We will write out the formula for the special case when, corresponding to λ_σ, there is a single elementary divisor whose degree is equal to the multiplicity of the root, $e_\sigma = q_\sigma$. We obtain

$$\begin{aligned} f(\lambda)^{-1} &= (\lambda E - B)^{-1} \\ &= \sum_\sigma{}' (k_{s+1}\varkappa_{s+1} + k_{s+2}\varkappa_{s+2} + \dots + k_{s+q_\sigma}\varkappa_{s+q_\sigma}) \frac{1}{\lambda - \lambda_\sigma} \\ &+ (k_{s+2}\varkappa_{s+1} + k_{s+3}\varkappa_{s+2} + \dots + k_{s+q_\sigma}\varkappa_{s+q_\sigma-1}) \frac{1}{(\lambda - \lambda_\sigma)^2} \\ &\dots + k_{s+q_\sigma}\varkappa_{s+1} \frac{1}{(\lambda - \lambda_\sigma)^{q_\sigma}} \end{aligned} \tag{2.63}$$

In this formula we again write s instead of $s^{(\sigma)}$. In this case $s^{(\sigma)}$ is equal to the sum of the multiplicities of all the preceding eigenvalues – see (2.52).

Comparing (2.56) with (2.61), and (2.57) with (2.63), and with account of (2.55), we obtain some matrix identities which are very important in practice. For simple eigenvalues λ_i, we have

$$k_i\varkappa_i = \frac{F(\lambda_i)}{\Delta'(\lambda_i)} \qquad i = 1, 2, \ldots, n \tag{2.64}$$

For the case when λ_σ is a characteristic value of multiplicity q_σ with a single elementary divisor of degree $e_\sigma = q_\sigma$, we obtain

$$\left.\begin{aligned}
&k_{s+1}\varkappa_{s+1} + k_{s+2}\varkappa_{s+2} + \ldots + k_{s+q_\sigma}\varkappa_{s+q_\sigma} \\
&\qquad = \frac{1}{(q_\sigma - 1)!}\left[\frac{F(\lambda_\sigma)}{\Delta_\sigma(\lambda_\sigma)}\right]^{(q_\sigma-1)} \\
&k_{s+2}\varkappa_{s+1} + k_{s+3}\varkappa_{s+2} + \ldots + k_{s+q_\sigma}\varkappa_{s+q_\sigma-1} \\
&\qquad = \frac{1}{(q_\sigma - 2)!}\left[\frac{F(\lambda_\sigma)}{\Delta_\sigma(\lambda_\sigma)}\right]^{(q_\sigma-2)} \\
&\ldots\ldots\ldots\ldots\ldots\ldots\ldots\ldots \\
&k_{s+\rho}\varkappa_{s+1} + k_{s+\rho+1}\varkappa_{s+2} + \ldots + k_{s+q_\sigma}\varkappa_{s+q_\sigma-\rho+1} \\
&\qquad = \frac{1}{(q_\sigma - \rho)!}\left[\frac{F(\lambda_\sigma)}{\Delta_\sigma(\lambda_\sigma)}\right]^{(q_\sigma-\rho)} \\
&\ldots\ldots\ldots\ldots\ldots\ldots\ldots\ldots \\
&k_{s+q_\sigma}\varkappa_{s+1} = \frac{F(\lambda_\sigma)}{\Delta_\sigma(\lambda_\sigma)}
\end{aligned}\right\} \tag{2.65}$$

In particular, for a binary root $\lambda_{n-1} = \lambda_n$ with a single elementary divisor of the second degree, we have $s = n - 2, \sigma = n - 1$, $q_\sigma = 2$ and $e_\sigma = 2$. Then, according to (2.65), we will have

$$\left.\begin{aligned}
k_{n-1}\varkappa_{n-1} + k_n\varkappa_n &= \left[\frac{F(\lambda_{n-1})}{\Delta_{n-1}(\lambda_{n-1})}\right]^{(1)} \\
k_n\varkappa_{n-1} &= \frac{F(\lambda_{n-1})}{\Delta_{n-1}(\lambda_{n-1})}
\end{aligned}\right\} \tag{2.66}$$

The description of the left-hand sides of (2.65) undergoes a change if several elementary divisors correspond to λ_σ. The general formulae are cumbersome and difficult to interpret. We shall not, therefore, write them out but confine ourselves to making a remark about their basic character.

Suppose several elementary divisors correspond to the root λ_σ of multiplicity q_σ. Let e_σ denote the power of the biggest divisor, $e_\sigma < q_\sigma$. Then, from an analysis of the structure of (2.54), it is easy to show that those terms $(\lambda - \lambda_\sigma)^{-\rho}$ for which $\rho > e_\sigma$ drop out of (2.63). From this it immediately follows that

$$\frac{1}{(q_\sigma - \rho)!}\left[\frac{F(\lambda_\sigma)}{\Delta_\sigma(\lambda_\sigma)}\right]^{(q_\sigma-\rho)} = 0, \quad \rho = e_\sigma + 1, \quad e_\sigma + 2, \ldots, q_\sigma \tag{2.67}$$

Using the matrix identities (2.64) and (2.65), and (2.51), we represent (2.46) and (2.53) in the form

$$B^m = \sum_{i=1}^{n} \lambda_i^m \frac{F(\lambda_i)}{\Delta'(\lambda_i)} \tag{2.68}$$

$$B^m = \sum_{\sigma}{}' \sum_{\rho=0}^{q_\sigma - 1} \lambda_\sigma^{m-\rho} \frac{m!}{(m-\rho)!\,\rho!} \, \frac{1}{(q_\sigma - \rho - 1)!} \left[\frac{F(\lambda_\sigma)}{\Delta_\sigma(\lambda_\sigma)}\right]^{(q_\sigma - \rho - 1)} \tag{2.69}$$

which are valid for simple λ_i and multiple λ_σ characteristic values of the matrix B, respectively. (This way of writing it does not depend on the number and power of the elementary divisor corresponding to the repeated root.) The last equality may be rewritten in the form

$$B^m = \sum_{\sigma}{}' \sum_{\rho=0}^{q_\sigma - 1} \frac{\left(\lambda_\sigma^m\right)^{(\rho)}}{\rho!} \cdot \frac{1}{(q_\sigma - \rho - 1)!} \left[\frac{F(\lambda_\sigma)}{\Delta_\sigma(\lambda_\sigma)}\right]^{(q_\sigma - \rho - 1)} \tag{2.70}$$

where

$$\left(\lambda_\sigma^m\right)^{(\rho)} = \left.\frac{d^\rho(\lambda^m)}{d\lambda^\rho}\right|_{\lambda=\lambda_\sigma} = \frac{m!\,\lambda_\sigma^{m-\rho}}{(m-\rho)!}$$

This notation is convenient in that it is easily generalised to the case of any polynomial of matrix B with matrix or scalar coefficients [5].

In many cases it is convenient to operate with canonical forms of the matrices. For the inverse transition it is necessary to know K and its inverse K^{-1}. The matrix identities permit one to make such a transition, since they express definite combinations of columns k_i and rows $\varkappa_j$ of the matrices K and K^{-1} in terms of the adjoint matrix $F(\lambda)$ and, in general, its first few derivatives. The adjoint matrix is constructed from the elements of the original matrix in accordance with the rules indicated above. Various methods exist for systematising and simplifying the process of constructing an adjoint matrix (see, for example, [6]).

Finally, in all the operations with canonical matrices which we shall be carrying out, we can assume that a single elementary divisor corresponds to λ_σ since, after the conversion to the adjoint matrix $F(\lambda)$, the calculation of the change of structure of the corresponding expansions is carried out, as it were, automatically. This is a consequence of (2.67).

In conclusion, let us look at some of the properties of the adjoint matrix $F(\lambda)$. We note that $\varkappa_i$ and k_j are the rows and columns of the mutually inverse matrices K^{-1} and K Then, in accordance with (2.15) and (2.25), we will have in all cases

$$\varkappa_i k_j = \begin{cases} 1 & \text{for} \quad i = j \\ 0 & \text{for} \quad i \neq j \end{cases} \tag{2.71}$$

and

$$KK^{-1}=\sum_{i=1}^{n} k_i\varkappa_i=E \tag{2.72}$$

From (2.64) and the formulae just given, it is easy to show that the equalities

$$\left.\begin{aligned} \frac{F(\lambda_i)}{\Delta'(\lambda_i)}\frac{F(\lambda_j)}{\Delta'(\lambda_j)}&=0 \qquad (i\neq j)\\ \left(\frac{F(\lambda_i)}{\Delta'(\lambda_i)}\right)^m&=\frac{F(\lambda_i)}{\Delta'(\lambda_i)}\\ \sum_{i=1}^{n}\frac{F(\lambda_i)}{\Delta'(\lambda_i)}&=E \end{aligned}\right\} \tag{2.73}$$

are true for simple eigenvalues λ_i.

In the case of multiple roots λ_σ, it is possible to deduce similar properties for the matrix $F(\lambda)/\Delta_\sigma(\lambda)$ and for its derivatives, which we can formulate in the following way: the coefficients of the expansion (2.57) into elementary fractions standing in front of $(\lambda-\lambda_\sigma)^{-1}$ satisfy (2.73) provided that, in the third equality, the sum is taken over the various roots λ_σ. The product, taken in arbitrary order, of any two coefficients of (2.57) corresponding to the roots λ_σ and λ_τ is equal to zero. The product in arbitrary order of the two coefficients standing in front of $(\lambda-\lambda_\sigma)^{-\rho}$ and $(\lambda-\lambda_\sigma)^{-\rho'}$ is equal to the coefficient standing in front of $(\lambda-\lambda_\sigma)^{-(\rho+\rho'-1)}$ or equal to zero if this power is absent from the expansion. The last possibility always occurs when $\rho+\rho'-1>q_\sigma$ or, by virtue of (2.67), when $\rho+\rho'-1>e_\sigma$, where e_σ is the highest power of the elementary divisor corresponding to the root λ_σ.

2.4 SOLUTION OF LINEAR FINITE-DIFFERENCE EQUATIONS WITH CONSTANT COEFFICIENTS

Consider a system of finite-difference equations of the form

$$\left.\begin{aligned} x_1(m)&=p^*_{11}x_1(m-1)+p^*_{12}x_2(m-1)\\ &\quad\dots+p^*_{1n}x_n(m-1)+g^*_1(m-1)\\ x_2(m)&=p^*_{21}x_1(m-1)+p^*_{22}x_2(m-1)\\ &\quad\dots+p^*_{2n}x_n(m-1)+g^*_2(m-1)\\ &\dots\dots\dots\dots\dots\dots\dots\\ x_n(m)&=p^*_{n1}x_1(m-1)+p^*_{n2}x_2(m-1)\\ &\quad\dots+p^*_{nn}x_n(m-1)+g^*_n(m-1) \end{aligned}\right\} \tag{2.74}$$

where the p^*_{ij} are constant real numbers and $x_i(m)$ and $g^*_i(m)$ are real functions of a variable integer m with values that run over all the natural numbers 0, 1, 2, ... etc.

The integral variable m will be called a discrete argument, and (2.74) will be called difference equations with a discrete argument. (Starting in Chapter 4, the role of the discrete argument will be played by equally spaced instants of time whose interval is T. The discrete argument mT will then run over the values 0, T, $2T$ ... etc. In Chapters 2 and 3 we will be using an abbreviated notation for the discrete argument, i.e. we will assume $T=1$). The functions $g^*_i(m)$ are considered to be known and capable of being specified by various methods.

In particular, they can be obtained from continuous functions for integral values of the independent variable. In that case the continuous functions are called generating functions. According to the form of the generating functions, one can have the discrete exponential e^{am}, the discrete sine $\sin \omega m$ and cosine $\cos \omega m$ etc. System (2.74) will be called the normal form for finite-difference equations.

We will now rewrite (2.74) in matrix form

$$\begin{Vmatrix} x_1(m) \\ x_2(m) \\ \vdots \\ x_n(m) \end{Vmatrix} = \begin{Vmatrix} p^*_{12} & \cdots & p^*_{1n} \\ p^*_{22} & \cdots & p^*_{2n} \\ \vdots & \cdots & \vdots \\ p^*_{n2} & \cdots & p^*_{nn} \end{Vmatrix} \begin{Vmatrix} x_1(m-1) \\ x_2(m-1) \\ \vdots \\ x_n(m-1) \end{Vmatrix} + \begin{Vmatrix} g^*_1(m-1) \\ g^*_2(m-1) \\ \vdots \\ g^*_n(m-1) \end{Vmatrix} \tag{2.75}$$

or in the more compact form

$$x(m) = P^*x(m-1) + g^*(m-1), \tag{2.76}$$

where P^* is a square matrix of the elements p^*_{ij} of order n, and $x(m)$ and $g^*(m)$ are column matrices of the elements $x_i(m)$ and $g^*_i(m)$ $(i=1, 2, \ldots, n)$, respectively. The matrix equation (2.76) is a recurrence relation which permits one to obtain $x(m)$ successively for any value of the discrete argument m with an arbitrarily specified initial value $x(0)$.

Thus, we have

$$\left.\begin{aligned} x(1) &= P^*x(0) + g^*(0) \\ x(2) &= P^{*2}x(0) + P^*g^*(0) + g^*(1) \\ &\cdots\cdots\cdots\cdots\cdots\cdots \\ x(m) &= P^{*m}x(0) + P^{*m-1}g^*(0) \\ &\quad + P^{*m-2}g^*(1) + \ldots + g^*(m-1) \end{aligned}\right\} \tag{2.77}$$

and the general solution of (2.76) which satisfies the initial condition

$x(0)$ may be written in the form

$$x(m)=P^{*m}x(0)+\sum_{m'=0}^{m-1}P^{*m-m'-1}g^*(m') \tag{2.78}$$

For this equality to be true when $m=0$, we will assume that discrete functions for a negative discrete argument are equal to zero. Equality (2.78) can be written:

$$x(m)=P^{*m}x(0)+\sum_{m'=0}^{m-1}P^{*m'}g^*(m-m'-1) \tag{2.79}$$

The solution (2.78) may be transformed by making use of (2.68) and (2.70) where, instead of the matrix B, its characteristic polynomial $\Delta(\lambda)$, eigenvalues λ_i and the adjoint matrix $F(\lambda)$ for $(\lambda E-B)$, we will write, respectively, P^*, $\Delta^*(\lambda)$, λ_i^* and $F^*(\lambda)$.

Let us substitute into (2.78) or (2.79) the quantities P^{*m} and $P^{*m-m'-1}$ found from (2.68) and (2.70), and let us reverse the order of summation. Then, for simple eigenvalues λ_i^* of the matrix P^*, we will have

$$x(m)=\sum_{i=1}^{n}\frac{F^*(\lambda_i^*)}{\Delta^{*\prime}(\lambda_i^*)}\left\{\lambda_i^{*m}x(0)+\sum_{m'=0}^{m-1}\lambda_i^{*m-m'-1}g^*(m')\right\} \tag{2.80}$$

or

$$x(m)=\sum_{i=1}^{n}\frac{F^*(\lambda_i^*)}{\Delta^{*\prime}(\lambda_i^*)}\left\{\lambda_i^{*m}x(0)+\sum_{m'=0}^{m-1}\lambda_i^{*m'}g^*(m-m'-1)\right\} \tag{2.81}$$

For multiple eigenvalues λ_σ, we obtain in a similar way

$$x(m)=\sum_{\sigma}{}'\sum_{\rho=0}^{q_\sigma-1}\frac{1}{(q_\sigma-\rho-1)!}\left[\frac{F^*(\lambda_\sigma^*)}{\Delta_\sigma^*(\lambda_\sigma^*)}\right]^{(q_\sigma-\rho-1)}\left\{\frac{\left(\lambda_\sigma^{*m}\right)^{(\rho)}}{\rho!}x(0)+\sum_{m'=0}^{m-1}\frac{\left(\lambda_\sigma^{*m-m'-1}\right)^{(\rho)}}{\rho!}g^*(m')\right\} \tag{2.82}$$

or

$$x(m)=\sum_{\sigma}{}'\sum_{\rho=0}^{q_\sigma-1}\frac{1}{(q_\sigma-\rho-1)!}\left[\frac{F^*(\lambda_\sigma^*)}{\Delta_\sigma^*(\lambda_\sigma^*)}\right]^{(q_\sigma-\rho-1)}\left\{\frac{\left(\lambda_\sigma^{*m}\right)^{(\rho)}}{\rho!}x(0)+\sum_{m'=0}^{m-1}\frac{\left(\lambda_\sigma^{*m'}\right)^{(\rho)}}{\rho!}g^*(m-m'-1)\right\} \tag{2.83}$$

We recall that – see (2.68) –

$$\frac{\left(\lambda_\sigma^{*m}\right)^{(\rho)}}{\rho!} = \frac{m(m-1)\ldots(m-\rho+1)}{\rho!}\lambda_\sigma^{*m-\rho} \tag{2.84}$$

(when $\rho=0$ the right-hand side of (2.84) is equal to λ^{*m}).

The choice of the form of the solution of the difference equation depends on the nature of the problem and on the technical feasibility of doing the calculations. If we put $g^*(m)\equiv 0$ in (2.80)-(2.83), we obtain a general homogeneous system. From the structure of the solution of a homogeneous system it is easy to establish the following: regardless of how $x(0)$ is represented, the matrix variable $x(m)$ tends to zero as the time m is increased without limit if and only if the moduli of all the eigenvalues $\lambda_1^*, \lambda_2^*, \ldots, \lambda_n^*$ of the matrix P^* are less than unity. Under these same conditions, one of the variables $x_1(m), x_2(m), \ldots, x_n(m)$ in column $x(m)$ will certainly tend to infinity if the modulus of at least one characteristic value is greater than unity.

We have discussed finite-difference equations in their normal form. We shall encounter such equations in the analysis of relay and discontinuous control systems. However, in the majority of cases, one can reduce an arbitrary linear system to the normal form. The procedure for carrying out this reduction and the conditions under which it is possible are similar to those used in the theory of differential equations and which we applied in Section 1.4.

2.5 SOLUTION OF LINEAR DIFFERENTIAL EQUATIONS WITH CONSTANT COEFFICIENTS

2.5.1 Matrix exponential

A system of linear differential equations with constant coefficients given in the normal form can be solved particularly easily with the help of the matrix exponential. The matrix exponential e^{Pt} is defined as the sum of the power series

$$e^{Pt} = E + P\frac{t}{1!} + P^2\frac{t^2}{2!} + \ldots + P^m\frac{t^m}{m!} + \ldots \tag{2.85}$$

where P is an arbitrary square matrix of order n, and t is a scalar parameter. It is not difficult to show that the scalar power series (2.85) corresponding to the matrix elements e^{Pt} converge absolutely and uniformly for all values of t, i. e. they have an infinite radius of convergence. If, in place of P, we substitute the scalar quantity λ (which may be real or complex) into (2.85) we obtain a power series for the exponential $e^{\lambda t}$.

The principal properties of the scalar exponential $e^{\lambda t}$ hold also for the matrix exponential e^{Pt}. Thus, for example, if two matrices P' and P'' commute, i. e. if $P'P''=P''P'$, then

$$e^{P't}e^{P''t} = e^{P''t}e^{P't} = e^{(P'+P'')t} \tag{2.86}$$

Whence, in particular, we find

$$e^{Pt}e^{-Pt} = e^{-Pt}e^{Pt} = E \tag{2.87}$$

where the matrices e^{Pt} and e^{-Pt} are mutually inverse to each other. These properties stem from the fact that the arithmetical operations carried out on commuting matrices and scalar quantities formally obey the same rules.

Consider the canonical form of the matrix exponential. Let $P = KJK^{-1}$. Then, with account of (2.43) and the distributive law of multiplication, we can put (2.85) in the form

$$e^{Pt} = K\left(E + J\frac{t}{1!} + J^2\frac{t^2}{2!} + \ldots + J^m\frac{t^m}{m!} + \ldots\right)K^{-1} \tag{2.88}$$

But according to the definition of the matrix exponential

$$e^{Jt} = E + J\frac{t}{1!} + J^2\frac{t^2}{2!} + \ldots + J^m\frac{t^m}{m!} + \ldots \tag{2.89}$$

and consequently

$$e^{Pt} = Ke^{Jt}K^{-1} \tag{2.90}$$

Let us introduce the special notation

$$M(t) = e^{Jt} \tag{2.91}$$

The matrix $M(t)$ can be regarded as the canonical form of the exponential e^{Pt}. Thus, the matrices P and e^{Pt} can be reduced to the canonical form by one and the same transforming matrix. The two matrices Jt_1 and Jt_2 therefore commute so that, in accordance with the general formulae (2.86) and (2.87), we have

$$M(t_1)\,M(t_2) = M(t_2)\,M(t_1) = M(t_1 + t_2) \tag{2.92}$$

and

$$M(t)\,M(-t) = M(-t)\,M(t) = E \tag{2.93}$$

The form of the matrix $M(t)$ depends on the type of root λ_i of the characteristic equation $\Delta(\lambda) = \det(\lambda E - P) = 0$, or, using the other terminology, of the eigenvalues of matrix P. For simple roots J is the diagonal matrix Λ (2.44). However, from (2.89) one can immediately show that $M(t)$ is a diagonal matrix of form

$$M(t) = \left\| \begin{array}{cccc} e^{\lambda_1 t} & 0 & \ldots & 0 \\ 0 & e^{\lambda_2 t} & \ldots & 0 \\ \cdot & \cdot & \cdot & \cdot \\ 0 & 0 & \ldots & e^{\lambda_n t} \end{array} \right\| \tag{2.94}$$

For the case of repeated roots λ_σ, the matrix J is the quasidiagonal matrix (2.36) with Jordan cells (blocks) J_σ. In a similar way we find that $M(t)$ is a quasidiagonal matrix

$$M(t) = \begin{Vmatrix} e^{J_1 t} & & & & \\ & e^{J_2 t} & & & \\ & & \cdot & & \\ & & & \cdot & \\ & & & & \cdot \\ & & & & & e^{J_s t} \end{Vmatrix} \tag{2.95}$$

whose blocks are the matrix exponentials $e^{J_\sigma t}$ corresponding to Jordan cells J_σ of order e_σ, and the index s indicates the number of roots λ_σ having different elementary divisors $(\lambda - \lambda_\sigma)^{e_\sigma}$. Since for $m < e_\sigma$ and $m \geqslant e_\sigma$ the m th power of the Jordan cell J_σ^m is given by (2.50) and (2.51), one can find the sum of series (2.89) for $J = J_\sigma$ by direct calculation in the form of a left triangular matrix of order e_σ with equal elements $d_\rho^{(\sigma)}$ in the oblique rows

$$e^{J_\sigma t} = \begin{Vmatrix} d_0 & 0 & \dots & 0 \\ d_1 & d_0 & \dots & 0 \\ \cdot & \cdot & \cdot & \cdot \\ d_{e_\sigma - 1} & d_{e_\sigma - 2} & \dots & d_0 \end{Vmatrix} \tag{2.96}$$

where

$$d_\rho^{(\sigma)} = \frac{t^\rho e^{\lambda_\sigma t}}{\rho!} \qquad \rho = 0, 1, 2, \dots, e_\sigma - 1 \tag{2.97}$$

and in the matrix (2.96) the notation has been simplified by omitting the upper index σ from $d_\rho^{(\sigma)}$. In particular, if the roots $\lambda_1, \lambda_2, \dots, \lambda_{n-2}$ are simple and $\lambda_{n-1} = \lambda_n = 0$ with a single elementary divisor, then from (2.94)-(2.97) it follows that

$$M(t) = \begin{Vmatrix} e^{\lambda_1 t} & 0 & \dots & 0 & 0 & 0 \\ 0 & e^{\lambda_2 t} & \dots & 0 & 0 & 0 \\ \cdot & \cdot & \cdot & \cdot & \cdot & \cdot \\ 0 & 0 & \dots & e^{\lambda_{n-2} t} & 0 & 0 \\ 0 & 0 & \dots & 0 & 1 & 0 \\ 0 & 0 & \dots & 0 & t & 1 \end{Vmatrix} \tag{2.98}$$

It is appropriate here to note that the eigenvalues of the matrix e^{Pt} are equal to $e^{\lambda_i t}$, where λ_i are the eigenvalues of the matrix P regardless of whether they are simple or multiple.

Let us consider the rules for differentiating and integrating matrix exponentials. Power series may be differentiated and integrated term by term within their circles of convergence. This property holds for the series (2.85). Differentiating it term by term, we obtain

$$\frac{d}{dt} e^{Pt} = P + P^2 \frac{t}{1!} + \ldots + P^m \frac{t^{m-1}}{(m-1)!} + \ldots \tag{2.99}$$

whence we immediately find that

$$\frac{d}{dt} e^{Pt} = Pe^{Pt} = e^{Pt}P \tag{2.100}$$

or, by virtue of (2.42), (2.90) and (2.91), we obtain

$$\frac{d}{dt} e^{Pt} = KJM(t)K^{-1} = KM(t)JK^{-1} \tag{2.101}$$

In particular

$$\frac{d}{dt} M(t) = \dot{M}(t) = JM(t) = M(t)J \tag{2.102}$$

[Here, and in what follows (unless the contrary is specifically stated), a dot above a symbol indicates differentiation with respect to t.] Further, integrating series (2.85) term by term, we obtain

$$\int_0^t e^{Pt}\,dt = Et + P\frac{t^2}{2!} + \ldots + P^m \frac{t^{m+1}}{(m+1)!} + \ldots \tag{2.103}$$

Assuming that the matrix P is non-singular, we put (2.103) in the form

$$\int_0^t e^{Pt}\,dt = P^{-1}\left\{-E + E + P\frac{t}{1!} + P^2\frac{t^2}{2!} + \ldots + P^m\frac{t^m}{m!} + \ldots\right\} \tag{2.104}$$

or, what amounts to the same thing, in the form

$$\int_0^t e^{Pt}\,dt = \left\{-E + E + P\frac{t}{1!} + P^2\frac{t^2}{2!} + \ldots + P^m\frac{t^m}{m!} + \ldots\right\}P^{-1} \tag{2.105}$$

whence we obtain

$$\int_0^t e^{Pt}\,dt = P^{-1}\left(e^{Pt} - E\right) = \left(e^{Pt} - E\right)P^{-1} \tag{2.106}$$

In this case we write

$$\int_0^t e^{Pt}\,dt = \frac{e^{Pt} - E}{P} \tag{2.107}$$

[Here the division is subject to the rule of division of scalar numbers in the sense that the result does not depend on the order in which the multiplication of the numerator by P^{-1} is carried out (from the right or from the left).] If we make use of (2.42), (2.90), (2.91) and (2.43) with $m=-1$, (2.107) can be put in the form

$$\int_0^t e^{Pt} = K \frac{M(t)-E}{J} K^{-1} \tag{2.108}$$

from which, in particular, we have

$$\int_0^t M(t)\,dt = \frac{M(t)-E}{J} \tag{2.109}$$

Let us introduce the special notation

$$N(t) = \frac{M(t)-E}{J} = J^{-1}\{M(t)-E\} = \{M(t-E)\}\,J^{-1} \tag{2.110}$$

In the general case, when the roots λ_σ are repeated, the matrix $N(t)$ is quasidiagonal, having the form

$$N(t) = \begin{Vmatrix} \frac{e^{J_1 t}-Ee_1}{J_1} & & & \\ & \frac{e^{J_2 t}-Ee_2}{J_2} & & \\ & & \ddots & \\ & & & \frac{e^{J_s t}-Ee_s}{J_s} \end{Vmatrix} \tag{2.111}$$

where, as before, J_σ is a Jordan cell of order e_σ and E_{e_σ} is a unit matrix of the same order. Let us determine the blocks of matrix (2.111) for index σ. In accordance with (2.33) and (2.49) we will have

$$J_\sigma^{-1} = \lambda_\sigma^{-1}E - \lambda_\sigma^{-2}H^1 + \ldots + (-1)^\rho \lambda_\sigma^{-\rho-1}H^\rho \\ \ldots + (-1)^{e_\sigma-1}\lambda_\sigma^{-e_\sigma}H^{e_\sigma-1} \tag{2.112}$$

With account of (2.96), we represent the matrix $e^{J_\sigma t}-E_{e_\sigma}$ in terms of the unit oblique series H^ρ in the form

$$e^{J_\sigma t} - E_{e_\sigma} = (d_0^{(\sigma)}-1)E + d_1^{(\sigma)}H^1 \ldots + d_\rho^{(\sigma)}H^\delta + \ldots + d_{e_\sigma-1}^{(\sigma)}H^{e_\sigma-1} \tag{2.113}$$

The lower index e_σ has been omitted in the right-hand sides of (2.112) and (2.113) for E and H^ρ. Multiplying out both sides of (2.112) and (2.113) using (2.27), we obtain

$$J_\sigma^{-1}\{e^{J_\sigma t}-E_{e_\sigma}\} = c_0^{(\sigma)}E + c_1^{(\sigma)}H^1 \\ \ldots + c_\rho^{(\sigma)}H^\rho + \ldots + c_{e_\sigma-1}^{(\sigma)}H^{e_\sigma-1} \tag{2.114}$$

where

$$c_\rho^{(\sigma)} = \lambda_\sigma^{-1} d_\rho^{(\sigma)} - \lambda_\sigma^{-2} d_{\rho-1}^{(\sigma)} + \ldots + (-1)^\rho \lambda_\sigma^{-\rho-1} \left(d_0^{(\sigma)} - 1\right) \qquad (2.115)$$

By virtue of (2.97), we can put $c_\rho^{(\sigma)}$ in the form

$$c_\rho^{(\sigma)} = \frac{t^\rho}{\rho!} \frac{e^{\lambda_\sigma t}}{\lambda_\sigma} - \frac{t^{\rho-1}}{(\rho-1)!} \cdot \frac{e^{\lambda_\sigma t}}{\lambda_\sigma^2} + \ldots + (-1)^\rho \frac{e^{\lambda_\sigma t} - 1}{\lambda_\sigma^{\rho+1}} \qquad (2.116)$$

$$(\rho = 0,\ 1,\ 2,\ \ldots,\ e_0 - 1)$$

Equations (2.111) and (2.114) were derived on the assumption that the matrix P is non-singular or, what amounts to the same, on the assumptions that all the roots λ_σ of its characteristic equation are non-zero. However, the integral of e^{Pt} exists in any case and it can be evaluated in accordance with the formulae we have found also in the case when one or more of the roots λ_σ are equal to zero by a limiting process. For this purpose we rewrite (2.116) in the form

$$c_\rho^{(\sigma)} = (-1)^\rho \frac{e^{\lambda_\sigma t}}{\lambda_\sigma^{\rho+1}} \left[1 - \frac{\lambda_\sigma t}{1!} + \ldots + (-1)^\rho \frac{(\lambda_\sigma \cdot t)^\rho}{\rho!}\right] - (-1)^\rho \frac{1}{\lambda_\sigma^{\rho+1}} \qquad (2.117)$$

In this equality the polynomial with respect to $\lambda_\sigma t$ represents the first ρ terms of the series expansion of the exponential (2.117). Taking this into account, we rewrite (2.117) in the form

$$c_\rho^{(\sigma)} = (-1)^\rho \frac{e^{\lambda_\sigma t}}{\lambda_\sigma^{\rho+1}} \left[e^{-\lambda_\sigma t} - (-1)^{\rho+1} \frac{(\lambda_\sigma t)^{\rho+1}}{(\rho+1)!} - (-1)^{\rho+2} \frac{(\lambda_\sigma t)^{\rho+2}}{(\rho+2)!} - \ldots\right] - (-1)^\rho \frac{1}{\lambda_\sigma^{\rho+1}}$$

or, after reduction,

$$c_\rho^{(\sigma)} = e^{\lambda_\sigma t}\left[\frac{t^{\rho+1}}{(\rho+1)!} - \frac{\lambda_\sigma t^{\rho+2}}{(\rho+2)!} + \ldots\right]$$

In the limiting case when $\lambda_\sigma = 0$, we have

$$c_\rho^{(\sigma)} = \frac{t^{\rho+1}}{(\rho+1)!}, \qquad \rho = 0,\ 1,\ 2,\ \ldots,\ e_\sigma - 1 \qquad (2.118)$$

Thus the matrix (2.114), which is a block element of the quasi-diagonal matrix $N(t)$ (2.111), is a left triangular matrix

$$\frac{e^{J_\sigma t} - E_{e_\sigma}}{J_\sigma} = \begin{Vmatrix} c_0 & 0 & \ldots & 0 \\ c_1 & c_0 & \ldots & 0 \\ \cdot & \cdot & \cdot & \cdot \\ c_{e_\sigma - 1} & c_{e_\sigma - 2} & \ldots & c_0 \end{Vmatrix} \qquad (2.119)$$

with equal elements $c_\rho^{(\sigma)}$ in the oblique rows. These elements are found from (2.116) for $\lambda_\sigma \neq 0$ and from (2.118) for $\lambda_\sigma = 0$. (Here, for simplicity of presentation, the upper index in the coefficients $c_\rho^{(\sigma)}$ has been omitted.)

Let us now find the form of the matrix $N(t)$ (2.111) in certain special cases. In the case when all the roots λ_i are simple and non-zero, matrix $N(t)$ can be represented in the form

$$N(t) = \begin{Vmatrix} \frac{e^{\lambda_1 t} - 1}{\lambda_1} & 0 & \dots & 0 \\ 0 & \frac{e^{\lambda_2 t} - 1}{\lambda_2} & \dots & 0 \\ \dots & \dots & \dots & \dots \\ 0 & 0 & \dots & \frac{e^{\lambda_n t} - 1}{\lambda_n} \end{Vmatrix} \tag{2.120}$$

Whence, for $\lambda_n = 0$ we obtain

$$N(t) = \begin{Vmatrix} \frac{e^{\lambda_1 t} - 1}{\lambda_1} & \dots & 0 & 0 \\ \dots & \dots & \dots & \dots \\ 0 & \dots & \frac{e^{\lambda_{n-1} t} - 1}{\lambda_{n-1}} & 0 \\ 0 & \dots & 0 & t \end{Vmatrix} \tag{2.121}$$

If the roots $\lambda_1, \lambda_2, \dots, \lambda_{n-2}$ are simple, and the root $\lambda_{n-1} = \lambda_n = 0$, with the corresponding elementary divisor of second degree $(e_{n-1} = q_{n-1} = 2)$, then

$$N(t) = \begin{Vmatrix} \frac{e^{\lambda_1 t} - 1}{\lambda_1} & \dots & 0 & 0 & 0 \\ \dots & \dots & \dots & \dots & \dots \\ 0 & \dots & \frac{e^{\lambda_{n-2} t} - 1}{\lambda_{n-2}} & 0 & 0 \\ 0 & \dots & 0 & t & 0 \\ 0 & \dots & 0 & \frac{t^2}{2!} & t \end{Vmatrix} \tag{2.122}$$

With account of (2.110), we can finally put (2.108) in the form

$$\int_0^t e^{Pt}\, dt = KN(t)K^{-1} \tag{2.123}$$

where the matrix $N(t)$ is determined by (2.111) and (2.119). This formula is convenient in that it is valid also when $\det P = 0$. In that case one should determine only the coefficients of matrix (2.119) in accordance with (2.118) and not (2.116), which is valid for $\det P \neq 0$.

Consider the inhomogeneous linear system of differential equations in the normal form

$$\left.\begin{aligned} \dot{x}_1 &= p_{11}x_1 + p_{12}x_2 + \dots + p_{1n}x_n + g_1(t) \\ \dot{x}_2 &= p_{21}x_1 + p_{22}x_2 + \dots + p_{2n}x_n + g_2(t) \\ &\dots\dots\dots\dots\dots\dots \\ \dot{x}_n &= p_{n1}x_1 + p_{n2}x_2 + \dots + p_{nn}x_n + g_n(t) \end{aligned}\right\} \tag{2.124}$$

where the p_{ij} are constants and $g_i(t)$ are known functions of time which may admit of discontinuities of the first kind. The system (2.125) may be represented in matrix form

$$\left\|\begin{matrix} \dot{x}_1 \\ \dot{x}_2 \\ \cdot \\ \cdot \\ \cdot \\ \dot{x}_n \end{matrix}\right\| = \left\|\begin{matrix} p_{11} & p_{12} & \dots & p_{1n} \\ p_{21} & p_{22} & \dots & p_{2n} \\ \dots & \dots & \dots & \dots \\ \dots & \dots & \dots & \dots \\ \dots & \dots & \dots & \dots \\ p_{n1} & p_{n2} & \dots & p_{nn} \end{matrix}\right\| \left\|\begin{matrix} x_1 \\ x_2 \\ \cdot \\ \cdot \\ \cdot \\ x_n \end{matrix}\right\| + \left\|\begin{matrix} g_1 t \\ g_2 t \\ \cdot \\ \cdot \\ \cdot \\ g_n t \end{matrix}\right\| \tag{2.125}$$

or, more compactly, in the form

$$\dot{x} = Px + g(t) \tag{2.126}$$

where P is a square matrix of the elements p_{ij}, x is a column matrix of the variables x_i, and $g(t)$ is a column matrix of the known functions $g_i(t)$. For $t = t_0$, let the matrix coordinate have the value $x(t_0)$. Then the solution of (2.126) satisfying this initial condition can be represented in the form

$$x(t) = e^{P(t-t_0)}x(t_0) + \int_{t_0}^{t} e^{P(t-\xi')}g(\xi')\,d\xi' \tag{2.127}$$

as can easily be seen by a simple substitution of (2.127) into (2.126). Outwardly, (2.127) has the same form as it would have if all the elements of (2.126) had been scalar quantities. As in the case of scalar equations, the first term of (2.127) determines the solution of the corresponding homogeneous equation which, for $t = t_0$, satisfies the initial condition $x(t_0)$, and the second term is a particular solution of the inhomogeneous equation satisfying the zero initial condition.

Having put $t - t_0 = \tau$ and $\xi' = \xi + t_0$, we convert (2.127) to the form

$$x(t_0 + \tau) = e^{P\tau}x(t_0) + \int_{0}^{\tau} e^{P(\tau-\xi)}g(t_0 + \xi)\,d\xi \tag{2.127'}$$

which, using the notation introduced in (2.90) and (2.91), can be rewritten in the form

$$x(t_0+\tau)=KM(\tau)K^{-1}x(t_0)+\int_0^\tau KM(\tau-\xi)K^{-1}g(t_0+\xi)\,d\xi \qquad (2.128)$$

Putting $g(t)=h$ with elements $h_i=$ constant, and bearing in mind (2.109), (2.110) and rule (2.92) for multiplying out the matrices $M(t_1)$ and $M(t_2)$, we can obtain from (2.128) the expression

$$x(t_0+\tau)=KM(\tau)K^{-1}x(t_0)+KN(\tau)K^{-1}h \qquad (2.129)$$

which we shall frequently have occasion to use in the theory of pulse systems. The right-hand side of (2.129) can be expanded with respect to the elements of matrices $M(t)$ and $N(t)$ by making use of (2.24). For the case when the roots λ_i of the characteristic equation $\Delta(\lambda)=\det(\lambda E-P)=0$, matrices $M(t)$ and $N(t)$ are defined by (2.94) and (2.120), and the expansion will have the form

$$x(t_0+\tau)=\sum_{i=1}^{n} e^{\lambda_i\tau}k_i\varkappa_i x(t_0)+\sum_{i=1}^{n}\frac{e^{\lambda_i\tau}-1}{\lambda_i}k_i\varkappa_i h \qquad (2.130)$$

or

$$x(t_0+\tau)=\sum_{i=1}^{n} e^{\lambda_i\tau}\frac{F(\lambda_i)}{\Delta'(\lambda_i)}x(t_0)+\sum_{i=1}^{n}\frac{e^{\lambda_i\tau}-1}{\lambda_i}\frac{F(\lambda_i)}{\Delta'(\lambda_i)}h \qquad (2.131)$$

if use is made of the matrix in (2.64). In the analysis of discontinuous control systems one comes across the case when the roots $\lambda_1, \lambda_2, \ldots, \lambda_{n-2}$ are simple and $\lambda_{n-1}=\lambda_n=0$. In that event, the matrices $M(t)$ and $N(t)$ are determined by (2.98) and (2.122). According to (2.24), we obtain the expansion

$$\begin{aligned}x(t_0+\tau)=&\sum_{i=1}^{n-2} e^{\lambda_i\tau}k_i\varkappa_i x(t_0)+(k_{n-1}\varkappa_{n-1}+k_n\varkappa_n)x(t_0)\\&+\tau k_n\varkappa_{n-1}x(t_0)+\sum_{i=1}^{n-2}\frac{e^{\lambda_i\tau}-1}{\lambda_i}k_i\varkappa_i h\\&+\tau(k_{n-1}\varkappa_{n-1}+k_n\varkappa_n)h+\frac{\tau^2}{2!}k_n\varkappa_{n-1}h\end{aligned} \qquad (2.132)$$

[(2.132) corresponds to the case when one elementary divisor of the second degree corresponds to the root $\lambda_{n-1}=\lambda_n=0$. If two elementary divisors of the first degree correspond to the root, then there is no term in (2.122) containing t^2.] With the help of (2.66) with $\lambda_{n-1}=0$, (2.132) can be represented in the form

$$x(t_0+\tau)=\sum_{i=1}^{n-2} e^{\lambda_i\tau}\frac{F(\lambda_i)}{\Delta'(\lambda_i)}x(t_0)+\left[\frac{F(0)}{\Delta_{n-1}(0)}\right]^{(1)}x(t_0)$$

$$+\tau\left[\frac{F(0)}{\Delta_{n-1}(0)}\right]x(t_0)+\sum_{i=1}^{n-2}\frac{e^{\lambda_i\tau}-1}{\Delta'(\lambda_i)}\frac{F(\lambda_i)}{\Delta'(\lambda_i)}h$$

$$+\tau\left[\frac{F(0)}{\Delta_{n-1}(0)}\right]^{(1)}h+\frac{\tau^2}{2!}\left[\frac{F(0)}{\Delta_{n-1}(0)}\right]h \tag{2.133}$$

Expansion (2.133) is valid for all cases since, according to (2.67), the term containing t^2 drops out automatically if $e_{n-1}=1$, i.e. if two elementary divisors of degree unity correspond to the double zero root. We shall call (2.129) [and expansions of type (2.130) and (2.132) which correspond to it] the classical form of the solution of the system of differential equations (2.125)-(2.126), and expansions of type (2.131) and (2.133) we shall call the spectral form of the solution. In the given case the spectral solution was obtained from the classical form with the help of the matrix identities (2.64) and (2.65). The spectral solution can be obtained independently by applying operational calculus methods based on the Laplace transformation.

2.6 QUADRATIC FORMS

Consider the real quadratic form

$$U=\sum_{i,j=1}^{n}a_{ij}x_ix_j \qquad (a_{ij}=a_{ji}) \tag{2.134}$$

of the real variables $x_1, x_2, \ldots, x_n$. In matrix notation we can represent it in the form

$$U=\|x_1x_2\ldots x_n\|\begin{Vmatrix}a_{11} & a_{12} & \ldots & a_{1n}\\ a_{21} & a_{22} & \ldots & a_{2n}\\ \cdot & \cdot & \cdot & \cdot\\ \cdot & \cdot & \cdot & \cdot\\ \cdot & \cdot & \cdot & \cdot\\ a_{n1} & a_{n2} & \ldots & a_{nn}\end{Vmatrix}\begin{Vmatrix}x_1\\ x_2\\ \cdot\\ \cdot\\ \cdot\\ x_n\end{Vmatrix} \tag{2.135}$$

or, more compactly,

$$U=x^TAx \tag{2.136}$$

where x is a column matrix of the variable x_i and A is a symmetric square matrix of the elements a_{ij}, i.e. $A=A^T$. The determinant of matrix A is called the discriminant of the form. The characteristic equation det $[\lambda E-A]=0$ is called the secular equation. The roots $\lambda_1, \lambda_2, \ldots, \lambda_n$ of a secular equation are always real.

There is always an orthogonal transforming matrix O $(O^T=O^{-1})$ which will reduce a symmetric matrix A to the canonical form

$$O^{-1}AO = \Lambda = \begin{Vmatrix} \lambda_1 & 0 & \dots & 0 \\ 0 & \lambda_2 & \dots & 0 \\ \cdot & \cdot & \cdot & \cdot \\ 0 & 0 & \dots & \lambda^n \end{Vmatrix} \tag{2.137}$$

and the form of the canonical matrix (2.137) is independent of the multiplicity of the roots λ_i of the secular equation. If one makes use of the transforming matrix O in order to effect the transition to the new matrix variable, i.e. to the column y, with the help of the linear transformation

$$x = Oy, \quad y = O^{-1}x = O^T x \tag{2.138}$$

then the quadratic form U (2.136) is converted to the form

$$U = y^T O^T A O y = y^T \Lambda y = \sum_{i=1}^{n} \lambda_i y_i^2 \tag{2.139}$$

i.e. it is represented in the form of a sum of the squares of the new independent variables $y_1, y_2, \dots, y_n$ with coefficients $\lambda_1 \lambda_2, \dots, \lambda_n$ which are equal to the roots of the secular equation.

The quadratic form U is said to be positive definite if it vanishes only when $x_1 = x_2 = \dots = x_n = 0$ or, in the matrix notation, when $x = 0$, and for all the other values of the variables $x_1, x_2, \dots, x_n$, or for the matrix variable x, it is positive. From (2.138), it follows that x and y vanish simultaneously. From this, and from (2.139), we can show that the quadratic form U is positive definite if and only if all the roots $\lambda_1, \lambda_2, \dots, \lambda_n$ of its secular equation are positive. In particular, the discriminant of the positive definite form

$$\det A = \lambda_1 \lambda_2 \dots \lambda_n \tag{2.140}$$

is greater than zero. From this one can conclude that all the principal minors of the discriminant of a positive definite form U are positive.

In fact these principal minors are, in their turn, discriminants of the positive definite special forms which are obtained from U by equating the side of the variables to zero.

2.6.1 Silvester's theorem

For a quadratic form U to be positive definite it is necessary and sufficient that the principal minors of its discriminant be positive, i.e. that the following inequalities should hold

$$D_1 = a_{11} > 0$$

$$D_2=\begin{vmatrix} a_{11} & a_{12} \\ a_{21} & a_{22} \end{vmatrix}>0, \ \ldots, \ D_n=\begin{vmatrix} a_{11} & a_{12} & \ldots & a_{1n} \\ a_{21} & a_{22} & \ldots & a_{2n} \\ \cdot & \cdot & \cdot & \cdot \\ a_{n1} & a_{n2} & \ldots & a_{nn} \end{vmatrix}>0 \tag{2.141}$$

The necessity of these conditions has already been demonstrated; their sufficiency can be deduced from the following considerations. Consider the right triangular matrix

$$\tilde{A}=\begin{Vmatrix} \tilde{a}_{11} & \tilde{a}_{12} & \ldots & \tilde{a}_{1n} \\ 0 & \tilde{a}_{22} & \ldots & \tilde{a}_{2n} \\ \cdot & \cdot & \cdot & \cdot \\ 0 & 0 & \ldots & \tilde{a}_{nn} \end{Vmatrix} \tag{2.142}$$

with elements

$$\tilde{a}_{ij}=\frac{\tilde{A}_{ij}}{\tilde{A}_{jj}} \qquad i\leqslant j \tag{2.143}$$

where the $\tilde{A}_{ij}$ are the cofactors (minors with the sign included) for the elements $a_{ij}=a_{ji}$ of the determinants D_j featuring in the inequalities (2.141). It is clear that $\tilde{A}_{jj}=D_{j-1}$. The determinant of the triangular matrix (2.142) is equal to unity since all its diagonal elements $\tilde{a}_{ii}$ are equal to unity. Using the rule for determinants (2.12), we find by direct calculation that

$$\tilde{A}^T A \tilde{A}=\begin{Vmatrix} D_1 & 0 & 0 & \ldots & 0 \\ 0 & \frac{D_2}{D_1} & 0 & \ldots & 0 \\ 0 & 0 & \frac{D_3}{D_1} & \ldots & 0 \\ \cdot & \cdot & \cdot & \cdot & \cdot \\ 0 & 0 & 0 & \ldots & \frac{D_n}{D_{n-1}} \end{Vmatrix} \tag{2.144}$$

Consider the linear transformation

$$x=\tilde{A}z \tag{2.145}$$

or in expanded form

$$\left.\begin{aligned} x_1 &= z_1+\tilde{a}_{12}z_2+\tilde{a}_{13}z_3+\ldots+\tilde{a}_{1n}z_n \\ x_2 &= \qquad z_2+\tilde{a}_{23}z_3+\ldots+\tilde{a}_{2n}z_n \\ x_3 &= \qquad\qquad z_3+\ldots+a_{3n}z_n \\ &\cdots\cdots\cdots\cdots\cdots \\ x_n &= \qquad\qquad\qquad\qquad z_n \end{aligned}\right\} \tag{2.146}$$

Noting (2.136), (2.144) and (2.145), we put the quadratic form U into the form of the sum of the squares of the new independent variables $z_1, z_2, \ldots, z_n$, i.e.

$$U = z^T \tilde{A}^T A \tilde{A} z = D_1 z_1^2 + \frac{D_2}{D_1} z_2^2 + \ldots + \frac{D_n}{D_{n-1}} z_n^2 \qquad (2.147)$$

This form has an independent and very important meaning – and incidentally, it determines the sufficiency of (2.141).

Along with (2.134)-(2.136), consider the second real quadratic form

$$V = \sum_{i\,j=1}^{n} b_{ij} x_i x_j \qquad (b_{ij} = b_{ji}) \qquad (2.148)$$

or, in matrix notation,

$$V = x^T B x \qquad (B = B^T) \qquad (2.149)$$

The two quadratic forms U and V define the form

$$\rho U - V = x^T (\rho A - B) x \qquad (2.150)$$

where ρ is a parameter. Form (2.150) is said to be regular if form U is positive definite. For a regular form there is always a real matrix L which simultaneously reduces matrices A and B to the form

$$L^T A L = \begin{Vmatrix} 1 & 0 & \ldots & 0 \\ 0 & 1 & \ldots & 0 \\ \cdot & \cdot & \cdot & \cdot \\ 0 & 0 & \ldots & 1 \end{Vmatrix} = E \qquad (2.151)$$

and

$$L^T B L = \begin{Vmatrix} \rho_1 & 0 & \ldots & 0 \\ 0 & \rho_2 & \ldots & 0 \\ \cdot & \cdot & \cdot & \cdot \\ 0 & 0 & \ldots & \rho_n \end{Vmatrix} = R \qquad (2.152)$$

where $\rho_1, \rho_2, \ldots, \rho_n$ are roots of the characteristic equation

$$\det [\rho A - B] = 0 \qquad (2.153)$$

of the form (2.150). The characteristic equation (2.153) of a regular form always has n real roots. The linear transformation of the variables

$$x = Ly \qquad (2.154)$$

simultaneously reduces the quadratic forms U and V to the form

$$U = \sum_{i=1}^{n} y_i^2 \qquad (2.155)$$

$$Y = \sum_{i=1}^{n} \rho_i y_i^2 \qquad (2.156)$$

From (2.155) and (2.156) it follows that a quadratic form V will be positive definite if, and only if, all the roots $\rho_1, \rho_2, \ldots, \rho_n$ of the characteristic equation of the regular form are positive. If $U = \sum_{i=1}^{n} x_i^2$ then $A = E$ [and the characteristic equation (2.153) of the regular form] is transformed to the secular equation for the form V.

Chapter 3

STABILITY OF A MOTION DEFINED BY DIFFERENCE EQUATIONS

3.1 GENERAL THEOREMS ON THE DIRECT METHOD OF LYAPUNOV

In the study of discontinuous and relay control systems the problem arises of determining the stability about the equilibrium point or stationary state of a motion when the states of the system at discrete, equally spaced instants of time are known. This problem is closely connected with the properties of the solutions of systems of difference equations, and so it is desirable to study such systems in their general form. The problem was discussed by the author in 1953 from the point of view of Lyapunov functions (see [30]). In 1919, Cotton [33] applied Lyapunov's theory of characteristic values to the analysis of the solutions of nonlinear difference equations.

Consider a system of nonlinear difference equations specified in the normal form

$$x_i(m) = X_i[x_1(m-1),\ x_2(m-1),\ \ldots,\ x_n(m-1)] \qquad (3.1)$$
$$i = 1,\ 2,\ \ldots,\ n$$

which establish a connection between the variables $x_1, x_2, \ldots, x_n$ for two successive, equally spaced values of the independent variable m. Without destroying the generality of our argument, we can assume that m assumes only integral values 0, 1, 2 ... . As before, we shall designate the independent variable m the discrete time.

The right-hand sides of (3.1) are single-valued, continuous functions of the variables $x_1, x_2, \ldots, x_n$. Under these conditions, any system of values of the variables $x_1(0), x_2(0), \ldots, x_n(0)$ which can be accepted as the initial conditions when $m=0$ determines a unique solution $x_1(m)$, $x_2(m)$, ..., $x_n(m)$ of (3.1) which, for $m=0$, satisfies these initial conditions.

Let us assume that there exists a region

$$\sum_{i=1}^{n} x_i^2 \leqslant b > 0 \tag{3.2}$$

of variation of the variables $x_1, x_2, \ldots, x_n$ inside which the right-hand sides of (3.1) are continuous and do not vanish simultaneously, apart from those values of the variables

$$x_1 = x_2 = \ldots = x_n = 0 \tag{3.3}$$

for which

$$X_i(0, 0, \ldots, 0) = 0, \qquad i = 1, 2, \ldots, n \tag{3.4}$$

Geometrically, the variables $x_1, x_2, \ldots, x_n$ can be regarded as coordinates of an n-dimensional state space and the solution of the system of difference equations (3.1) can be regarded as representing the motion of an image point. The system of equations (3.1) can then be called a discrete dynamical system. (This designation is in accord with the terminology accepted in the qualitative theory of differential equations.) A discrete dynamical system brings about a point-wise transformation in the state space. The point $x_1 = x_2 = \ldots = x_n = 0$ is a fixed, or invariant, point of this transformation.

The equalities (3.3) determine the trivial solution of (3.1). In accordance with Lyapunov's terminology – see [16, 28 and 17] – $x_1 = x_2 = \ldots = x_n = 0$ will be called the equilibrium point of the system, (3.1) will be called the equations of the perturbed motion, and their solutions will be called the perturbed motions of the system.

The right-hand sides of (3.1) do not depend explicitly on the discrete time m so that their solution determines the stability of the equilibrium point. However, as will be demonstrated below, the problem of the stability of the periodic modes of oscillation of relay-operated control systems reduces to an analysis of the solutions of difference equations of this sort.

For discrete systems we can formulate the definition of Lyapunov stability.

If, for every arbitrarily specified number $a > 0$ no matter how small, one can find a positive number λ such that, for all perturbed motions $x_1(m)$, $x_2(m)$, ..., $x_n(m)$ which satisfy at zero time $m=0$ the inequalities

$$\sum_{i=1}^{n} x_i^2(0) \leqslant \lambda \tag{3.5}$$

there will hold the inequalities

$$\sum_{i=1}^{n} x_i^2(m) < a \tag{3.6}$$

for any value of $m > 0$, then equilibrium point is said to be stable. If, in addition, the limiting equalities

$$\lim_{m=\infty} x_i(m) = 0, \qquad i = 1, 2, \ldots, n \tag{3.7}$$

hold, then this is termed asymptotic stability. Otherwise the motion is unstable.

A natural extension of the theorems pertaining to the direct method of Lyapunov provides a solution to the problem of the stability of discrete systems in the case we are considering. Let there exist a real function $V(x_1, x_2, \ldots, x_n)$ with respect to which we shall suppose that there exists a sufficiently small region

$$\sum_{i=1}^{n} x_i^2 \leqslant c \tag{3.8}$$

inside which the function is single-valued, continuous and vanishes when all the coordinates x_i are zero. We shall call the function V positive definite (negative definite) if, for all values of the coordinates x_i satisfying the inequality (3.8) apart from $x_1 = x_2 = \ldots = x_n = 0$, we find that $V > 0$ $(V < 0)$; the function is termed positive semi-definite (negative semi-definite) if under the above conditions, we find that $V \geqslant 0$ $(V \leqslant 0)$.

If the function V is neither definite nor semi-definite in any neighbourhood of the point $x_1 = x_2 = \ldots = x_n = 0$ which may be as small as one pleases, then the function V is said to be sign indefinite.

Let us introduce the notation

$$V[x_1(m), x_2(m), \ldots, x_n(m)] = V_m \tag{3.9}$$

and say that the first difference $V_m - V_{m-1}$ of function V is taken in accordance with (3.1) if the variables $x_1(m)$ $x_2(m), \ldots, x_n(m)$ in this function are expressed in terms of $x_1(m-1), x_2(m-1)$ $\ldots, x_n(m-1)$ through these equations.

First stability theorem

If the difference equations (3.1) of the perturbed motion are such that one can find a sign definite function V whose first difference according to these equations is a semi-definite function of sign opposite to that of V, or is identically equal to zero, then the equilibrium point is stable.

This theorem can be proved simply by repeating almost word for word the corresponding Lyapunov theorem. Let $c < b$ and let V be a positive definite function and let its first difference, taken in accordance with (3.1), satisfy the equation

$$V_m - V_{m-1} = -U_{m-1} \tag{3.10}$$

where U is a positive semi-definite function.

We obtain by summation

$$V_m - V_0 = -\sum_{i=0}^{m-1} U_i \tag{3.11}$$

and, by virtue of the fact that $U_i \geqslant 0$, we have

$$V_m \leqslant V_0 \tag{3.12}$$

Furthermore, let a be an arbitrarily small positive number (which we shall suppose is in any case less than c); let l be a lower boundary point of the function V on the sphere a

$$\sum_{i=1}^{n} x_i^2 = a \tag{3.13}$$

i.e. for all points of the sphere a the following inequality holds:

$$V(x_1, x_2, \ldots, x_n) \geqslant l \tag{3.14}$$

The number l is non-zero and positive since V is a positive definite function. On the other hand, for l it is possible to find a $\lambda > 0$ such that, for values of the variables x_i satisfying the condition

$$\sum_{i=1}^{n} x_i^2 \leqslant \lambda \tag{3.15}$$

the values of the function V will satisfy the inequality

$$V(x_1, x_2, \ldots, x_n) < l \tag{3.16}$$

We know that such a value of $\lambda > 0$ must necessarily exist by virtue of properties of the function V stipulated above. If for $m = 0$ the initial values of the coordinates $x_1(0), x_2(0), \ldots, x_n(0)$ are chosen in such a way that they satisfy (3.15), then, in accordance with (3.9), (3.12) and (3.16), for any instant of discrete time the inequality

$$V_m \leqslant V_0 < l \tag{3.17}$$

will hold.

In view of the fact that, in the process of motion, V_m does not reach the value l then, by virtue of (3.13) and (3.14), the image point never reaches the sphere a, i.e. for any instant of time m the coordinates x_i satisfy (3.6).

Second stability theorem

If the difference equations (3.1) of a perturbed motion are such that it is possible to find a positive definite function V whose first difference according to these equations is a negative definite function, then the unperturbed motion is asymptotically stable.

In this case it is obvious that the conditions of the first theorem are fulfilled, and that consequently the equilibrium point is stable. We shall show that here the limiting equalities (3.7) are valid. Let the function V be positive definite. Then in (3.10) $V_m \geqslant 0$ and $U_m \geqslant 0$ and the equality holds for $x_1 = x_2 = \ldots = x_n = 0$. From (3.10) it follows that, with a change in the time m the function V_m forms a monotonically decreasing sequence

$$V_m < V_{m-1} < \cdots < V_0 \tag{3.18}$$

which is bounded from below by zero. Consequently, this sequence tends to a limit if m increases without limit. This limit is equal to zero and, by virtue of the fact that V is a positive definite function, the coordinates $x_i(m)$ are also equal to zero in the limit.

This statement can be proved by contradiction. Let this limit be $l' > 0$; then there exists a neighbourhood bounded by the sphere a' into which an image point with coordinates $x_i(m)$ is unable to penetrate for all m. As a consequence of this definition, a positive function U_m will, for any m, be greater than a certain non-zero positive number l''. However, from (3.11), it is easy to derive the inequality

$$V_m < V_0 - ml'' \tag{3.19}$$

according to which the function V_m must certainly become negative, which is impossible.

The following two instability theorems are valid:

First instability theorem

If the difference equations (3.1) of a perturbed motion are such that it is possible to find a function V which, by virtue of these equations, possesses a first difference which is a sign definite function and is such that, by a suitable choice of the quantities $x_1, x_2, \ldots, x_n$ (which may be numerically as small as one pleases), one can make its sign the same as that of its first difference, then the equilibrium point is unstable.

Second instability theorem

If the difference equations (3.11) of a perturbed motion are such that it is possible to find a function V whose first difference can, by virtue of these equations, be reduced to the form

$$\dot{V}_m - V_{m-1} = \alpha V_{m-1} + W \tag{3.20}$$

where α is a positive constant and W is either identically equal to zero or is a semi-definite function, and if at the same time the V function found is such that, by a suitable choice of $x_1, x_2, \ldots, x_n$ (which may be numerically as small as one pleases) one can make its sign the same as that of W, then the equilibrium point is unstable.

In particular, the conditions of these instability theorems will certainly be fulfilled if the function V is semi-definite or is a definite function identical to that of the first difference (first theorem) or to that of function W (second theorem).

These theorems are easily proved if one follows the appropriate arguments given by Lyapunov introducing the simplifications and changes which were made clear in the proofs of the stability theorems given earlier, and will not be given here.

The Lyapunov stability theorems given above have a very straightforward geometrical interpretation [17]. Let V be a positive definite function. Consider in state space the non-parameter family of surfaces

$$V(x_1, x_2, \ldots, x_n) = d \tag{3.21}$$

where the parameter d is a positive number. We shall call them surfaces of equal level. Clearly, $V(0)$ degenerates into a point. There exists a sufficiently small neighbourhood (3.8) around the zero point $x_1 = x_2 = \ldots = x_n = 0$ in which the level surfaces will be closed and in which they will enclose the zero point. When d changes from zero to a sufficiently small value, we obtain a one-parameter family of closed surfaces of equal level enclosing the zero point contracting to this point when $d = 0$. In Fig 3.1 these surfaces are represented in the form of curves of equal level for the case of a two-dimensional state space, or the so-called phase plane.

Each surface (3.21) envelopes all the surfaces of the family corresponding to a lower level of the parameter d. The quantities V_m calculated for the running values $x_i(m)$ of the coordinates of the image point always coincide with definite values of d. This means that, at a given instant of time, an image point with coordinates $x_i(m)$ falls onto the surface of a given level. Thus, in a discrete motion, the image point jumps from one level surface to another and, if the first difference $V_m - V_{m-1}$ is a sign definite function opposite to that of V (that is, in this case, a negative definite function), the image point always jumps from the surface

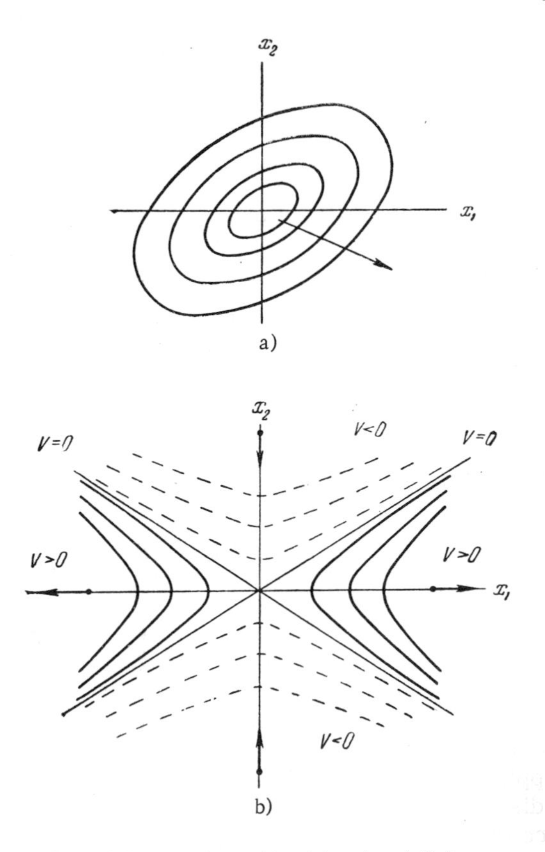

Fig 3.1 Curves of equal level for sign definite functions and indefinite functions V.

of the given level to the surface of a lower level and must therefore asymptotically approach the zero point. If the first difference $V_m - V_{m-1}$ taken in accordance with (3.1) is identically equal to zero, or is a semi-definite function opposite to that of V, then the image point may get trapped on the surface of a certain level. If this happens, it cannot indefinitely approach the limit point.

If now V is an indefinite function, the level surfaces (3.21) will not be closed. In this case the parameter d can assume positive and negative values so that, in the neighbourhood of the zero point, there will be regions $V > 0$ and $V < 0$ separated from each other by the surface $V = 0$, which passes through the point $x_1 = x_2 = \ldots = x_n = 0$. In the region $V > 0$ the surfaces (3.21) of equal level do not intersect, by virtue of the uniqueness of function V, and with an increase of the positive parameter d in the region $V < 0$ they move away from the zero point; with an increase of the negative parameter d on the other hand, the surfaces of equal level approach the zero point in the dividing surface $V = 0$. Figure 3.1b represents such a situation for state space. In any neighbourhood of the zero point, however small, there exist regions $V > 0$ and $V < 0$.

Let the first difference $V_m - V_{m-1}$ calculated in accordance with (3.1) be a positive definite function. Then, if the motion starts from a point in the region $V > 0$ which is in any neighbourhood of the origin, the image point will, during the course of its discrete motion, jump from a lower level surface to a higher level surface. Without leaving the region $V > 0$, the image point must necessarily go outside the limits of any pre-assigned region (3.6), as the equilibrium point will be unstable. We note that to prove that an equilibrium point is unstable it is sufficient to find just one phase trajectory which passes outside the limits of region (3.6).

In the above discussion we did not make full use of the fact that the first difference is a positive definite function. To prove this it is sufficient to know that, in any neighbourhood of the zero point, there is a region $V > 0$, and that the first difference $V_m - V_{m-1}$ taken in accordance with (3.1) is positive in the region.

Such a generalisation of Lyapunov's instability theorems has been given by Chetaev [28].

General theorems concerning the stability of the equilibrium point of discrete dynamical systems should be regarded as a natural extension of the corresponding Lyapunov theorems. A function V possessing the properties laid down in one of the four theorems is usually called a Lyapunov function.

The application of the general theorems to the study of the stability of discrete linear systems requires a special approach, and it is here that the methods of matrix analysis are found to be particularly useful. The special techniques of matrix analysis are such that they enable such problems to be solved in a rather simple way.

3.2 STABILITY OF A MOTION DEFINED BY CONSTANT-COEFFICIENT LINEAR DIFFERENCE EQUATIONS

Consider the system of constant-coefficient linear difference equations

$$\left.\begin{aligned} x_1(m) &= p^*_{11}x_1(m-1) + p^*_{12}x_2(m-1) \\ &\qquad \dots + p^*_{1n}x_n(m-1), \\ x_2(m) &= p^*_{21}x_1(m-1) + p^*_{22}x_2(m-1) \\ &\qquad \dots + p^*_{2n}x_n(m-1), \\ &\dots\dots\dots\dots\dots\dots \\ x_n(m) &= p^*_{n1}x_1(m-1) + p^*_{n2}x_2(m-1) \\ &\qquad \dots + p^*_{nn}x_n(m-1). \end{aligned}\right\} \quad (3.22)$$

Equalities (3.3) in this case determine the trivial solution of (3.22) or the equilibrium point. Thus, (3.22) is a particular form of the general equations (3.1) of the perturbed motion of the system. Clearly, the general theorems of the direct Lyapunov method set out in the preceeding section are applicable to (3.22), and in the given case the solution of the stability problem is easily reduced to an algorithm provided one constructs the function V as a quadratic form of the variables $x_1, x_2, \ldots, x_n$.

We will represent the system of linear difference equations (3.22) in matrix notation

$$x(m) = P^* x(m-1) \tag{3.23}$$

where P^* is a square matrix of nth order with elements p^*_{ij}, and x is a column matrix of the variables $x_1, x_2, \ldots, x_n$. Let us examine the problem of the stability of equilibrium point $x_1 = x_2 = \ldots = x_n = 0$. For this purpose we will seek a quadratic form

$$V = \sum_{i,\, j=1}^{n} b_{ij} x_i x_j \qquad (b_{ij} = b_{ji}) \tag{3.24}$$

or, in matrix notation (see Section 2.6)

$$V = x^T B x \qquad (B = B^T) \tag{3.25}$$

which, in accordance with (3.22) and (3.23), solves the difference equation

$$V_m - V_{m-1} = -U_{m-1} \tag{3.26}$$

where

$$U = \sum_{i,\, j=1}^{n} a_{ij} x_i x_j \qquad (a_{ij} = a_{ji}) \tag{3.27}$$

or, in matrix notation

$$U = x^T A x \qquad (A = A^T) \tag{3.28}$$

is any pre-assigned positive definite quadratic form.

Let us consider this problem in more detail. In (3.26) we will replace V and U by expressions (3.25) and (3.28). Equation (3.26) can then be rewritten in the form

$$x^T(m) B x(m) - x^T(m-1) B x(m-1) = -x^T(m-1) A x(m-1) \tag{3.29}$$

We now substitute in here the column $x(m)$ given by (3.23) and the row $x^T(m)$ obtained from the same formula by transposition. We then obtain the equation

$$x^T(m-1) P^{*T} B P^* x(m-1) - x^T(m-1) B x(m-1) = -x^T(m-1) A x(m-1) \tag{3.30}$$

which, after some obvious transformations, we can represent in the form

$$x^T(m-1)\left(P^{*T}BP^* - B + A\right)x(m-1) = 0 \tag{3.31}$$

The left-hand side of (3.31) determines a quadratic form since the matrix inside the brackets is symmetric. The quadratic form vanishes for arbitrary values of the matrix variable x if, and only if, all its coefficients are equal to zero. Thus, from (3.31) we obtain the matrix equation

$$P^{*T}BP^* - B + A = 0 \tag{3.32}$$

which is equivalent to $\frac{n(n+1)}{2}$ scalar algebraic equations of the form

$$\sum_{r,s=1}^{n} p^*_{ri}p^*_{sj}b_{rs} - b_{ij} + a_{ij} = 0, \quad i \leqslant j = 1, 2, \ldots, n \tag{3.33}$$

In these equations p^*_{ij} and a_{ij} are known quantities, and the coefficients $b_{ij} = b_{ji}$ of the quadratic form V (3.24), (3.25) that we are seeking are unknown. The number of equations (3.33) and the number of unknowns b_{ij} are the same, and equal to $\frac{n(n+1)}{2}$.

Let us denote the determinant of the system of algebraic equations (3.33) by $D(p^*_{ij})$, which emphasises its dependence on the coefficients p^*_{ij} of the assumed system of difference equations (3.22). If the determinant $D(p^*_{ij})$ is non-zero, then there exists a unique system of values b_{ij} which are a solution of (3.33). Thus, when $D(p^*_{ij}) \neq 0$, one can always construct (and, moreover, in a unique way) a quadratic form V (3.24), (3.25) which, by virtue of (3.22), (3.23), satisfies the difference equation (3.26). If, in addition, the quadratic form V that is found is positive definite, i.e. if in accordance with the Silvester criterion (2.141) its coefficients b_{ij} satisfy the inequalities

$$D_1 = b_{11} > 0$$

$$D_2 = \begin{vmatrix} b_{11} & b_{12} \\ b_{21} & b_{22} \end{vmatrix} > 0, \ \ldots, \ D_n = \begin{vmatrix} b_{11} & b_{12} & \ldots & b_{1n} \\ b_{21} & b_{22} & \ldots & b_{2n} \\ \cdot & \cdot & \cdot & \cdot \\ b_{n1} & b_{n2} & \ldots & b_{nn} \end{vmatrix} > 0 \tag{3.34}$$

then, by virtue of the general theorems of Lyapunov, the perturbed motion will not only be stable but also asymptotically stable.

The general theorems of Lyapunov define only the sufficient conditions of stability. However, for the linear case, the conditions that have been found are also necessary conditions.

This assertion can be proved in the following way. We establish first a connection between the condition $D(p^*_{ij}) \neq 0$ for the solvability of the algebraic equations (3.33) and the roots $\lambda^*_1, \lambda^*_2, \ldots, \lambda^*_n$ of the characteristic equation of the matrix P^* appearing in (3.23).

We note that $D(p^*_{ij}) = 0$ if, and only if, the system of homogeneous algebraic equations obtained from (3.33) by equating all the coefficients a_{ij}, to zero has a zero solution.

In matrix notation this is equivalent to the requirement that there should exist a non-zero symmetric matrix B that will satisfy the matrix equation

$$P^{*T} B P^* = B \tag{3.35}$$

obtained from (3.32) for $A = 0$. For simplicity, let us consider the case of simple roots λ^*_i. On the basis of (2.40)-(2.44) and the rules for transposing matrix products, we put (3.35) in the form

$$\left(K^{*T}\right)^{-1} \Lambda^{*T} K^{*T} B K^* \Lambda^* K^{*-1} = B \tag{3.36}$$

where P^* denotes the canonical matrix for Λ^*. The matrix Λ^* is a diagonal matrix of type (2.44) with elements λ^*_i. For this matrix we have $\Lambda^{*T} = \Lambda^*$.

We multiply both sides of (3.36) from the left by K^{*T} and from the right by K^*, obtaining

$$\Lambda^* C \Lambda^* = C \tag{3.37}$$

where

$$C = K^{*T} B K^* \tag{3.38}$$

together with B, is a symmetric matrix. Since, in multiplying matrix C from the left by the diagonal matrix Λ^*, all the elements of the ith row are multiplied by λ^*_i, and in multiplying from the right by Λ^*, all the elements of the jth column are multiplied by λ^*_j, it is clear that the matrix equation (3.37) is equivalent to n^2 scalar equations

$$\lambda^*_i \lambda^*_j c_{ij} = c_{ij}, \qquad i, j = 1, 2, \ldots, n \tag{3.39}$$

in which c_{ij} are the elements of matrix (3.38). It is clear that for $\lambda^*_i \lambda^*_j \neq 1$ we have $c_{ij} = c_{ji} = 0$ and for $\lambda^*_i \lambda^*_j = 1$, the coefficient $c_{ij} = c_{ji}$ may assume a non-zero value.

Thus, the requirement that $D(p^*_{ij}) \neq 0$ is equivalent to the requirement that the characteristic values λ^*_i of matrix P^* shall not satisfy any of the relations

$$\lambda^*_i \lambda^*_j = 1 \qquad i, j = 1, 2, \ldots, n \tag{3.40}$$

Let us now suppose that the motion of the system (3.22), (3.23) is asymptotically stable. In that case, on the basis of the results of Section 2.4, it follows that the absolute values of all the characteristic values λ_i^* of the matrix P^* are such that

$$|\lambda_i^*| < 1, \qquad i = 1, 2, \ldots, n \tag{3.41}$$

When conditions (3.41) are fulfilled, not one of the relations in (3.40) are fulfilled, so that in this case one can construct, and in a unique way, a quadratic form V (3.24), (3.25) which solves (3.26) in the above sense.

Let us show that the form V must be positive definite, i.e. that its sign is opposite to the sign of its first difference $V_m - V_{m-1}$. Actually, the function V_m defined in accordance with the difference equations (3.22), (3.23) forms a monotonically decreasing sequence (3.18) since the right-hand side of (3.26) is less than zero by definition and vanishes only when $x_1 = x_2 = \ldots = x_n = 0$. However, when $m \to \infty$ on the other hand, the quadratic form V_∞ must vanish in the limit together with the variables $x_1(m), x_2(m), \ldots, x_n(m)$ since the system is, according to supposition, asymptotically stable. Consequently, regardless of the order in which the quantities in (3.18) are taken, V_m cannot assume negative values since, if this were possible, there would be certain instants of time when V_m must increase in order to reach zero in the limit, and this is impossible.

Thus, the fact that it is possible to construct a quadratic form V (3.24), (3.25) that will, in accordance with (3.22), (3.23), solve (3.26) for an arbitrary positive definite quadratic form U (3.27), (3.28), and one can construct coefficients b_{ij} which will satisfy (3.34), is a necessary and sufficient condition for a discrete linear system to be asymptotically stable.

Consider now the properties of function V and the problem of constructing it when the modulus of at least one characteristic value λ_i^* of matrix P^* is greater than unity. In that case the equilibrium point will be unstable. Let us suppose first that the values λ_i^* do not satisfy any of the relations in (3.40). According to the usual procedure, we must in this case find a quadratic form V which, by virtue of (3.22) and (3.23), satisfies (3.26) for a positive definite form U. The form V that is found cannot be positive definite since an equilibrium point would be asymptotically stable, which contradicts the condition. The quadratic form V cannot be of fixed positive sign; in fact in this case one can always find initial conditions $x_1(0), x_2(0), \ldots, x_n(0)$ *)(not simultaneously equal to zero) for which $V_0 = 0$. However, in the next discrete instant of time, V becomes negative since, by virtue of (3.26) and the description of the form U, the first difference $V_m - V_{m-1}$ is negative. Thus the form V that is found must be such that in any neighbourhood of the origin it can assume values having the same sign as its first difference. This corresponds to the conditions laid down in the first Lyapunov instability theorem.

If amongst the characteristic values λ_i^* there are some that satisfy (3.40), then the corresponding Lyapunov arguments can be extended to include this case also. Let us introduce the new matrix variable

$$x(m) = \chi^{-m} x'(m) \qquad (3.42)$$

where χ is a number satisfying the inequality $0 < \chi < 1$. For (3.42), the elements of the column $x(m)$ will tend to infinity together with the corresponding elements of the column $x'(m)$. In the new variables, (3.23) appears in the form

$$x'(m) = \chi P^* x'(m-1) \qquad (3.43)$$

The roots λ_i' of the characteristic equation

$$\det(\lambda E - \chi P^*) = 0 \qquad (3.44)$$

of the matrix representing the change system will be expressed through λ_i^* in the form

$$\lambda_i' = \chi \lambda_i^*, \qquad i = 1, 2, \ldots, n \qquad (3.45)$$

Let us choose a positive number χ which is so close to unity that the roots λ_i' satisfy none of the relations in (3.40) and, in addition, let this number be chosen so that to the roots λ_i^* with $|\lambda_i^*| \lesssim 1$ there correspond roots λ_i' with $|\lambda_i'| \lesssim 1$. This is always possible in the case we are considering. For the transformed equation (3.43), by virtue of the above choice of the number χ, it is always possible to construct a quadratic form

$$V' = (x')^T Bx' \qquad (3.46)$$

whose first difference by virtue of these equations

$$V'_m - V'_{m-1} = [x'(m-1)]^T \{\chi^2 P^{*T} BP^* - B\} x'(m-1)$$

$$= -[x'(m-1)]^T Ax'(m-1) \qquad (3.47)$$

will, for example, be a negative definite form. Moreover, in accordance with what has been said above, for a suitable choice of the elements of the column x' (which numerically may be as small as one pleases) this form will have the same sign as that of its first difference.

Consider now the quadratic form

$$V = \chi^2 x^T Bx \qquad (3.47')$$

By virtue of (3.23), the first difference V can be put in the form

$$V_m - V_{m-1} = x^T(m-1)\chi^2\left[P^{*T}BP^* - B\right]x(m-1) \tag{3.48}$$

or, with account of equality (3.47), in the form

$$V_m - V_{m-1} = \alpha V_{m-1} + W_{m-1} \tag{3.49}$$

where

$$\alpha = \frac{1-\chi^2}{\chi^2} > 0$$

and $W = -x^T Ax$ is a negative definite form. Thus, in the present case, the conditions of the second instability theorem of Lyapunov hold and, for roots λ_i^* having the properties indicated above, the question of instability is completely solved within the freamework of Lyapunov's direct method.

Finally, we must consider the case when the modulus of at least one root is equal to unity and the moduli of the remaining roots λ_i^* are less than unity. Clearly, if only one of the relations in (3.40) is fulfilled, it is not possible in the present case to construct a quadratic form V in accordance with (3.23)-(3.28). Nor can the procedure for constructing the form V associated with the application of the transformation (3.42) be carried out successfully.

In the critical cases we are considering, the theorems of the direct Lyapunov method do not decide the question of the stability of the unperturbed motion of a discrete linear dynamic system (3.22)-(3.23). In the critical case, stability depends on the fine structure of the solution of a system of linear difference equations. From (2.80)-(2.83), and for $g^*(m) \equiv 0$, one can show that in the critical case an equilibrium point may be simply stable, or stable, depending on the power e_σ of the elementary divisors corresponding to the roots λ_σ^* with moduli equal to unity: for $e_\sigma = 1$ the equilibrium point is stable, for $e_\sigma > 1$ it is unstable.

3.3 STABILITY WITH RESPECT TO THE FIRST APPROXIMATION

Let us suppose that the right-hand sides of (3.1) are such that the difference equations can be put in the form

$$\left.\begin{aligned}
x_1(m) &= p_{11}^* x_1(m-1) + p_{12}^* x_2(m-1) \\
&\quad \dots + p_{1n}^* x_n(m-1) + \tilde{X}_1\left[x_1(m-1), \dots, x_n(m-1)\right] \\
x_2(m) &= p_{21}^* x_1(m-1) + p_{22}^* x_2(m-1) \\
&\quad \dots + p_{2n}^* x_n(m-1) + \tilde{X}_2\left[x_1(m-1), \dots, x_n(m-1)\right] \\
&\cdots\cdots\cdots\cdots\cdots\cdots \\
x_n(m) &= p_{n1}^* x_1(m-1) + p_{n2}^* x_2(m-1) \\
&\quad \dots + p_{nn}^* x_n(m-1) + \tilde{X}_n\left[x_1(m-1), \dots, x_n(m-1)\right]
\end{aligned}\right\} \tag{3.50}$$

where $\tilde{X}_i(x_1, x_2, \ldots, x_n)$ are absolutely converging series in powers of $x_1, x_2, \ldots, x_n$ which start with terms not lower than second order; the p^*_{ij} are constants. In matrix form these equations can be written

$$x(m) = P^* x(m-1) + \tilde{X}[x_1(m-1), x_2(m-1), \ldots, x_n(m-1)] \quad (3.51)$$

where $\tilde{X}$ is a column matrix of the functions $\tilde{X}_i$ and the rest of the notation is the same as that used for (3.23). The linear part of the matrix equation, which formally coincides with (3.23), will be called the equations of the first approximation. The following Lyapunov theorem holds.

Theorem

If the moduli of all the roots λ^*_i of the characteristic equation of the first approximation, i.e. the moduli of the characteristic values of the matrix P^*, are less than unity, then the unperturbed motion is asymptotically stable independently of the higher order terms $\tilde{X}_i$ in the difference equations (3.50).

In view of the fact that $|\lambda^*_i| < 1$, $i = 1, 2, \ldots, n$, and in accordance with the results of the preceding section for an equation of the first approximation which formally coincides with (3.23), one can construct a positive definite quadratic form V (3.24), (3.25) and, moreover, one can construct it in a unique way, which will solve (3.26) for a positive definite form U (3.27), (3.28). For the quadratic form V thus constructed, we form the first difference according to the assumed system of nonlinear difference equations (3.51). We then obtain

$$V_m - V_{m-1} = -U_{m-1} + \tilde{W}_{m-1} \quad (3.52)$$

where

$$\tilde{W}_{m-1} = x^T(m-1) P^{*T} B \tilde{X}_{m-1} + \tilde{X}^T_{m-1} B P^* x(m-1) + \tilde{X}^T_{m-1} B \tilde{X}_{m-1} \text{ *)} \quad (3.53)$$

(here, for brevity, we have written $\tilde{X}_{m-1}$ for $\tilde{X}[x_1(m-1), x_2(m-1), \ldots, x_n(m-1)]$). Since the functions $\tilde{X}_i$ are in the form of an absolutely converging series containing terms which are not lower than second order, the last three terms on the right-hand side of (3.52) can be written in the form of an absolutely converging series which will start with terms not lower than third order with respect to the variables $x_1, x_2, \ldots, x_n$. There will then certainly exist a sufficiently small region (3.8) inside which the first difference

$V_m - V_{m-1}$ will be negative definite and, by virtue of the general theorems of Section 3.1, the equilibrium point of a nonlinear discrete system (3.42), (3.43) will be asymptotically stable.

Theorem

If among the roots λ_i^* of the characteristic equation of the first approximation, i.e. among the characteristic values of the matrix P^*, there is at least one root with a modulus greater than unity, then the unperturbed motion is stable independently of the higher order terms $\tilde{X}_i$ in the difference equations (3.50).

In this case, for the equations of the first approximation [which formally coincide with (3.22) and (3.23)], it will always be possible to construct a quadratic form V which will, for example, by virtue of the equations of the first approximation, satisfy (3.49). We will consider here only this more general case.

Let us, in accordance with the nonlinear matrix equation (3.51), construct the first difference of the form V that has been found. With account of (3.49), we obtain

$$V_m - V_{m-1} = \alpha V_{m-1} + W_{m-1} + \tilde{W}_{m-1} \tag{3.54}$$

where $\tilde{W}_{m-1}$ is a form defined by (3.53) representing absolutely converging series containing terms of order higher than the second with respect to the variables $x_1, x_2, \ldots, x_n$, and where W is a negative definite quadratic form. Therefore, again, there exists a small region (3.8) inside which $(W + \tilde{W})$ will be a negative definite function.

In accordance with the general instability theorems of Lyapunov, the unperturbed motion for the nonlinear system (3.50)-(3.51) will be unstable outside the dependence of the nonlinear terms $\tilde{X}_i$ which possess the properties indicated above.

In the critical case, the equations of the first approximation do not enable us to come to a conclusion about the stability of the system because, in this case, stability depends upon the nature of the nonlinear terms $\tilde{X}_i$ of (3.50)-(3.51) which are thrown away. A simple example will make this clear. Consider the difference equation

$$x_1(m) = x_1(m-1) + \mu x_1^3(m-1) \tag{3.55}$$

which is a special case of (3.50). We have here a critical case since $\lambda_1^* = 1$ for the equation of the first approximation $x_1(m) = x_1(m-1)$. We choose now a positive definite quadratic form V in the form

$$V = x_1^2 \tag{3.56}$$

The first difference $V_m - V_{m-1}$, defined in accordance with (3.55), can then be represented in the form

$$V_m - V_{m-1} = 2\mu x_1^4(m-1) + \mu^2 x_1^6(m-1) \tag{3.57}$$

or

$$V_m - V_{m-1} = \mu x_1^4(m-1)[2 + \mu x_1^2(m-1)] \tag{3.58}$$

There exists a region of the type (3.8)

$$x_1^2 < \frac{2}{|\mu|} \tag{3.59}$$

inside which the first difference is a negative definite or a positive definite function according to whether μ is a negative or positive number. However, by virtue of the general theorems of Lyapunov, we obtain the result that for $\mu < 0$ the unperturbed motion will be asymptotically stable, and for $\mu > 0$ it will be unstable.

Chapter 4

DISCONTINUOUS CONTROL SYSTEMS

Two basic methods have been developed for the theoretical investigation of discontinuous automatic control systems. One of these methods is based on mathematical techniques associated with the discrete Laplace transform and certain of its modifications. This method enables us to introduce the concepts of a transfer function and frequency characteristics. We can also apply methods of investigation which are, in a certain sense, analogous to those characteristic of the frequency analysis of continuous automatic control systems. The main credit for the development of this particular approach to the theory of discontinuous systems belongs to Tsipkin [26] and Jury [7].

The second approach involves the application of a method in which the initial conditions are included in such a way that the problem is reduced to one of solving linear or nonlinear finite difference equations. This approach goes back to the beginning of this century when Zhukovskii [9] used it in control system studies.

In this book we shall make use of this second method in combination with matrix analysis techniques. It is interesting to note that, at a certain stage in our analysis of the stability of linear discontinuous control systems, we approach quite closely – though only formally – the procedures typical of frequency analysis. Matrix methods can, in general, also be used in conjunction with the methods of operational calculus; in certain cases such a combination of techniques can lead more rapidly and more simply to the required results, depending on the experience of the investigator.

4.1 CONTINUOUS AND DISCRETE MOTIONS OF LINEAR DISCONTINUOUS SYSTEMS

Consider a linear discontinuous control system. The perturbed motion of such a system is described by (1.12) or (1.15), containing a control function $\psi(\sigma)$ defined by (1.10), which is graphically represented by a discontinuous curve of the kind illustrated in Fig. 1.7. Let us consider first the equations of motion in their normal form (1.15). We represent the system of equations in matrix form

$$\left.\begin{aligned}
\begin{Vmatrix} \dot{x}_1 \\ \dot{x}_2 \\ \vdots \\ \dot{x}_n \end{Vmatrix} &= \begin{Vmatrix} p_{11} & p_{12} & \cdots & p_{1n} \\ p_{21} & p_{22} & \cdots & p_{2n} \\ \cdot & \cdot & \cdot & \cdot \\ p_{n1} & p_{n2} & \cdots & p_{nn} \end{Vmatrix} \begin{Vmatrix} x_1 \\ x_2 \\ \vdots \\ x_n \end{Vmatrix} + \begin{Vmatrix} h_1 \\ h_2 \\ \vdots \\ h_n \end{Vmatrix} \psi(\sigma) + \begin{Vmatrix} g_1 \\ g_2 \\ \vdots \\ g_n \end{Vmatrix} \\
\sigma &= \begin{Vmatrix} \gamma_1 & \gamma_2 & \cdots & \gamma_n \end{Vmatrix} \begin{Vmatrix} x_1 \\ x_2 \\ \vdots \\ x_n \end{Vmatrix}
\end{aligned}\right\} \tag{4.1}$$

or in the more compact form

$$\left.\begin{aligned} \dot{x} &= Px + h\psi(\sigma) + g \\ \sigma &= \gamma x \end{aligned}\right\} \tag{4.2}$$

The square matrix P will sometimes be called the matrix of the open loop control system.

We will investigate first the free-motion of a pulsed system arising from the action of instantaneous perturbations defined mathematically by specifying an initial value for the matrix variable x for $t=0$. The free-motions are described by

$$\left.\begin{aligned} \dot{x} &= Px + h\psi(\sigma) \\ \sigma &= \gamma x \end{aligned}\right\} \tag{4.3}$$

which are obtained from (4.2) with $g \equiv 0$, i. e. for the case when there is no external forcing action. Two circumstances simplify the problem of finding a solution to (4.3). In the first place the control function $\psi(\sigma)$ is a step-function with discontinuities of the first kind, so that during the discrete intervals of time where $\psi(\sigma)$ is constant, (4. 3) is a system of linear differential equations with constant coefficients with a right-hand side which is constant. Secondly, the interval of time between the start of two successive values of the function $\psi(\sigma)$, the so-called recurrence interval T, is constant. It is therefore sufficient to establish the relation between the states of the system at the ends of the first cycle in order to be able, by applying the normal methods of the matrix analysis, to determine the state of the system at the end of any repetition cycle, and also at any intermediate point.

As we have already mentioned, in solving (4.3) we shall make use of the method of incorporating initial conditions; i.e. as the points of discontinuity of the function $\psi(\sigma)$ pass, the initial value of the matrix variable x in the approaching interval of time will be taken as the final value in the interval that has just passed. In this way we can construct a continuous solution of (4.3) which will be uniquely determined by specifying an arbitrary initial value for the matrix variable x at $t=0$, or at any other instant of time which we choose to take as the initial instant.

For convenience, and to simplify later discussion, we will choose the initial instant of time $t=0$ so that it coincides with the start of the first pulse of function $\psi(\sigma)$. One can always move the time origin t to make it coincide with the start of the first cycle, if necessary, without loss of generality. The beginning of any recurrence interval will then coincide with the start of the corresponding pulse. Consider the mth recurrence interval which starts at time $(m-1)T$. In addition, let the state of the system for $t=0$ be determined by the initial condition $x=x(0)$; at $t=(m-1)T$ we have $x=x[(m-1)T]$.

Let us take $x=x[(m-1)T]$ for the initial value of the matrix coordinate x for the mth recurrence interval, i.e. for $(m-1)T\leqslant t \leqslant mT$. It will sometimes be convenient to extract from t an integral number of repetition periods, supposing that $t=(m-1)T+\tau$ for $0\leqslant\tau\leqslant T_1$ and $t=(m-1)T+T_1+\tau$ for $0\leqslant\tau\leqslant T_2$. Then, according to (1.10) and the second of equations (4.3), we will have

$$\psi(\sigma)=\sigma[(m-1)T]=\gamma x[(m-1)T] \tag{4.4}$$

With account of (4.4) the first equation of (4.3) can be rewritten in the form

$$\dot{x}=Px+h\gamma x[(m-1)T],\quad (m-1)T\leqslant t\leqslant(m-1)T+T_1 \tag{4.5}$$

Since $\gamma x[(m-1)T]$ is a scalar constant and h is a column matrix of constants, then on the basis of (2.129), the solution of (4.5) – which reduces to $t=[(m-1)T]$ when $x[(m-1)T]$ – can be represented in the form

$$\begin{aligned} x[(m-1)T+\tau]&=KM(\tau)K^{-1}x[(m-1)T]\\ &+KN(\tau)K^{-1}h\gamma x[(m-1)T] \end{aligned} \tag{4.6}$$

which is clearly valid in the interval $0\leqslant\tau\leqslant T_1$. In (4.6), K should be the transition matrix for the matrix P. Solution (4.6) can be rewritten in the form

$$x[(m-1)T+\tau]=Q'(\tau)x[(m-1)T],\quad 0\leqslant\tau\leqslant T_1 \tag{4.7}$$

where

$$Q'(t) = KM(t)K^{-1} + KN(t)K^{-1}h\gamma \tag{4.8}$$

On completion of the control pulse $\psi(\sigma)$, i. e. immediately after the instant of time $t=(m-1)T+T_1$, there is a pause which lasts for a time T_2. During this interval of time the control loop is effectively open, i. e. $\psi(\sigma)=0$. The first equation of (4.3) can then be written in the form

$$\dot{x} = Px, \quad (m-1)T+T_1 \leqslant t \leqslant mT \tag{4.9}$$

We will integrate (4.9), taking as our initial value of the matrix coordinate x its final value in the preceding time interval – which can be obtained from (4.7) and (4.8) with $t=(m-1)T+T_1$. The solution of (4.9) can be put in the form

$$x[(m-1)T+T_1+\tau] = KM(\tau)K^{-1}x[(m-1)T+T_1] \tag{4.10}$$

which will be valid over the interval $0 \leqslant \tau \leqslant T_2$. In (4.10) we will replace $x[(m-1)T+T_1]$ by the value obtained from (4.7) with $t=(m-1)T+T_1$, and we will make use of (2.92). We then obtain

$$x[(m-1)T+T_1+\tau] = Q''(\tau)x[(m-1)T], \quad 0 \leqslant \tau \leqslant T_2 \tag{4.11}$$

where

$$Q''(t) = KM(t+T_1)K^{-1} + KM(t)N(T_1)K^{-1}h\gamma \tag{4.12}$$

at the end of the mth period, i. e. for $\tau = T_2$, we obtain from (4.11) the relation

$$x(mT) = Q''(T_2)x[(m-1)T] \tag{4.13}$$

For the matrix $Q''(T_2)$ we will introduce the special notation

$$P^* = Q''(T_2) \tag{4.14}$$

Matrix P^* can be written in the form

$$P^* = KM(T)K^{-1} + KM(T_2)N(T_1)K^{-1}h\gamma \tag{4.15}$$

and, in the new notation, (4.13) can be rewritten in the form

$$x(mT) = P^*x[(m-1)T] \tag{4.16}$$

Expression (4.16) is a recurrence relation which establishes a connection between the values of the variable x at the beginning and end of an arbitrary recurrence period. Matrix P^* will be referred to as the matrix operator for the system.

Relation (4.16) can be regarded as a matrix difference equation with a discrete argument represented in normal form. The solution of the difference equation (4.16) is uniquely determined by the initial condition $x=x(0)$ for $t=0$. The solution of the initial differential equation (4.3) is also uniquely determined by the same initial condition. The solutions of the differential equation (4.3) and the difference equation (4.16), which satisfy the same initial condition, coincide at discrete, equally spaced instants of time mT. The difference equation determines the discrete motion of the discontinuous system and it can be used independently to solve problems concerned with the dynamics of discontinuous control systems.

The solution of (4.16) satisfying the value $x=x_0$ for $t=0$ can be represented in the form (see Section 2.4)

$$x(mT)=P^{*m}x(0) \tag{4.17}$$

The value of the matrix variable for an arbitrary instant of time can be obtained from (4.7) and (4.11) if one substitutes in these equations the value of $x[(m-1)T]$ obtained from (4.17). Thus, we will have

$$x[(m-1)T+\tau]=Q'(\tau)P^{*m-1}x(0), \quad 0\leqslant\tau\leqslant T_1 \tag{4.18}$$

and

$$x[(m-1)T+T_1+\tau]=Q''(\tau)P^{*m-1}x(0), \quad 0\leqslant\tau\leqslant T_2 \tag{4.19}$$

Since $Q'(0)=E$ and $Q''(T_2)=P^*$, then, for $\tau=0$ and $\tau=T_2$, (4.18) and (4.19) determine, respectively, $x[(m-1)T]$ and $x(mT)$ in the same form as that obtained from (4.17). Thus, (4.18) and (4.19) determine the continuous motion of the system as a function of the initial value $x(0)$ or, to use another terminology, these formulae give the continuous trajectory of motion as a function of the initial point $x(0)$.

As has been shown above, the coordinates of the continuous trajectory determined by (4.18) and (4.19) coincide with the coordinates of the discrete phase trajectory determined by (4.17) at the discrete, equally spaced instants of time mT.

4.2 STABILITY AND THE CHARACTERISTIC EQUATION

We can approach the problem of the stability of a linear discontinuous system from two points of view. The first approach is based on the study of the structure of the general solution of a homogeneous system of equations of motion; the second is based on the construction of Lyapunov functions. In this section we shall be concerned with the first point of view.

The differential equation (4.3) and the difference equation (4.16) are equivalent to each other in the sense specified in Section 4.2. The stability of a linear discontinuous system may therefore be defined in terms of the characteristic solution of (4.16). A system will be stable and, moreover, asymptotically stable, if the absolute magnitude of all the roots $\lambda_i^*, \lambda_2^*, \ldots, \lambda_n^*$ of the characteristic equation of matrix P^* [see (4.16)] is less then unity, and it will be unstable if the absolute magnitude of at least one of the roots λ_i^* is greater than unity. The borderline is represented by the case when some of the roots have absolute magnitude equal to unity, and the moduli of the others is less than unity.

At the boundary of asymptotic stability, a discrete system can be stable or unstable according to the power e_σ of the elementary divisors corresponding to the roots with moduli equal to unity. For $e_\sigma = 1$ the system is stable; for $e_\sigma > 1$ it is unstable.

The last circumstance can readily be established from (2.82) which, for $g^*(m') = 0$, provides a general solution of the homogeneous system of multiple roots λ_σ^*, by reference to (2.84). It is clear that, when the roots λ_i^* are simple, ordinary stability always holds on the boundary.

The behaviour of a linear discontinuous system on the boundary of asymptotic stability, when the conditions for ordinary stability are fulfilled, depends on the form of the root whose modulus is equal to unity. If this root can be represented in the form $e^{i2\pi\frac{p}{q}}$, where p and q are integers, then, regardless of the initial conditions, periodic oscillations with period qT establish themselves in the systems. In the remaining cases the system will execute limited motions which never repeat themselves.

In a linear continuous system, motions of a similar type are quasiperiodic; they can arise only in those cases when the boundary of oscillatory stability is determined by two or more purely imaginary roots which are unequal.

Let us consider the problem of constructing the characteristic equation for the matrix P^* appearing in (4.16) and determined by (4.15). As we showed in Chapter 2, similar matrices have identical characteristic equations. Consequently, it is convenient to replace the above matrix by a similar matrix since it will be generally easier to find the characteristic equation for a similar matrix. Having made use of this device, we represent matrix (4.15) in the form

$$P^* = KP^{**}K^{-1}$$

where

$$P^{**} = M(T) + M(T_2)\,N(T_1)\,K^{-1}h\gamma K \qquad (4.20)$$

We should remember that K is the transition matrix for P, the matrix of the open control system appearing in (4.2).

To abbreviate still further, we introduce the notation

$$u = K^{-1}h, \qquad \beta = \gamma K \qquad (4.21)$$

The elements $u_1, u_2, \ldots, u_n$ of the column matrix u and the elements $\beta_1, \beta_2, \ldots, \beta_n$ of the row matrix β are found from

$$u_i = \varkappa_i h; \qquad \beta_i = \gamma k_i, \qquad i = 1, 2, \ldots, n \qquad (4.22)$$

where $\varkappa_i$ and k_i are the rows and columns of the matrix K^{-1} and K, respectively. In the new notation, matrix (4.20) appears in the form

$$P^{**} = M(T) + M(T_2) N(T_1) u\beta \tag{4.23}$$

or in the simpler form

$$P^{**} = M(T) + u'\beta \tag{4.24}$$

if we use u' to denote the column matrix

$$u' = M(T_2) N(T_1) u \tag{4.25}$$

(This introduction by stages of new notation will be found to be more convenient later on.)

The matrix P^{**} has a very simple form since, on the right-hand side (4.24), $M(T)$ is a diagonal or, to be more general, a quasidiagonal matrix, and $u'\beta$ is a matrix of rank unity [see (2.11)]. Thus, the characteristic equation of the matrix P^* can be represented in the form

$$\begin{aligned}\Delta^*(\lambda) &= \det(\lambda E - P^*) = \det(\lambda E - P^{**}) \\ &= \det(\lambda E - M(T) - u'\beta) = 0\end{aligned} \tag{4.26}$$

a) Case of simple characteristic values $\lambda_1, \lambda_2, \ldots, \lambda_n$ *of the matrix* P

Bearing in mind (2.11), (2.94) and (4.26), let us represent the characteristic equation for the matrix P^* in the form

$$\Delta^*(\lambda) = \begin{vmatrix} \lambda - e^{\lambda_1 T} - u_1'\beta_1 & -u_1'\beta_2 & \cdots & -u_1'\beta_n \\ -u_2'\beta_1 & \lambda - e^{\lambda_2 T} - u_2'\beta_2 & \cdots & -u_2'\beta_n \\ \cdot & \cdot & \cdot & \cdot \\ -u_n'\beta_1 & -u_n'\beta_2 & \cdots & \lambda - e^{\lambda_n T} - u_n'\beta_n \end{vmatrix} = 0 \tag{4.27}$$

Now the determinant

$$D = \begin{vmatrix} z_1 + c_{11} & c_{12} & \cdots & c_{1n} \\ c_{21} & z_2 + c_{22} & \cdots & c_{2n} \\ \cdot & \cdot & \cdot & \cdot \\ c_{n1} & c_{n2} & \cdots & z_n + c_{nn} \end{vmatrix} \tag{4.28}$$

can be expanded about the diagonal elements z_i and represented in the form

$$D = \prod_{j=1}^{n} z_j \left\{ 1 + \sum_{i=1}^{n} \frac{c_{ii}}{z_i} + \sum_{\substack{i,k=1 \\ i<k}}^{n} \frac{\begin{vmatrix} c_{ii} & c_{ik} \\ c_{ki} & c_{kk} \end{vmatrix}}{z_i z_k} + \ldots + \frac{\begin{vmatrix} c_{11} & c_{12} & \cdots & c_{1n} \\ c_{21} & c_{22} & \cdots & c_{2n} \\ \cdot & \cdot & \cdot & \cdot \\ c_{n1} & c_{n2} & \cdots & c_{nn} \end{vmatrix}}{z_1 z_2 \ldots z_n} \right\} \tag{4.29}$$

In the determinant (4.27) the role of the terms z_i is played by the expressions $(\lambda - e^{\lambda_i T})$, and the matrix with elements c_{ij} is $(-u'\beta)$. But a matrix which is a product of a column u' and a row β has rank unity. All its minors of order above the first are equal to zero. Therefore the characteristic equation (4.27) can be represented in expanded form as:

$$\Delta^*(\lambda) = \prod_{j=1}^{n} (\lambda - e^{\lambda_j T}) \left\{ 1 - \sum_{i=1}^{n} \frac{u'_i \beta_i}{\lambda - e^{\lambda_i T}} \right\} = 0 \qquad (4.30)$$

In this equation the elements u'_i of column u' can be expressed in terms of the elements u_i of column u with the help of (4.25). Bearing in mind the fact that, for the case of simple roots λ_i, the matrices $M(t)$ and $N(t)$ are diagonal and can be determined from (2.94) and (2.120), the matrix $M(T_2)N(T_1)$ is also diagonal with elements $e^{\lambda_i T_2}(e^{\lambda_i T_1} - 1)/\lambda_i$. Thus from (4.25) we obtain

$$u'_i = \frac{e^{\lambda_i T_2}(e^{\lambda_i T_1} - 1)}{\lambda_i} u_i, \qquad i = 1, 2, \ldots, n \qquad (4.31)$$

Substituting u'_i from (4.31) into (4.30), we have

$$\Delta^*(\lambda) = \prod_{j=1}^{n} (\lambda - e^{\lambda_j T}) \left\{ 1 - \sum_{i=1}^{n} \frac{e^{\lambda_i T_2}(e^{\lambda_i T_1} - 1)}{\lambda_i (\lambda - e^{\lambda_i T})} u_i \beta_i \right\} = 0 \qquad (4.32)$$

According to (4.22), the product of the two scalar quantities u_i and β_i can conveniently be written in the form

$$u_i \beta_i = \beta_i u_i = \gamma k_i \varkappa_i h \qquad (4.33)$$

Using the associative law of multiplication of matrices and the matrices (2.64), we represent (4.33) in the form

$$\beta_i u_i = \frac{\gamma F(\lambda_i) h}{\Delta'(\lambda_i)}, \qquad i = 1, 2, \ldots, n \qquad (4.34)$$

Substituting the $u_i \beta_i$ from here into (4.32), we finally represent the characteristic equation we are seeking in the form

$$\Delta^*(\lambda) = \prod_{j=1}^{n} (\lambda - e^{\lambda_j T}) \left\{ 1 - \sum_{i=1}^{n} \frac{e^{\lambda_i T_2}(e^{\lambda_i T} - 1)}{\lambda_i (\lambda - e^{\lambda_i T})} \frac{\gamma F(\lambda_i) h}{\Delta'(\lambda_i)} \right\} = 0 \qquad (4.35)$$

The characteristic equation (4.35) for the matrix (4.15) is expressed in terms of the parameters of the initial matrix differential equation (4.3), since λ_i are the roots of the characteristic equation $\Delta(\lambda) = \det(\lambda E - P)$ and $F(\lambda)$ is the adjoint matrix for $(\lambda E - P)$.

In many cases of practical importance, the structure of matrix P of the open loop system is such that $\Delta(\lambda)$ appears in the form of a product of several polynomials in λ. In these cases the roots λ_i can easily be determined.

b) Case of multiple characteristic values $\lambda_1, \lambda_2, \ldots, \lambda_n$ of the matrix P

Consider now an important case that frequently arises in practical applications when two characteristic values of the matrix P are equal to zero. For definiteness we shall assume that $\lambda_{n-1}=\lambda_n=0$, and that to this repeated root there corresponds a single elementary divisor of second degree, i.e. $e_{n-1}=2$.

For the case under consideration, the matrix $M(T)$ is given by (2.98) with $t=T$. Using (2.98), we can rewrite the characteristic equation (4.26) in the form

$$\begin{vmatrix} \lambda-e^{\lambda_1 T}-u_1'\beta_1 & -u_1'\beta_2 & \ldots & -u_1'\beta_{n-1} & -u_1'\beta_n \\ -u_2\beta_1 & \lambda-e^{\lambda_2 T}-u_2'\beta_2 & \ldots & -u_2'\beta_{n-1} & -u_2'\beta_n \\ \cdot & \cdot & \cdot & \cdot & \cdot \\ -u_{n-1}'\beta_1 & -u_{n-1}^1\beta_2 & \ldots & \lambda-1-u_{n-1}'\beta_{n-1} & -u_{n-1}'\beta_n \\ -u_n'\beta_1 & -u_n'\beta_2 & \ldots & -T-u_n'\beta_{n-1} & \lambda-1-u_n'\beta_n \end{vmatrix}=0 \tag{4.36}$$

The determinant of the left-hand side of (4.36) can be written in the form of a sum of two determinants, one of which coincides with the determinant in (4.27) – this determinant is represented in expanded form by the left-hand side of (4.30) – and the other of which can be written in the form of the $(n-1)$-th order determinant

$$T\begin{vmatrix} \lambda-e^{\lambda_1 T}-u_1'\beta_1 & -u_2'\beta_1 & \ldots & -u_1'\beta_{n-2} & -u_1'\beta_n \\ -u_2'\beta_1 & \lambda-e^{\lambda_2 T}-u_2'\beta_2 & \ldots & -u_2'\beta_{n-2} & -u_2'\beta_n \\ \cdot & \cdot & \cdot & \cdot & \cdot \\ -u_{n-2}'\beta_1 & -u_{n-2}'\beta_2 & \ldots & \lambda-e^{\lambda_{n-2}T}-u_{n-2}'\beta_{n-2} & -u_{n-2}'\beta_n \\ -u_{n-1}'\beta_1 & -u_{n-1}'\beta_2 & \ldots & -u_{n-1}'\beta_{n-2} & -u_{n-1}'\beta_n \end{vmatrix} \tag{4.37}$$

In multiplying out this determinant we again use (4.28) and (4.29). Here one should remember that the term in the last diagonal element z_{n-1} is equal to zero and the matrix of the elements u_i' and β_i' in this case also has rank unity. Taking this into account, we can put (4.37) in the form

$$-Tu_{n-1}'\beta_n \prod_{i=1}^{n-2}(\lambda-e^{\lambda_i T}) \tag{4.38}$$

Taking the sum in the left-hand side of (4.30) [where one should assume only that $\lambda_{n-1}=\lambda_n=0$], using expression (4.38) and equating the sum of zero, we obtain the characteristic equation (4.46) in expanded form

$$\Delta^*(\lambda)=(\lambda-1)^2\prod_{j=1}^{n-2}(\lambda-e^{\lambda_j T})\left\{1-\sum_{i=1}^{n-2}\frac{u_i'\beta_i}{\lambda-e^{\lambda_j T}}\right.$$

$$\left.-\frac{u_{n-1}'\beta_{n-1}+u_n'\beta_n}{\lambda-1}\right\}-Tu_{n-1}'\beta_n\prod_{j=1}^{n-2}(\lambda-e^{\lambda_j T})=0 \tag{4.39}$$

which can be represented as

$$\Delta^*(\lambda)=(\lambda-1)^2\prod_{j=1}^{n-1}(\lambda-e^{\lambda_j T})\left\{1-\sum_{i=1}^{n-2}\frac{u_i'\beta_i}{\lambda-e^{\lambda_i T}}\right.$$

$$\left.-\frac{u_{n+1}'\beta_{n-1}+u_n'\beta_n}{\lambda-1}-T\frac{u_{n-1}'\beta_n}{(\lambda-1)^2}\right\}=0 \tag{4.40}$$

Let us express the quantities u_i' through u_i. In (4.25) the matrices $M(T_2)$ and $N(T_1)$ are determined, in this case, by (2.98) and (2.122) by putting $t=T_2$ and $t=T_1$, respectively, into these formulae. Let us expand the matrices $M(T_2)$ and $M(T_1)$ into blocks and represent them in the form of the quasidiagonal matrices

$$M(T_2)=\left\|\begin{array}{cccc|cc} e^{\lambda_1 T_2} & 0 & \dots & 0 & 0 & 0\\ \cdot & \cdot & \cdot & \cdot & \cdot & \cdot\\ 0 & 0 & \dots & e^{\lambda_{n-2}T_2} & 0 & 0\\ \hline 0 & 0 & \dots & 0 & 1 & 0\\ 0 & 0 & \dots & 0 & T_2 & 1 \end{array}\right\|=\left\|\begin{array}{cc} M_1 & 0\\ 0 & M_2\end{array}\right\| \tag{4.41}$$

$$M(T_1)=\left\|\begin{array}{cccc|cc} \frac{e^{\lambda_i T_1}-1}{\lambda} & 0 & \dots & 0 & 0 & 0\\ \cdot & \cdot & \cdot & \cdot & \cdot & \cdot\\ 0 & 0 & \dots & \frac{e^{\lambda_{n-2}T_1}-1}{\lambda_{n-2}} & 0 & 0\\ \hline 0 & 0 & \dots & 0 & T_1 & 0\\ 0 & 0 & \dots & 0 & \frac{T_1^2}{2} & T_1 \end{array}\right\|=\left\|\begin{array}{cc} N_1 & 0\\ 0 & N_2\end{array}\right\| \tag{4.42}$$

Multiplying out the quasidiagonal matrices (4.41) and (4.42), we obtain

$$M(T_2)N(T_1)=\left\|\begin{array}{cc} M_1N_1 & 0\\ 0 & M_2N_2\end{array}\right\| \tag{4.43}$$

The sub-matrices M_1 and N_1 are diagonal, so that their product M_1N_1 can be found without difficulty. It is obvious that M_1N_1 is a diagonal matrix of order $(n-2)$ with the common element $\frac{e^{\lambda_i T_2}(e^{\lambda_i T_i}-1)}{\lambda_i}$.

The product of the sub-matrices M_2N_2 in this case can be readily obtained by direct calculation. However, bearing in mind more complex cases, we will apply the following procedure for calculating them. The sub-matrices M_2 and N_2 are left triangular matrices of second order, with equal elements in the oblique rows; consequently, with the help of unit oblique rows [see (2.26)], they can be put in the form

$$\left.\begin{aligned} M_2 &= E + T_2H^1 \\ N_2 &= T_1E + \frac{T_1^2}{2}H^1 \end{aligned}\right\} \tag{4.44}$$

Multiplying out the left- and right-hand sides of (4.44), and using (2.27), we obtain

$$M_2N_2 = T_1E + \left(T_1T_2 + \frac{T_1^2}{2}\right)H^1 \tag{4.45}$$

or, in the usual notation,

$$M_2N_2 = \begin{Vmatrix} T_1 & 0 \\ T_1T_2 + \frac{T_1^2}{2} & T_1 \end{Vmatrix} \tag{4.46}$$

Thus, we have finally

$$M(T_2)N(T_1) = \begin{Vmatrix} \frac{e^{\lambda_i T_2}(e^{\lambda_1 T_1}-1)}{\lambda_1} & \cdots & 0 & 0 & 0 \\ \cdot\;\cdot\;\cdot & \cdot\;\cdot\;\cdot & \cdot\;\cdot\;\cdot & \cdot\;\cdot\;\cdot & \cdot\;\cdot\;\cdot \\ 0 & \cdots & \frac{e^{\lambda_{n-2} T_2}(e^{\lambda_{n-2} T_1}-1)}{\lambda_{n-2}} & 0 & 0 \\ 0 & \cdots & 0 & T_1 & 0 \\ 0 & \cdots & 0 & T_1T_2 + \frac{T_1^2}{2} & T_1 \end{Vmatrix} \tag{4.47}$$

Using (4.47), we find from (4.25)

$$\left.\begin{aligned} u_i' &= \frac{e^{\lambda_i T_2}(e^{\lambda_i T_1}-1)}{\lambda_i}u_i, \quad i = 1, 2, \ldots, n-2 \\ u_{n-1}' &= T_1u_{n-1} \\ u_n' &= \left(T_1T_2 + \frac{T_1^2}{2}\right)u_{n-1} + T_1u_n \end{aligned}\right\} \tag{4.48}$$

Substituting u_i' from (4.48) into (4.40), we reduce the characteristic equation to the form

$$\Delta^*(\lambda)=(\lambda-1)^2\prod_{j=1}^{n-2}\left(\lambda-e^{\lambda_j T}\right)\left\{1-\sum_{i=1}^{n-2}\frac{e^{\lambda_i T_2}\left(e^{\lambda_i T_1}-1\right)}{\lambda_i\left(\lambda-e^{\lambda_i T}\right)}u_i\beta_i\right.$$

$$-\frac{T_1}{\lambda-1}(u_{n-1}\beta_{n-1}+u_n\beta_n)$$

$$\left.-\left(T_2+\frac{T_1}{2}+\frac{T}{\lambda-1}\right)\frac{T_1}{\lambda-1}u_{n-1}\beta_n\right\}=0 \quad (4.49)$$

Moreover, according to (4.22), we have

$$\left.\begin{aligned} u_i\beta_i&=\gamma k_i\varkappa_i h, \qquad i=1,\ 2,\ \ldots,\ n-2\\ u_{n-1}\beta_{n-1}+u_n\beta_n&=\gamma(k_{n-1}\varkappa_{n-1}+k_n\varkappa_n)\,h\\ u_{n-1}\beta_n&=\gamma k_n\varkappa_{n+1}h \end{aligned}\right\} \quad (4.50)$$

For the first $(n-2)$ equalities (4.50), we will make use of the matrices (2.64), and for the two last we will use (2.66). We will then have

$$\left.\begin{aligned} \beta_i u_i&=\frac{\gamma F(\lambda_i)\,h}{\Delta'(\lambda_i)}, \qquad i=1,\ 2,\ \ldots,\ n-2\\ \beta_{n-1}u_{n-1}+\beta_n u_n&=\gamma\left[\frac{F(0)}{\Delta_{n-1}(0)}\right]^{(1)}h\\ \beta_n u_{n-1}&=\gamma\left[\frac{F(0)}{\Delta_{n-1}(0)}\right]h \end{aligned}\right\} \quad (4.51)$$

Using (4.51) we finally rewrite the characteristic equation (4.49) in the form

$$\Delta^*(\lambda)=(\lambda-1)^2\prod_{j=1}^{n-2}\left(\lambda-e^{\lambda_j T}\right)\left\{1-\sum_{i=1}^{n-2}\frac{e^{\lambda_i T_2}\left(e^{\lambda_i T_1}-1\right)}{\lambda_i\left(\lambda-e^{\lambda_i T}\right)}\frac{\gamma F(\lambda_i)\,h}{\Delta'(\lambda_i)}\right.$$

$$-\frac{T_1}{\lambda-1}\gamma\left[\frac{F(0)}{\Delta_{n-1}(0)}\right]^{(1)}h$$

$$\left.-\left(T_2+\frac{T_1}{2}+\frac{T}{\lambda-1}\right)\frac{T_1}{\lambda-1}\gamma\left[\frac{F(0)}{\Delta_{n-1}(0)}\right]h\right\}=0 \quad (4.52)$$

The characteristic equation (4.52) is expressed in final form in terms of the parameters of the original matrix equation (4.3).

4.3 STABILITY CRITERIA

The stability criteria for a linear discontinuous system establish the fact that the moduli of all the roots λ_i^* of the characteristic equation of the matrix P^* (4.15) are less than unity or, expressing it geometrically, all the roots λ_i^* lie inside a circle of unit radius with the coordinate origin as centre.

Various methods exist for defining the necessary and sufficient conditions for stability. We shall discuss here two criteria: one criterion reduces to the Hurwitz problem; the other is based on the application of the theory of residues to contour integration. The second criterion is more general, since it is applicable to those cases when the characteristic equation of the matrix P^* of a closed loop system is a transcendental equation.

a) Reduction to the Hurwitz problem

In the characteristic equation let us replace λ by the new variable ν using the substitution (see, for example, the book by Oldenbourg and Sartorius [20]):

$$\lambda = \frac{\nu + 1}{\nu - 1} \tag{4.53}$$

and, conversely

$$\nu = \frac{\lambda + 1}{\lambda - 1} \tag{4.54}$$

The substitution (4.53) is a conformal transformation which establishes a relationship between the points of the complex plane λ lying inside a circle of unit radius with the coordinate origin as centre and the points of the complex plane ν lying to the left of the imaginary axis. In fact, if the point λ describes a counter-clockwise unit circle $\lambda = e^{i\alpha}$ with $(0 \leqslant \alpha \leqslant 2\pi)$, then the point $\nu = -i \cot \frac{\alpha}{2}$ moves along the imaginary axis from minus infinity to plus infinity. With the stipulated direction of movement around the circle, the internal points of the unit circle on the plane λ and the points of the left-hand half-plane ν lie to the left of their boundary curves so that after the conformal transformation (4.53) they turn out to be reciprocal.

Let us consider first the case of simple characteristic values λ_i of the matrix P and let us choose the characteristic equation of the closed discontinuous system in the form (4.30) corresponding to this case. We substitute into (4.30) λ from (4.53). Then, remembering that

$$\lambda - e^{\lambda_i T} = \frac{(\nu + 1) - (\nu - 1)\, e^{\lambda_i T}}{\nu - 1} = \frac{\nu\left(1 - e^{\lambda_i T}\right) + \left(1 + e^{\lambda_i T}\right)}{\nu - 1} \tag{4.55}$$

we obtain

$$\prod_{j=1}^{n} \left[\frac{\nu\left(1 - e^{\lambda_j T}\right) + \left(1 + e^{\lambda_j T}\right)}{\nu - 1} \right] \times \left\{ 1 - (\nu - 1) \sum_{i=1}^{n} \frac{u_i' \beta_i}{\nu\left(1 - e^{\lambda_i T}\right) + \left(1 + e^{\lambda_i T}\right)} \right\} = 0 \tag{4.56}$$

or, after cancelling the non-zero factor $(\nu-1)^{-n}$, we can write the transformed characteristic equation in the form

$$\prod_{j=1}^{n}[\nu(1-e^{\lambda_j T})+(1+e^{\lambda_j T})]$$

$$\times\left\{1-(\nu-1)\sum_{i=1}^{n}\frac{u'_i\beta_i}{\nu(1-e^{\lambda_i T})+(1+e^{\lambda_i T})}\right\}=0 \tag{4.57}$$

We should note that when $\nu=1$ it follows from (4.53) that $\lambda=\infty$ but, since we are only interested in the region where $|\lambda|<1$, this case is excluded.

In the case of a double zero root $\lambda_{n-1}=\lambda_n=0$ we should take the characteristic equation (4.40). Repeating the procedure just given, we obtain for this case the transformed characteristic equation in the form

$$\prod_{j=1}^{n-2}[\nu(1-e^{\lambda_j T})+(1+e^{\lambda_j T})]$$

$$\times\left\{1-(\nu-1)\sum_{i=1}^{n-2}\frac{u'_i\beta_i}{\nu(1-e^{\lambda_i T})+(1+e^{\lambda_i T})}\right.$$

$$\left.-\frac{\nu-1}{2}(u'_{n-1}\beta_{n-1}+u'_n\beta_n)-T\frac{(\nu-1)^2}{4}u'_{n-1}\beta_n\right\}=0 \tag{4.58}$$

To express the transformed characteristic equations (4.57) and (4.58) in terms of the initial parameters of (4.3) it is necessary to apply successively (4.31) and (4.34) or (4.48) and (4.51) to (4.57) and (4.58), respectively. Such a successive conversion was carried out above for the untransformed characteristic equation. We will leave the transformed equations in the shortened form (4.57) and (4.58), bearing in mind that a similar conversion can be simply performed using the formulae indicated above. The transformed characteristic equations are equations of the nth degree.

$$p'_0\nu^n+p'_1\nu^{n-1}+\dots+p'_n=0 \tag{4.59}$$

with real coefficients p'_i. For a linear system to be asymptotically stable it is necessary and sufficient that all the roots of (4.59) shall lie to the left of the imaginary axis and, for this, it is necessary and sufficient that the following inequalities be fulfilled:

$$\Delta_1>0,\ \Delta_2>0,\ \dots,\ \Delta_{n-1}>0,\ p'_n>0 \quad \text{for} \quad p'_0>0 \tag{4.60}$$

or

$$\Delta_1<0,\ \Delta_2>0,\ \dots,\ (-1)^{n-1}\Delta_{n-1}>0,\ p'_n<0 \quad \text{for} \quad p'_0<0 \tag{4.61}$$

where Δ_i are the Hurwitz determinants. The determinants Δ_i are main diagonal minors of the Hurwitz matrix

$$\left\| \begin{matrix} p'_1 & p'_0 & 0 & 0 & 0 & 0 & \dots \\ p'_3 & p'_2 & p'_1 & p'_0 & 0 & 0 & \dots \\ p'_5 & p'_4 & p'_3 & p'_2 & p'_1 & p_0 & \dots \\ p'_7 & p'_6 & p'_5 & p'_4 & p'_3 & p'_2 & \dots \\ \cdot & \cdot & \cdot & \cdot & \cdot & \cdot & \dots \\ \cdot & \cdot & \cdot & \cdot & \cdot & \cdot & \dots \end{matrix} \right\| \tag{4.62}$$

In the Hurwitz matrix one must put a zero instead of p'_i wherever $i > n$. Thus, the substitution (4.53) enables us to reduce the solution of the problem of the stability of a linear discontinuous system to the Hurwitz criterion, a criterion widely applied in the theory of linear continuous control systems.

b) Application of the argument principle

Let us consider the characteristic equation $\Delta^*(\lambda)$ of a linear discontinuous system in the form (4.30) and (4.40), regarding this form as a shortened notation for (4.35) and (4.52) which are expressed in terms of the original parameters of the system. We will introduce a special notation for the rational function appearing in (4.30) and (4.40). We shall put

$$\left.\begin{aligned} W^*(\lambda) &= \frac{e^*(\lambda)}{\Delta_p^*(\lambda)} = -\sum_{i=1}^{n} \frac{u'_i\beta_i}{\lambda - e^{\lambda_i T}}, \\ W^*(\lambda) &= \frac{e^*(\lambda)}{\Delta_p^*(\lambda)} = -\sum_{i=1}^{n-2} \frac{u'_i\beta_i}{\lambda - e^{\lambda_i T}} - \frac{u'_{n-1}\beta_{n-1} + u'_n\beta_n}{\lambda - 1} \\ &\quad - T\frac{u'_{n-1}\beta_n}{(\lambda - 1)^2} \end{aligned}\right\} \tag{4.63}$$

The denominator $\Delta_p^*(\lambda)$ is a polynomial of degree n which can always be written in the form

$$\Delta_p^*(\lambda) = \prod_{i=1}^{n} (\lambda - e^{\lambda_i T}) \tag{4.64}$$

if we assume that the roots λ_i can be repeated. From a comparison of (4.30) and (4.40) with (4.63) and (4.64), we find

$$1 + W^*(\lambda) = \frac{\Delta^*(\lambda)}{\Delta_p^*(\lambda)} \tag{4.65}$$

from which it follows that the zero of the function $1 + W^*(\lambda)$ coincide with the roots λ_i^* of the characteristic polynomial $\Delta^*(\lambda)$ and its poles are the roots $e^{\lambda_i T}$ of the polynomial $\Delta_p^*(\lambda)$. Since the power of the numerator $e^*(\lambda)$ of the rational function (4.63) is less than the power of the denominator $\Delta_p^*(\lambda)$, the powers of the polynomials $\Delta^*(\lambda)$ and $\Delta_p^*(\lambda)$ are the same. Consequently, the function $1+W^*(\lambda)$ has an identical number of zeros and poles. The argument principle holds for the function $1+W^*(\lambda)$.

The argument principle

Suppose that on the complex plane λ we take a closed contour Γ which does not pass through zero nor through the poles of the function $1+W^*(\lambda)$. Then the difference between the number of zeros q and the number of poles p of the function $1+W^*(\lambda)$ which are enclosed by the contour Γ is equal to the variation of the argument $\delta \arg [1+W^*(\lambda)]$ of the function $1+W^*(\lambda)$ during a circuit of the point λ about the contour Γ in the positive direction divided by 2π. The number of times each zero and pole is counted is equal to the multiplicity, and the positive direction is assumed to be the counter-clockwise direction around the curve Γ.

Thus

$$\frac{\Psi_1 - \Psi_0}{2\pi} = q - p \tag{4.66}$$

where $\Psi_0 = \arg [1+W^*(\lambda)]$ at the initial point λ_0, and Ψ_1 is the argument of the same point λ_0 after a circuit around the curve Γ in the positive direction.

To prove the argument principle we will, using (4.65), represent the function $1+W^*(\lambda)$ in the form

$$1+W^*(\lambda) = \text{const}\,\frac{(\lambda-\lambda_1^*)(\lambda-\lambda_2^*)\ldots(\lambda-\lambda_n^*)}{(\lambda-e^{\lambda_1 T})(\lambda-e^{\lambda_2 T})\ldots(\lambda-e^{\lambda_n T})} \tag{4.67}$$

Here each elementary factor is written down the same number of times as the multiplicity of the corresponding root. This procedure permits us to apply the same reasoning as we did when discussing the case of multiple roots. Therefore, we have

$$\delta \arg [1+W^*(\lambda)] = \sum_{i=1}^{n} \delta \arg (\lambda - \lambda_i^*) - \sum_{k=1}^{n} \delta \arg (\lambda - e^{\lambda_k T}) \tag{4.68}$$

If root λ_i^* of the polynomial $\Delta^*(\lambda)$ or a root $e^{\lambda_k T}$ of the polynomial $\Delta_p^*(\lambda)$ lies inside the closed contour Γ then, as can be seen from Fig 4.1a, the vector represented by the difference $\lambda-\lambda_i^*$ or $\lambda - e^{\lambda_k T}$ rotates in a counter-clockwise direction through an angle equal to 2π when the point λ completes one circuit

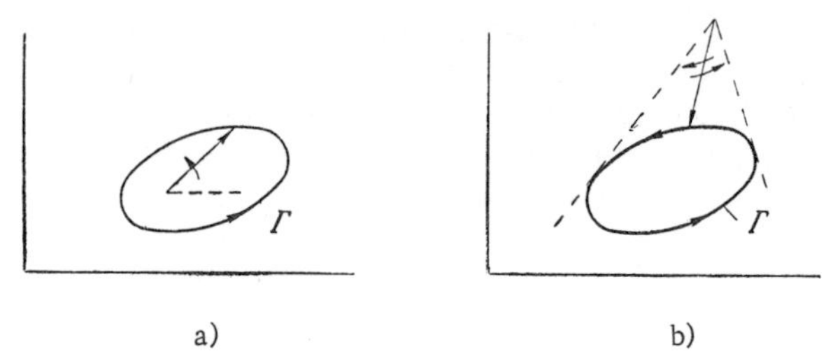

Fig 4.1 Change of argument during a circuit round a closed contour.

of the curve Γ in the positive direction. If the above roots lie outside the contour Γ, then from Fig 4.1b we can conclude that, under the same conditions, the corresponding vectors rotate through the same angle in both positive and negative directions so that the total angle through which they turn is equal to zero.

These considerations, together with (4.68), prove the validity of the argument principle for a rational function. In general, the argument principle is a consequence of the logarithmic remainder theorem and it is valid for any analytical function which has only isolated singular points (poles) inside a closed contour Γ – see, for example, Markushevich [18].

The argument principle has a geometrical interpretation. When the point λ makes a circuit around the closed contour Γ in the positive direction, the end of the vector $1+W^*(\lambda)$ describes a closed curve Γ′. In general, the vector $1+W^*(\lambda)$ makes a certain number of complete turns around the coordinate origin. Suppose we assign $+1$ to each complete turn in the positive direction and -1 if in the negative direction. Then the argument principle can be formulated in the following way: the difference between the number of zeros and poles of the function $1+W^*(\lambda)$ enclosed by the closed contour Γ is equal to the number of complete turns which the vector representing the function $1+W^*(\lambda)$ makes around the coordinate origin during the time that the point λ makes a circuit around the countour Γ in the positive direction.

In control theory it is usual to construct a closed curve which, under the conditions indicated above, describes a phasor representing the function $W^*(\lambda)$. The curve constructed in this way is sometimes called the hodograph of the phasor $W^*(\lambda)$. In this case it is obvious that, to find the difference between the number of zeros and the number of poles of the function $1+W^*(\lambda)$, one should count the number of complete turns made on the plane W^* by the corresponding phasor relative to the point -1.

The application of the argument principle to the study of stability is based on the following assumptions. The poles $e^{\lambda_i T}$ of the function $1+W^*(\lambda)$ are considered to be known, since they are given by the roots λ_i of the characteristic equation of the open loop control system. The distribution of poles $e^{\lambda_i T}$ with respect to a circle of unit radius with the coordinate origin as centre is found from the properties of the roots λ_i. Indeed, for any root λ_i it follows from the conditions $\operatorname{Re}\lambda_i \lesseqqgtr 0$ that $|e^{\lambda_i T}| \lesseqqgtr 1$.

In the terminology of control theory, this fact can be expressed in the following way. If a control system is stable in the open state, then all the poles of the function $1+W^*(\lambda)$ (4.63) which are roots of the polynomial $\Delta_p^*(\lambda)$ (4.64) lie inside a unit circle with the coordinate origin as centre; in the opposite case one or more poles are located outside this circle. Finally, if an open system finds itself on the boundary of stability, then one or more of the poles is located on a bounding circumference of unit radius.

When a linear discontinuous system is stable, all the roots λ_i^* of the function $1+W^*(\lambda)$ which are roots of the characteristic polynomial $\Delta^*(\lambda)$ lie inside a unit circle. Thus, for a stable linear discontinuous system, the difference between the number of zeros and poles inside a circle of unit radius is known in advance.

Consider now on the complex plane a unit circle $\lambda = e^{i\alpha}$ with $0 \leqslant \alpha \leqslant 2\pi$. This interval of variation corresponds to the point λ making a positive circuit around the unit circle. We can formulate the following stability criteria.

Stability criterion I

For a linear discontinuous system to be stable it is necessary and sufficient that the increment of the argument of the function $1+W^*(e^{i\alpha})$ should be given by Table 4.1 when α varies within the interval $0 \leqslant \alpha \leqslant 2\pi$.

Table 4.1

No.	Increment of $\arg[1+W^*(e^{i\alpha})]$	Characteristic of the roots λ_i of the open system
1	0	$\operatorname{Re}\lambda_i < 0,\ i=1, 2, \ldots, n$
2	$s\pi$	$\lambda_i = \lambda_2 = \ldots = \lambda_s = 0$ $\operatorname{Re}\lambda_k < 0,\ k = s+1,\ s+2, \ldots, n$
3	$2s\pi$	$\lambda_1 = \pm i\omega_1,\ \lambda_2 = \pm i\omega_2, \ldots, \lambda_s = \pm i\omega_s$ $\operatorname{Re}\lambda_k < 0,\ k = 2s+1,\ 2s+2, \ldots, n$
4	$2s\pi$	$\operatorname{Re}\lambda_i > 0,\ i=1, 2, \ldots, s$ $\operatorname{Re}\lambda_k < 0,\ k = s+1,\ s+2, \ldots, n$

The first and fourth cases are a consequence of the argument principle if the circle $\lambda = e^{i\alpha}$ $(0 \leqslant \alpha \leqslant 2\pi)$ is taken as the contour

(Fig 4.2a). In fact, in this case, $q=n$, $p=n$, and $p=n-s$, respectively, in the first and fourth cases.

For the second and third cases, a unit circle cannot serve as the contour Γ since that circle passes through certain poles of the function $1+W^*(\lambda)$. In these cases one must take the curves represented in Fig 4.2b and Fig 4.2c as the contour Γ. These curves consist of the arc $\lambda=e^{i\alpha}$ of the unit circle, and the arc $\lambda-e^{\lambda_j T}=\rho e^{i\alpha}$ of circles with small radii ρ whose centres are at poles $e^{\lambda_j T}$, for which $|e^{\lambda_j T}|=1$, i.e. for which either $\lambda_j=0$, or $\lambda_j=\pm i\omega$. These cases are illustrated in Figs 4.2b and 4.2c, respectively.

Within a small neighbourhood of the pole $e^{\lambda_j T}$, (4.67) can be written approximately in the form

$$1+W^*(\lambda)\approx \text{const}\frac{1}{\lambda-e^{\lambda_j T}} \qquad (4.69)$$

From this it follows that, as the point λ goes round the arc $\lambda-e^{\lambda_j T}=\rho e^{i\alpha}$ in the positive direction, the change in the argument of the function $1+W^*(\lambda)$ is the same as the increment of the angle α, but with its sign reversed. However, with $\rho\to 0$, the increment α tends to π. This, in conjunction with the fact that inside the contours Γ represented in Figs 4.2b and 4.2c the function $1+W^*(\lambda)$ has n zeros and n poles, proves that in the limit the change in $\arg[1+W^*(\lambda)]$ over the circumference of a unit circle as $\rho\to 0$ is that indicated for cases 2 and 3 of Table 4.1.

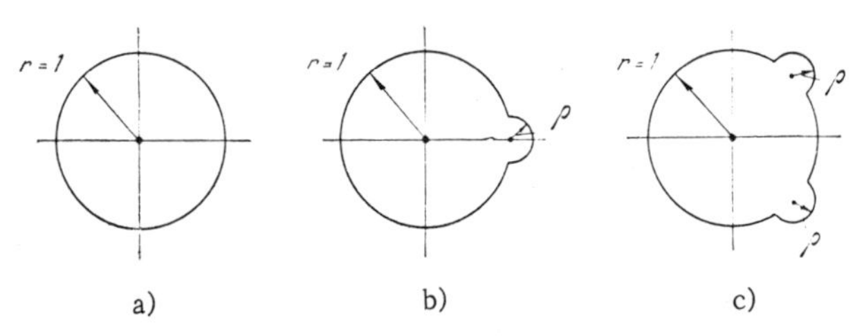

Fig 4.2 Closed contours Γ on the complex plane λ.

For discontinuous control systems which are stable in the open state, or which are at the boundary of stability (see cases 1 and 2 of Table 4.1), one can provide a unique formulation of the stability criterion by making use of the geometrical interpretation of the argument principle along with the supplementary remarks made in connection with the construction of the hodograph of the phasor $W^*(\lambda)$. Suppose we construct a hodograph of the phasor representing the function $W^*(\lambda)$ on the complex plane W^* when the point λ goes round one of the closed contours shown in Fig 4.2 in the positive direction. The stability criterion can now be formulated as follows.

Stability criterion II

For a linear discontinuous system (which is stable in the open state or which is at the boundary of stability) to be stable, it is necessary and sufficient that the overall rotation of the vector whose beginning is at the point -1 should be equal to zero when its end goes round the hodograph $W^*(\lambda)$ (allowing for infinite branches)

To make use of these stability criteria it is necessary, by means of the function $W^*(\lambda)$, to map the closed contours Γ (see Fig 4.2) of the complex plane λ on to the complex plane W^* Let us note some characteristic features of such a mapping.

The polynomials $e^*(\lambda)$ and $\Delta_p^*(\lambda)$ (4.63) have real coefficients, so that, for complex conjugate values λ, the function $W^*(\lambda)$ also has complex conjugate values. It follows that the hodograph of the phasor $W^*(\lambda)$ is symmetric with respect to the real axis of the complex plane W^*, since the closed curves Γ represented in Fig 4.2 are symmetric with respect to the real axis of the complex plane λ. In particular, the points of intersection of the curves Γ with the real axis correspond to analogous points of the hodograph $W^*(\lambda)$. To the arcs of the circles with small radii ρ there correspond large radius arcs (of the order $1/\rho$) on the plane W^* (4.69). The point W^* passes around these arcs in a negative direction, i.e. in a clockwise direction. When $\rho \to 0$, arcs of infinitely large radius on the plane W^* contain as many semicircles as the multiplicity of the corresponding pole $e^{\lambda_j T}$ for $|e^{\lambda_j T}| = 1$.

Figure 4.3 illustrates some typical hodographs of the phasor $W^*(\lambda)$. Figure 4.3a shows the hodograph $W^*(\lambda)$ for a linear system that is stable in the open state, and Figs 4.3b and 4.3c show the hodographs for a neutral-stable system in the open state when $\lambda_n = 0$ and $\lambda_{n-1} = \lambda_n = 0$, respectively. Finally, Fig 4.3d shows the case when $\lambda_{n-1,n} = \pm i\omega$. The end of the phasor whose beginning is at the point -1 represents the function $1 + W^*(\lambda)$ if -1 is taken as the coordinate origin. In all the diagrams of Fig 4.3, the position of the point -1 is such that all four cases correspond to stable systems.

Stability criterion II can be reformulated in the following way.

Stability criterion III

For a linear pulsed system (which is stable in the open state or is at the boundary of stability) to be stable, it is necessary and

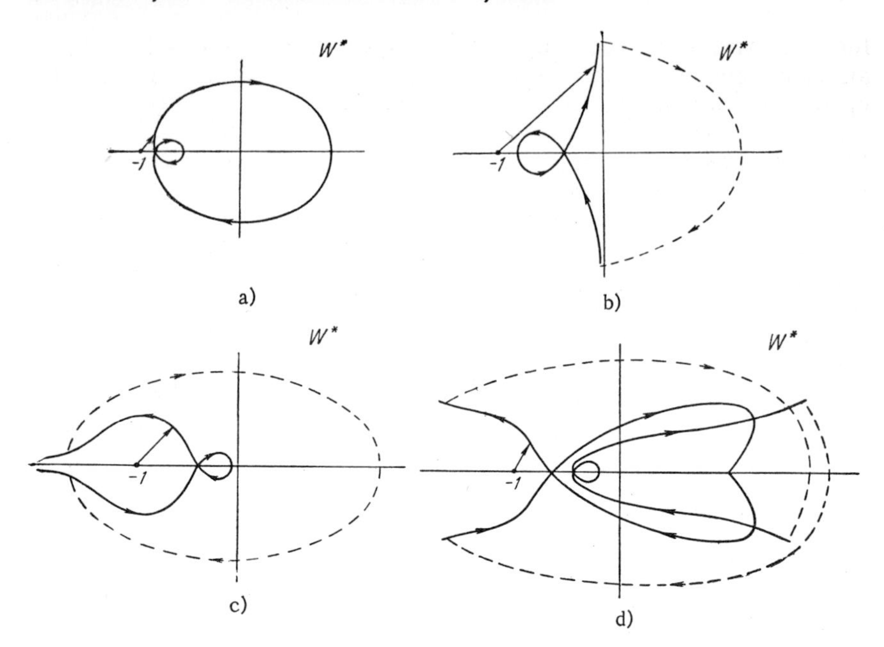

Fig 4.3 Typical hodographs of the vector $W^*(\lambda)$ on the complex plane W^*.

sufficient that the hodograph $W^*(\lambda)$ should intersect the real axis to the left of the point -1 the same number of times from below to above as from above to below when the point λ moves in a positive direction around the half-contour Γ (Fig 4.2) lying in the upper half-plane.

There is a special case when the application of a rational function $W^*(\lambda)$ can lead to incorrect deductions regarding the stability of a pulsed system. This case arises when the numerator $\Delta_p^*(\lambda)$ and the denominator $e^*(\lambda)$ of the rational function $W^*(\lambda)$ have identical roots $e^{\lambda_j T}$ for which $|e^{\lambda_j T}| > 1$. In this case it is obvious that $e^{\lambda_j T}$ will also be a root of the characteristic equation $\Delta^*(\lambda)$ and that the system will be unstable. However, in the formal application of the stability criteria and the function $W^*(\lambda)$, it may turn out that after the elementary factors $(\lambda - e^{\lambda_j T})$ have been cancelled the conditions for stability will be found to be fulfilled.

The stability criteria considered here are related to the Nyguist stability criterion in the theory of linear continuous control systems.

The first to obtain a stability criterion based on the application of the argument principle to a rational function was Tsipkin. In the theory of discontinuous systems which he developed on the basis of the discrete Laplace transform, the rational function $W^*(\lambda)$ for $\lambda = e^{sT}$ has a clear physical meaning since it is a transfer function within the meaning of the discrete Laplace transform for an open linear discontinuous system. We note that the idea of an open loop control system as used here is not the same as the open loop control system discussed by Tsipkin. In his book, Tsipkin [26] mentions certain peculiarities arising in the application of Stability Criterion III which are not discussed in this book.

4.3.1 Application of the argument principle to a characteristic polynomial

The characteristic polynomial $\Delta^*(\lambda)$ of a linear discontinuous system will be represented in terms of its roots λ_i^* in the form

$$\Delta^*(\lambda) = p_0^*(\lambda - \lambda_1^*)(\lambda - \lambda_2^*) \ldots (\lambda - \lambda_n^*) \tag{4.70}$$

where each elementary factor is repeated for a number of times equal to the multiplicity of the corresponding root λ_σ^*, and p_0^* is the coefficient of λ^n. The polynomial $\Delta^*(\lambda)$ may be regarded as a special case of the rational function $1 + W^*(\lambda)$, when $\Delta_p^*(\lambda) = 1$.

We shall apply the argument principle to the characteristic polynomial $\Delta^*(\lambda)$, having taken the circle of unit radius illustrated in Fig 4.2a as the contour Γ. The end of the phasor representing the function $\Delta^*(\lambda)$ for $\lambda = e^{i\alpha}$ describes a closed curve as α goes from 0 to 2π. This curve will be called the hodograph of the vector $\Delta^*(e^{i\alpha})$. The circle of unit radius and the hodograph of $\Delta^*(e^{i\alpha})$ are symmetric with respect to their real axes; consequently it is convenient to construct the hodograph $\Delta^*(e^{i\alpha})$ in the interval $0 \leqslant \alpha \leqslant \pi$ which corresponds to a circuit of the point λ round the upper semicircle in the positive direction. When $|\lambda_i^*| < 1$, $i = 1, 2, \ldots, n$, it follows from (4.70) that $\Delta^*(e^{i\alpha}) \neq 0$ in the interval $0 \leqslant \alpha \leqslant 2\pi$. The function $\Delta^*(e^{i\alpha})$ can vanish only when $|\lambda_j^*| = 1$ at the points α satisfying the equality $\lambda_j = e^{i\alpha}$ For $|\lambda_i^*| < 1$, $i = 1, 2, \ldots, n$, the sign of $\Delta^*(e^{i\alpha})$ for $\alpha = 0$ is the same as the sign of the leading coefficient p_0^* As α varies between 0 and 2π, the phasor representing the elementary factor $(e^{i\alpha} - \lambda_i^*)$ of the function $\Delta^*(e^{i\alpha})$, monotonically rotates in the positive direction (see Fig 4.1a) through an angle equal to 2π, if $|\lambda_i^*| < 1$.

Let all the roots λ_i^* lie inside a unit circle; then the increment of the argument of the functions $\Delta^*(e^{i\alpha})$ is equal to $2\pi n$ when α varies from 0 to 2π. This fact follows directly from (4.66) since $\Delta^*(\lambda)$ has no poles in any finite part of the complex plane λ. Starting from this fact one can formulate a stability criterion which is the analogue of Mikhailov's criterion stemming from the theory of linear continuous systems.

Stability criterion IV

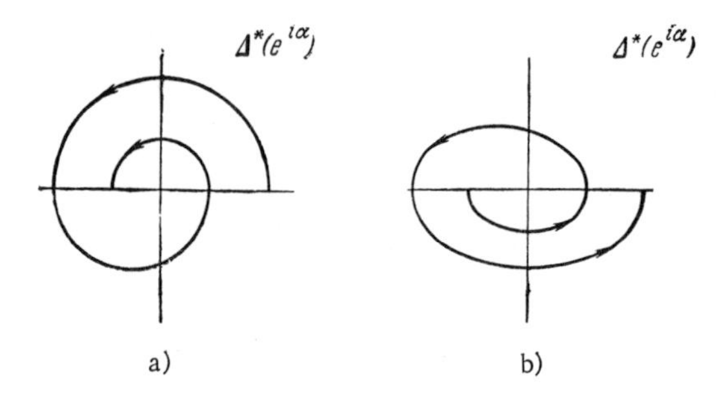

Fig 4.4 Hodograph of the phasor $\Delta^*(e^{i\lambda})$ with $0 \leqslant \alpha \leqslant \pi$ for (a) $p_0^* > 0$ and (b) $p_0^* < 0$.

For a linear discontinuous system to be stable it is necessary and sufficient that the vector of the hodograph $\Delta^*(e^{i\alpha})$ should rotate in a positive direction through an angle $n\pi$ when α varies from 0 to π. Here, n is the degree of the characteristic polynomial $\Delta^*(\lambda)$.

Figure 4.4 shows the hodograph of a polynomial $\Delta^*(\lambda)$ of third degree when the modulus of the roots λ_i^* are less than unity.

Tsipkin [26] was the first to generalise Mikhailov's criterion to the characteristic polynomial $\Delta^*(\lambda)$ of a linear discontinuous system.

4.4 SEPARATION OF THE STABILITY BOUNDARY

In the previous section we obtained the necessary and sufficient conditions for the stability of linear discontinuous systems. We will now discuss one particular method of investigating stability which is sometimes convenient in practical applications. The roots λ_i^* of the characteristic polynomial $\Delta^*(\lambda)$ are continuous functions of the parameters of the system. Suppose, in turn, that the parameters of the system are continuous functions of some parameter α; then the roots λ_i^* will also be continuous functions of this parameter. We shall refer to those values of the parameter α for which the modulus of at least one of the roots λ_i^* is equal to unity as the critical values. It is then clear that the stability can only change sign at critical values of the parameter α. Therefore, for all values of the parameter α lying between two successive critical values, the stability of the system has one and the same sign. The interval of variation of the parameter α inside which the system is stable is always limited by its critical values.

Thus one can apply the following method of the study of stability problems. It is first necessary to find all the critical values of the parameter α and then, in order to decide about the stability over the entire range of variation of the parameter α, it is sufficient to determine by some means or other (even if by direct calculation of the roots λ_i^*) the sign of the stability for only one value inside each interval of variation of α between two successive critical values. In the intervals mentioned above, one should choose those values of α for which the characteristic polynomial $\Delta^*(\lambda)$ assumes the simplest form. This will simplify the calculations necessary to determine the character of the roots of the polynomial.

One can in this way separate out from among all the critical values of the parameter α just those which are the boundaries of the region of stability of the system.

Let us apply the matrix method to the problem of finding the conditions which will ensure the existence of roots λ_i^* whose moduli are equal to unity. For this purpose we will represent the characteristic equation $\Delta^*(\lambda)=0$ in the form

$$\Delta^*(\lambda) = p_0^*\lambda^n + p_1^*\lambda^{n-1} + \dots + p_n^* = 0 \tag{4.71}$$

where, obviously, all the coefficients p_i^* are real numbers. Consider first the real roots of (4.71) with moduli equal to unity. These roots can be $\lambda^*=1$ and $\lambda^*=-1$. Substituting these values into (4.67), we obtain the necessary conditions for the existence of such roots.

$$\left.\begin{aligned} p_0^* + p_1^* + \dots + p_n^* = 0 \\ (-1)^n p_0^* + (-1)^{n-1} p_1^* + \dots + p_n^* = 0 \end{aligned}\right\} \tag{4.72}$$

It is not difficult to see that these necessary conditions are also sufficient conditions. We will investigate now the case of complex roots. Let λ_1^* and λ_2^* be a pair of complex conjugate roots of (4.71) with moduli equal to unity. In that case, when $\omega > 0$, one can write them in the form

$$\lambda_1^* = e^{i\omega}, \quad \lambda_2^* = e^{-i\omega} \tag{4.73}$$

Furthermore, we have

$$\lambda_1^*\lambda_2^* = 1; \quad \lambda_1^* + \lambda_2^* = 2\cos\omega = -\xi \tag{4.74}$$

We put the left-hand side of (4.71) in the form

$$\Delta^*(\lambda) = (\lambda^2 + \xi\lambda + 1)(q_0\lambda^{n-2} + q_1\lambda^{n-3} + \dots + q_{n-2}) \tag{4.75}$$

Multiplying out the brackets on the right-hand side and equating coefficients with identical powers of λ in (4.71) and (4.75), we obtain $(n+1)$ equations for determining the n unknowns $q_0, q_1, \ldots, q_{n-2}$ and ξ

$$\left.\begin{array}{l} q_0 = p_0^* \\ q_0\xi + q_1 = p_1^* \\ q_0 + q_1\xi + q_2 = p_2^* \\ \cdots\cdots\cdots\cdots\cdots\cdots \\ q_{s-2} + q_{s-1}\xi + q_s = p_s^* \\ \cdots\cdots\cdots\cdots\cdots\cdots \\ q_{n-4} + q_{n-3}\xi + q_{n-2} = p_{n-2}^* \\ q_{n-3} + q_{n-2}\xi = p_{n-1}^* \\ q_{n-2} = p_n^* \end{array}\right\} \tag{4.75}$$

Equations (4.75) are symmetrically described from above to below and vice versa. For perfect symmetry, and to simplify later calculations, we will assume for the present that n is odd, i.e. set $n = 2k+1$. We then have an even number $(2k+2)$ of (4.75). Let us introduce the left triangular matrix of the $(k+1)$ th order

$$\overbrace{\left\|\begin{array}{ccccccc} 1 & 0 & 0 & \ldots & 0 & 0 & 0 \\ \xi & 1 & 0 & \ldots & 0 & 0 & 0 \\ 1 & \xi & 1 & \ldots & 0 & 0 & 0 \\ \cdot & \cdot & \cdot & \cdot & \cdot & \cdot & \cdot \\ 0 & 0 & 0 & \ldots & 1 & 0 & 0 \\ 0 & 0 & 0 & \ldots & \xi & 1 & 0 \\ 0 & 0 & 0 & \ldots & 1 & \xi & 1 \end{array}\right\|}^{(k+1)\text{ columns}} \tag{4.76}$$

with equal elements in the oblique rows. The triangular matrix (4.76) can be represented in terms of the unit oblique rows (2.26) in the form

$$C = E + \xi H^1 + H^2 \tag{4.77}$$

It is easily seen that, with the help of matrix (4.76), the system of equations (4.75) can be written in the form of the two matrix equations

$$C\left\|\begin{array}{c} q_0 \\ q_1 \\ \vdots \\ q_{k-1} \\ q_k \end{array}\right\| = \left\|\begin{array}{c} p_0^* \\ p_1^* \\ \vdots \\ p_{k-1}^* \\ p_k^* \end{array}\right\| \tag{4.78}$$

and

$$\| q_{k-1} q_k \dots q_{2k-2} q_{2k-1} \| C = \| p^*_{k+1} p^*_{k+2} \dots p^*_{2k} p^*_{2k+1} \| \tag{4.79}$$

The solution of (4.78) and (4.79) can be represented in the form

$$\begin{Vmatrix} q_0 \\ q_1 \\ \vdots \\ q_{k-1} \\ q_k \end{Vmatrix} = C^{-1} \begin{Vmatrix} p^*_0 \\ p^*_1 \\ \vdots \\ p^*_{k-1} \\ p^*_k \end{Vmatrix} \tag{4.80}$$

and

$$\| q_{k-1} q_k \dots q_{2k-2} q_{2k-1} \| = \| p^*_{k+1} p^*_{k+2} \dots p^*_{2k} p^*_{2k+1} \| C^{-1} \tag{4.81a}$$

The inverse matrix C^{-1} in these equations can be obtained using the procedure for inverting a triangular matrix with equal elements in the oblique rows, which was given as a special example in Section 2.1 [see (2.33)].

From (4.77), quite formally, we obtain

$$C^{-1} = E - (\xi H^1 + H^2) + (\xi H^1 + H^2)^2 - \dots + (-1)^k (\xi H^1 + H^2)^k \tag{4.81b}$$

Thus, C^{-1} is a left triangular matrix with equal elements in the oblique rows, i. e.

$$C^{-1} = \begin{Vmatrix} c_0 & 0 & 0 & \dots & 0 & 0 & 0 \\ c_1 & c_0 & 0 & \dots & 0 & 0 & 0 \\ c_2 & c_1 & c_0 & \dots & 0 & 0 & 0 \\ \cdot & \cdot & \cdot & \cdot & \cdot & \cdot & \cdot \\ c_{k-2} & c_{k-3} & c_{k-4} & \dots & c_0 & 0 & 0 \\ c_{k-1} & c_{k-2} & c_{k-3} & \dots & c_1 & c_0 & 0 \\ c_k & c_{k-1} & c_{k-2} & \dots & c_2 & c_1 & c_0 \end{Vmatrix} \tag{4.82}$$

or, in the other notation,

$$C^{-1} = c_0 E + c_1 H^1 + \dots + c_k H^k \tag{4.83}$$

To find the coefficients c_ρ of this matrix we will expand the right-hand side of (4.81) according to Newton's rule. Then, remembering rule (2.27) for multiplying unit oblique rows, we obtain

$$C^{-1} = E - \xi H^1 - H^2 + \xi^2 H^2 + 2\xi H^3 + H^4 - \xi^3 H^3 - 3\xi^2 H^4$$
$$- 3\xi H^5 - H^6 + \xi^4 H^4 + 4\xi^3 H^5 + 6\xi^2 H^6 + 4\xi H^7 + H^8 - \dots \tag{4.84}$$

Equating coefficients of identical H^k in the (4.83) and (4.84), we find

$$\left.\begin{array}{l} c_0=1,\ c_1=-\xi,\ c_2=(\xi^2-1),\ c_3=-(\xi^2-2\xi) \\ c_4=(\xi^4-3\xi^2+1),\ \dots \end{array}\right\} \tag{4.85a}$$

A recurrence relation, expressed by

$$c_\rho=-(\xi c_{\rho-1}+c_{\rho-2}) \tag{4,85b}$$

holds between the coefficients (4.85) and with its help we can successively (starting with c_3) write out the values of any coefficeint c_ρ up to the highest coefficient c_k. The coefficients c_ρ are polynomials of degree ρ with respect to ξ.

In (4.80) and (4.81), two identical unknowns, q_{k-1} and q_k occur. By virtue of the form of the matrix C^{-1} defined by (4.82), we obtain from (4.80) and (4.81a) by direct calculation the scalar expressions

$$\left.\begin{array}{l} q_{k-1}=c_{k-1}p_0^*+c_{k-2}p_1^*+\dots+c_1p_{k-2}^*+c_0p_{k-1}^* \\ q_k=c_kp_0^*+c_{k-1}p_1^*+\dots+c_1p_{k-1}^*+c_0p_k^* \\ q_{k-1}=c_0p_{k+1}^*+c_1p_{k+2}^*+\dots+c_{k-1}p_{2k}^*+c_kp_{2k+1}^* \\ q_k=c_0p_{k+2}^*+c_1p_{k+3}^*+\dots+c_{k-1}p_{2k}^*+c_{k-1}p_{2k+1}^* \end{array}\right\} \tag{4.86}$$

which determine the unknown coefficients q_{k-1} and q_k. Eliminating q_{k-1} and q_k from (4.86), we obtain the relations

$$\left.\begin{array}{l} c_0(p_{k+1}^*-p_{k-1}^*)+c_1(p_{k+2}^*-p_{k-2}^*)+\dots \\ \qquad \dots+c_{k-1}(p_{2k}^*-p_0^*)+c_kp_{2k+1}^*=0 \\ c_0(p_k^*-p_{k+2}^*)+c_1(p_{k-1}^*-p_{k+3}^*)+\dots \\ \qquad \dots+c_{k-1}(p_1^*-p_{2k+1}^*)+c_kp_0^*=0 \end{array}\right\} \tag{4.87}$$

It is important to note the form in which (4.87) is written. If, for example, we write down the first equation (4.87), then the second equation is obtained from the first by replacing the coefficients p_j^* by p_{2k+1-j}^*.

After substituting c_ρ from (4.85a) and (4.85b), the left-hand sides of (4.87) can be represented in the form of polynomials of degree k with respect to the quantity ξ. Consequently, (4.87) may be regarded as simultaneous algebraic equations of the kth degree with respect to the quantity ξ.

According to the rules of algebra [29], two algebraic equations

$$\left.\begin{array}{l} a_0\xi^m+a_1\xi^{m-1}+\dots+a_m=0, \\ b_0\xi^n+b_1\xi^{n-1}+\dots+b_n=0 \end{array}\right\} \tag{4.88}$$

can have a common root if, and only if, their resultant vanishes.

This condition can be written in the form

$$\begin{vmatrix} a_0 & 0 & 0 & \dots & 0 & b_0 & 0 & 0 & \dots & 0 \\ a_1 & a_0 & 0 & \dots & 0 & b_1 & b_0 & 0 & \dots & 0 \\ a_2 & a_1 & a_0 & \dots & 0 & b_2 & b_1 & b_0 & \dots & 0 \\ \cdot & \cdot & \cdot & \cdot & \cdot & \cdot & \cdot & \cdot & \cdot & \cdot \\ 0 & 0 & 0 & \dots & a_{m-1} & 0 & 0 & 0 & \dots & b_{n-1} \\ 0 & 0 & 0 & \dots & a_m & 0 & 0 & 0 & \dots & b_n \end{vmatrix} = 0 \qquad (4.89)$$

(the first n columns contain the a's, the last m columns contain the b's: "n columns", "m columns")

Thus, if (4.71) has roots in the form (4.73), then its coefficients p_i^* must satisfy a condition of type (4.89). Consequently, (4.80) provides the necessary condition under which the characteristic equation (4.71) has two complex conjugate roots (4.73) whose moduli are equal to unity.

This condition, however, is not sufficient. Indeed, in our discussion we have made use only of the fact that $\lambda_1^*\lambda_2^* = 1$. But this property is possessed by any numbers that can be put in the form $\lambda_i^* = e^{\varepsilon \pm i\omega}$ and $\lambda_2^* = e^{-\varepsilon \mp i\omega}$ for arbitrary real numbers ε and ω. Therefore (4.89) is a necessary and sufficient condition for (4.71) to have two roots λ_1^* and λ_2^* whose product $\lambda_1^*\lambda_2^* = 1$. However, this fact does not prevent us from applying the above method of separating the regions of stability.

We shall now discuss the case when (4.71) is of even degree. It is not difficult to see that, when $n = 2k$, (4.75) can also be written in matrix form, (4.78) being left unaltered, and the order of the matrix in the second equation being reduced by unity, since there will be no coefficients p_{2k+1}^* and q_{2k-1} in it. If, therefore, we go through the same reasoning but with different orders of the matrix (4.76), we arrive again at the four equations of type (4.87), these equations being obtainable from (4. 87) as limiting cases if we insert $p_{2k+1}^* = 0$. It is not particularly difficult in actual cases to find the resultant (4.89).

In practice, one should keep to the following order. For definiteness, let n be odd. Then $n = 2k + 1$. We now write out the first equation (4.87) for the value of k that is found. Next, using (4.85a) and the recurrence relation (4.85b), we write out the coefficients c_ρ. We then represent the left-hand side of the first equation in (4.87) in the form of a polymial of degree k with respect to the parameter ξ from which, by replacing p_j^* by p_{2k+1-j}^*, we obtain the second equation in (4. 87). After this we compose the resultant (4.89).

As an example, consider a characteristic equation of the fifth degree. In the given case $k = 2$, so that the first equation in (4.87) is written in the form

$$c_0(p_3^* - p_1^*) + c_1(p_4^* - p_0^*) + c_2 p_5^* = 0 \qquad (4.90)$$

Substituting in here c_0, c_1, c_2 from (4.85) and collecting together similar terms in ξ, we obtain

$$p_5^*\xi^2+(-p_4^*+p_0^*)\xi+(-p_5^*+p_3^*-p_1^*)=0 \qquad (4.91)$$

Replacing in this equation p_j^* by p_{5-j}^*, we obtain

$$p_0^*\xi^2+(-p_1^*+p_5^*)\xi+(-p_0^*+p_2^*-p_4^*)=0 \qquad (4.92)$$

Transposing the columns, we write the resultant of (4.91) and (4.92) in the form [see (4.88) and (4.89)]:

$$\begin{vmatrix} p_0^* & p_5^* & 0 & 0 \\ p_5^*-p_1^* & p_0^*-p_4^* & p_0^* & p_5^* \\ -p_0^*+p_2^*-p_4^* & -p_5^*+p_3^*-p_1^* & p_5^*-p_1^* & -p_4^*+p_0^* \\ 0 & 0 & -p_0^*+p_2^*-p_4^* & -p_5^*+p_3^*-p_1^* \end{vmatrix}=0 \qquad (4.93)$$

Multiplying out the determinant, we put (4.93) in the form

$$[p_0^*(-p_4^*+p_0^*)-p_5^*(p_5^*-p_1^*)][(p_5^*-p_1^*)(-p_5^*+p_3^*-p_1^*)$$
$$-(-p_4^*+p_0^*)(-p_0^*+p_2^*-p_4^*)]$$
$$-[p_0^*(p_3^*-p_1^*)-p_5^*(p_2^*-p_4^*)]^2=0 \qquad (4.94)$$

From (4.94), by putting successively $p_5^*=0$, $p_5^*=p_4^*=0$, $p_5^*=p_4^*=p_3^*=0$, we obtain the resultant (4.89) for characteristic equations (4.71) of the fourth, third and second degree, as follows:

for a characteristic equation of the fourth degree in the form

$$(-p_4^*+p_0^*)[p_1^*(p_3^*-p_1^*)+(-p_4^*+p_0^*)(-p_0^*+p_2^*-p_4^*)]$$
$$+p_0^*(p_3^*-p_1^*)^2=0 \qquad (4.95)$$

for a characteristic equation of the third degree in the form

$$p_0^*(p_2^*-p_0^*)+p_3^*(p_3^*-p_1^*)=0 \qquad (4.96)$$

and for a characteristic equation of the second degree in the form

$$p_2^*-p_0^*=0 \qquad (4.97)$$

When relations (4.94)-(4.97) are fulfilled, the characteristic equation of corresponding degrees has at least two roots λ_1^* and λ_2^* for which $\lambda_1^*\lambda_2^*=1$.

Conditions (4.94)-(4.97) can also be obtained, although in a somewhat more devious way, by constructing the transformed characteristic equation (4.59) and equating its penultimate Hurwitz determinant to zero ($\Delta_{n-1}=0$).

4.5 STABILITY COEFFICIENTS AND USE OF THE LYAPUNOV FUNCTION TO ESTIMATE DAMPING

The stability of a linear discontinuous system can be investigated by constructing the corresponding Lyapunov function. The necessary and sufficient conditions for the stability of the solution of a system of linear difference equations with constant coefficients were obtained in Section 3.2 by the direct method of Lyapunov. Here we shall investigate the properties of the Lyapunov function more deeply, an exercise that will enable us not only to determine the qualitative state of the system, i.e. its stability, but also to establish certain quantitative characteristics of the process of transition from the perturbed motion of a system to its equilibrium stationary state.

4.5.1 Stability coefficients

The matrix difference equation (4.16) determines the discrete motion of a linear pulsed system about the equilibrium state which is defined by the matrix identity $x \equiv 0$. The Lyapunov function V corresponding to (4.16) is given as the quadratic form (3.25). By virtue of the matrix equation (4.16), the first difference of the form V satisfies (3.26) for an arbitrarily specified positive definite quadratic form U. The necessary and sufficient condition for stability is that the Lyapunov function, i.e. the quadratic form V, should be a positive definite function.

The quadratic forms U and V constitute a regular form (see Section 2.6) since U is chosen to be positive definite. As has already been remarked in Section 2.6, the form V will in this case be positive definite if, and only if, all the roots ρ_i of the characteristic equation (2.153) are greater than zero. Let us introduce the new quantities μ_i which are connected with the roots ρ_i through the relation

$$\mu_i = \frac{\rho_i - 1}{\rho_i} \qquad i = 1, 2, \ldots, n \tag{4.98}$$

If the root ρ_i runs from 1 to ∞, then μ_i runs through all values from 0 to 1. The quantities μ_i will be called stability coefficients. One can prove a series of propositions about these quantities.

Theorem I

For a linear discontinuous system to be stable (and also asymptotically stable) it is necessary and sufficient that the quantities shall satisfy the inequalities

$$0 \leqslant \mu_i < 1, \qquad i = 1, 2, \ldots, n \tag{4.99}$$

The Lyapunov function for a stable linear discontinuous system is a positive definite quadratic form V. Consider the forms U and V. With the help of the linear transformation (2.154) and a non-singular real matrix L we reduce simultaneously the forms U and V to the form (2.155) and (2.156). We substitute U and V in the form (2.155) and (2.156) into the difference equation (3.26). After some straight-forward multiplication, we obtain the equality

$$\sum_{i=1}^{n} \rho_i y_i^2 (mT) = \sum_{i=1}^{n} (\rho_i - 1) y_i^2 [(m-1)T] \tag{4.100}$$

which is valid for any instant of discrete time mT. In particular, when $m = 1$, we have

$$\sum_{i=1}^{n} \rho_i y_i^2 (T) = \sum_{i=1}^{n} (\rho_i - 1) y_i^2 (0) \tag{4.101}$$

The initial values $y_i(0)$ can be specified arbitrarily. This follows from the fact that the matrix of the L transformation (2.154) is non-singular so that for any set of values $y_1(0), y_2(0), \ldots, y_n(0)$ one can always find, and moreover find in a unique way, a corresponding set of values $x_1(0), x_2(0), \ldots, x_n(0)$ of the assumed coordinates. Then, by making each of the values $y_i(0)$ non-zero one after the other and assuming the remaining values are equal to zero, it is easy to establish that the quadratic form V can be positive definite and simultaneously satisfy (4.101) only if the roots ρ_i satisfy the inequalities

$$\rho_i \geqslant 1 \tag{4.102}$$

ρ_i being equal to unity at the end of the first interval T, we have already $y_i(T) = 0$. From (4.98) and (4.102) there follow the equalities (4.99).

Thus, for stable linear discontinuous systems, the positive stability coefficients μ_i, like the moduli of the roots λ_i^* of the characteristic equation $\Delta^*(\lambda) = 0$ (4.30)-(4.40), are less than unity.

In general, a rather closer connection exists between the stability coefficients μ_i and the roots of the characteristic equation $\Delta^*(\lambda) = 0$. This connection is established by the following theorems.

Theorem II

If a linear discontinuous system is asymptotically stable, the following equality is valid

$$\mu_{\min} < |\lambda_i^*|^2 < \mu_{\max}, \qquad i = 1, 2, \ldots, n \tag{4.103}$$

Consider a system of algebraic equations in the matrix notation (3.32) determining the Lyapunov function in terms of the form $V = x^T Bx$, when $U = x^T Ax$. The matrix P^* is the matrix operator of the linear discontinuous system.

In accordance with (2.151) and (2.152), we represent the matrices A and B in the form

$$A = (L^T)^{-1} EL^{-1}, \qquad B = (L^T)^{-1} RL^{-1} \tag{4.104}$$

We can do this because the real matrix L is non-singular. Let us substitute into (3.32) matrices A and B in the form (4.104). We then obtain

$$P^{*T}(L^T)^{-1} RL^{-1} P - (L^T)^{-1} RL^{-1} = (L^T)^{-1} EL^{-1}$$

We will multiply both sides of this equality from the right by L and from the left by L^T and transfer the second term from the left- to the right-hand side. We then have

$$L^T P^{*T}(L^T)^{-1} RL^{-1} P^* L = R - E \tag{4.105}$$

We introduce the new notation

$$Q = L^{-1} P^* L \tag{4.106}$$

and rewrite (4.105) in the form

$$Q^T RQ = R - E \tag{4.107}$$

The matrices P^* and Q are similar, so that λ_i^* are the characteristic values of matrix Q. For simplicity we shall suppose that all the characteristic values are simple. We can then write

$$Q = K\Lambda^* K^{-1} \tag{4.108}$$

where K is a matrix which transforms Q into the canonical Λ^*. With the assumptions we have made Λ^* is a diagonal matrix with elements λ_i^* [see (2.42) and (2.44)]. From (4.108) it is not difficult to see that the columns k_j of the matrix K satisfy the condition

$$Qk_j = \lambda_j^* k_j \tag{4.109}$$

Since the matrix Q is real, to the two complex conjugate roots λ_i^* and $\bar{\lambda}_j^*$ there correspond columns k_j and $\bar{k}_j$ with complex conjugate elements k_{ij} and $\bar{k}_{ij}$ for $i = 1, 2, \ldots, n$.

Let λ_j^* and $\bar{\lambda}_j^*$ be the conjugate roots. We will multiply both sides of (4.107) from the left by the row $\bar{k}_j^T$ and from the right by the column k_j. Then, with account of (4.109) and the relation obtained from it by transposing the matrix, we obtain the scalar expression

$$\bar{\lambda}_j\lambda_j\bar{k}_j^T R k_j = \bar{k}_j(R-E)k_j \tag{4.110}$$

which, by direct calculation, leads to the form

$$\bar{\lambda}_j\lambda_j \sum_{i=1}^{n} \rho_i \bar{k}_{ij} k_{ij} = \sum_{i=1}^{n} (\rho_i - 1)\bar{k}_{ij} k_{ij} \tag{4.111}$$

or, setting $\bar{\lambda}_j\lambda_j = |\lambda_j|^2$ and $\bar{k}_{ij}k_{ij} = |k_{ij}|^2$ and taking (4.98) into account, we can finally write (4.111) in the form

$$|\lambda_j|^2 = \frac{\mu_1\rho_1|k_{1j}|^2 + \mu_2\rho_2|k_{2j}|^2 + \dots + \mu_n\rho_n|k_{nj}|^2}{\rho_1|k_{ij}|^2 + \rho_2|k_{2j}|^2 + \dots + \rho_n|k_{nj}|^2} \tag{4.112}$$

$$j = 1, 2, \dots, n$$

If we regard the μ_i as coordinates of material points having negative masses $\rho_i|k_{ij}|^2$ distributed along a straight line, then $|\lambda_j|^2$ determines the coordinate of the centre of mass which lies between the extreme points. The theorem is therefore proved. It is not difficult to prove this theorem for the case where the roots λ_j^* are multiple.

The following corollary stems from Theorem II.

Corollary I

If only one of the roots λ_i^* of a discontinuous system is equal to zero, then the smallest stability coefficient μ_{min} also vanishes. This follows from the fact that, for a stable system, $\mu_{min} \geqslant 0$ and $\mu_{min} < |\lambda_j|^2$ for any number j.

Corollary II

If a linear discontinuous system is at the boundary of stability, then the largest stability coefficient μ_{max} is equal to unity.

Indeed, under the conditions stipulated above, the smallest modulus of the roots λ_j^* is equal to unity. However, for a stable discontinuous system, the inequalities $\mu_{max} < 1$ and $|\lambda_j|^2 < \mu_{max}$ hold for any number j. By proceeding to the limit $|\lambda_j| \to 1$, we can establish the validity of the condition we have formulated. Strictly speaking, this is not so much a proof as a logical extension to the definition of the quantity μ_{max} at the boundary of stability since, when $|\lambda_j| = 1$ a Lyapunov function does not exist. Finally, the following theorem holds.

Theorem III

If a linear discontinuous system is asymptotically stable, then the following equality holds

$$\prod_{i=1}^{n} \lambda_i^2 = \prod_{i=1}^{n} \mu_i \tag{4.113}$$

We equate the determinants of the matrices on the left- and right-hand sides of (4.107). Then, remembering the rule for forming the determinants of the product of matrices, we obtain

$$\det Q^T \det R \det Q = \det (R - E) \tag{4.114}$$

But, according to (2.152)

$$\det R = \prod_{i=1}^{n} \rho_i, \qquad \det (R - E) = \prod_{i=1}^{n} (\rho_i - 1)$$

and

$$\det Q^T = \det Q = \prod_{i=1}^{n} \lambda_i^*$$

since det Q is a free term of the characteristic polynomial of matrix Q. If we substitute into (4.114) the values found for the determinants, divide the left- and right-hand sides by the non-zero det R and take account of (4.98), we obtain (4.113).

4.5.2 Estimate of the damping

Let us now use the stability coefficients μ_i to estimate certain quantitative characteristics of the damping of the perturbed motion of a system towards its stationary state. We recall that we defined a perturbed motion as the free-motion of a system generated by instantaneous perturbations (instantaneous perturbations are defined mathematically by specifying the initial conditions).

Consider (4.100). By virtue of (4.98), we can rewrite this equation in the form

$$\sum_{i=1}^{n} \rho_i y_i^2 (mT) = \sum_{i=1}^{n} \mu_i \rho_i y_i^2 [(m - 1) T] \tag{4.115}$$

On the left-hand side of (4.115) stands the quadratic form V (the Lyapunov function) expressed in terms of the transformed coordinates y_i.

As before, μ_{min} and μ_{max} will denote the smallest and largest values of the stability coefficients $\mu_1, \mu_2, \ldots, \mu_n$ which satisfy

(4.99) for a stable system. Then, from (4.115), we establish the inequalities

$$\mu_{min} V_{m-1} \leqslant V_m \leqslant \mu_{max} V_{m-1} \qquad (4.116)$$

Successively applying (4.116), we obtain

$$V_0 \mu^m_{min} \leqslant V_m \leqslant V_0 \mu^m_{max} \qquad (4.117)$$

The inequalities (4.117) define curves which set upper and lower limits to the variation of a Lyapunov function in the form of V if the latter is treated as the function $x_1(mT)$ of the solutions of the difference equation (4.16). From the right-hand inequality (4.17) one can obtain the maximum curve for any coordinate $x_i(mT)$.

In accordance with (2.147), a positive definite quadratic form V (3.25) can be represented in the form

$$V = D_1 z_1^2 + \frac{D_2}{D_1} z_2^2 + \ldots + \frac{D_n}{D_{n-1}} z_n^2 \qquad (4.118)$$

where, by virtue of (2.141),

$$D_1 = b_{11} > 0, \quad D_2 = \begin{vmatrix} b_{11} & b_{12} \\ b_{21} & b_{22} \end{vmatrix} > 0, \; \ldots, \; D_n = \begin{vmatrix} b_{11} & b_{12} & \ldots & b_{1n} \\ b_{21} & b_{22} & \ldots & b_{2n} \\ \cdot & \cdot & \cdot & \cdot \\ b_{n1} & b_{n2} & \ldots & b_{nn} \end{vmatrix} > 0 \qquad (4.119)$$

According to (2.146), we have $z_n = x_n$. With account of (4.119) and $z_n = x_n$, we obtain from (4.118) the inequality

$$x_n^2 \leqslant \frac{D_{n-1}}{D_n} V \qquad (4.120)$$

Such an inequality can be obtained for any coordinate x_i having first given the coordinate the number n, so that we can write

$$x_i^2 \leqslant \frac{D_{i-1}}{D_i} V \qquad (i = 1, 2, \ldots, n) \qquad (4.121)$$

where $D_i = D_n$ is the discriminant of the quadratic form V and D_{i-1} is its chief minor obtained by eliminating the ith row and the ith column. From the right-hand side of (4.17) and from (4.120), we obtain the expressions

$$x_i^2(mT) \leqslant \frac{D_{i-1}}{D_i} V_0 \mu^m_{max} \qquad (4.122)$$

These expressions can be used to calculate the parameters of discontinuous control systems since they permit one to estimate, though admittedly it is an overestimate, the interval of time $m'T$ in a discrete motion which elapses before the coordinate x_i falls inside a previously specified zone characterising the precision of the control with respect to a given coordinate.

However, one can approach (4.122) from another point of view. It is not difficult to see that the smaller the quantity μ_{max} the faster will be the rate of fall of the curve representing the right-hand side of (4.122). Since the rate of decrease of the ordinates of the curve guarantees at least that the rate of decrease of all the coordinates x_i will not be smaller than that as the discrete time mT increases, the quantity μ_{max} plays the role of a stability coefficient characterising to a known degree the rate of damping of the perturbed motion of the system towards its equilibrium state. Consequently, in preliminary calculations we can use the maximum stability coefficient μ_{max} as the criterion of the degree of damping of the perturbed motion of the system, and we can carry out the caculation in such a way as to obtain the maximum possible value for this coefficient.

From (2.80) with $g(m)\equiv 0$ it follows that, in the perturbed motion defined by the initial perturbation $x(0)$, to each root λ_i^* of characteristic polynomial $\Delta^*(\lambda)$ of the operator matrix P^* there corresponds a partial perturbed motion, the modulus of the root λ_i^* determining the damping and the argument determining the oscillation of this motion. The over-all perturbed motion is given by a superposition of its partial motions. If the rate of damping of the perturbed motion is approximately determined from the rate of damping of that partial motion which dies away most slowly, then for the damping criterion one should take the greatest value of the modulus of all the roots λ_i^*, i. e. the quantity $|\lambda_i^*|_{max}$. This point of view is the one adopted by Tsipkin [26] (see also [27]) who introduced the notion of a degree of stability δ^* for a linear discontinuous system, having taken $\delta^* = |\lambda_i^*|_{max}$, and also gave a method of defining it.

From the right-hand side of (4.103) it follows that the degree of stability is always smaller than the maximum stability coefficient, i. e. $\delta^* = |\lambda_i^*|_{max} < \mu_{max}$. The fundamental reason for this stems from (4.122). In fact, according to the arguments just given, the maximising curve must decrease more slowly than any partial motion, but this is possible only when the condition $|\lambda_i^*|_{max} < \mu_{max}$ is fulfilled. Next, we note that according to the second corollary to Theorem II, at the boundary of stability the equality $|\lambda_i^*|^2_{max} = \mu_{max} = 1$ holds. Therefore, by virtue of their continuity, the difference between the quantities $|\lambda_i^*|^2_{max}$ and μ_{max} will be quite small close to the boundary of stability, and they will decrease simultaneously when one of the parameters is changed, at least up to the first minimum. Here, of course, it is assumed that the change in the selected parameter takes place in a direction away from the boundary and towards the internal part of the stability region. This remark lies at the basis of the method used to calculate the parameters of a system that has achieved maximum stability, i. e. a minimum value of $|\lambda_i^*|_{max}$. However, calculating the amount of damping from $(\mu_{max})_{min}$ has the advantage that it enables (4.122) to be used to determine the finite interval of time $m'T$ during which the modulus of any coordinate x_i will become smaller than a given value.

4.6 RESPONSE OF THE SYSTEM TO A DESIRED RESPONSE SIGNAL AND TO AN EXTERNAL PERTURBATION

Suppose the control system is acted upon by external perturbations, as a result of which the system takes up a certain position in state space by a certain instant of time. At this instant, which we shall assume to be the origin of the time axis, the external perturbation is removed and the system is left to itself. It will then execute a free-motion and, as we showed earlier, we determine the stability of the stationary state of the system according to the characteristics of this motion. In this sense the free-motion clearly does not depend on the nature of the previous perturbations but is entirely determined by the result of their action up to the instant of time being considered. It is therefore convenient to disregard the previous history of the perturbations acting on the system and to assume instead that the free-motion is due to instantaneous perturbations which immediately cause the system to go over to the deviated position. Mathematically, the action of the instantaneous perturbations is then determined by specifying the initial conditions, i. e. the values of the coordinates at $t = 0$. The free-motion of the system is given by the solution of homogeneous equations for the given initial conditions.

Let us consider the reverse situation. Suppose that up to a given instant of time the control system is in a stationary state, and suppose that at this instant a desired response signal is fed into the system or an external perturbation is superimposed. The subsequent motion of the system is referred to as its response to these external actions and perturbations.

Mathematically, the response of the system to desired response signals and to external actions is given by the solution of inhomogeneous equations satisfying the zero initial conditions. If the external actions and perturbations are such that when $t \to \infty$ the system goes over to an appropriate stationary state of motion of equilibrium, then the response of the system is regarded as consisting of steady-state and transient processes. In general, it is a complex problem to separate out the steady-state process. In certain special cases when the external perturbations are given by simple functions, this problem is completely soluble. In the theory of discontinuous control it is possible to consider the response of the system in either discrete or in continuous motion.

4.6.1 Motion of a system under the action of instantaneous perturbations

The discrete motion of a linear discontinuous system which is being acted upon by external perturbations is given by the solution of the matrix difference equation (4.16) for the initial condition $x(0)$. This solution can be expressed in terms of the power of the matrix operator P^* (4.15) or in terms of its characteristic polynomial $\Delta^*(\lambda)$, the roots λ_i^* and the adjoint matrix $F^*(\lambda)$. These forms of the solutions are given by (2.78), (2.80) and (2.82) with $g^*(m) \equiv 0$.

The first form of the solutions is convenient when digital computers are being used to calculate the motion, since in that case the calculation of the integral power of the matrix P^* is done in iterated cycles.

The second form of the solution penetrates more deeply into the internal structure and into the influence of that structure on the character of the free-motion, so that it is more suitable for general investigations of discontinuous control systems.

Our problem consists of expressing the matrices P^* and $F^*(\lambda)$ in terms of the assumed parameters of the system.

According to (4.15), we have

$$P^* = KM(T)K^{-1} + KM(T_2)N(T_1)K^{-1}h\gamma$$

We will expand the matrices $KM(T)K^{-1}$ and $KM(T_2)N(T_1)K^{-1}$ in the elements of the matrices $M(T)$ and $M(T_2)N(T_1)$ using (2.24). For the case of simple roots λ_i of the matrix P of the open loop system, the matrices $M(t)$ and $N(t)$ are diagonal and are given by (2.94) and (2.120), respectively. The matrix $M(T_2)N(T_1)$ is also diagonal with the common element

$$e^{\lambda_i T_2}\left(e^{\lambda_i T_1} - 1\right)/\lambda_i \, ^*)$$

[see also the discussion given just before (4.31)]. We then have

$$P^* = \sum_{i=1}^{n} e^{\lambda_i T} k_i \varkappa_i + \sum_{i=1}^{n} \frac{e^{\lambda_i T_2}\left(e^{\lambda_1 T_1} - 1\right)}{\lambda_i} k_i \varkappa_i h\gamma \qquad (4.123)$$

or, with account of the matrices (2.64), we obtain

$$P^* = \sum_{i=1}^{n} e^{\lambda_i T} \frac{F(\lambda_i)}{\Delta'(\lambda_i)} + \sum_{i=1}^{n} \frac{e^{\lambda_i T_2}\left(e^{\lambda_i T_1} - 1\right)}{\lambda_i} \frac{F(\lambda_i)}{\Delta'(\lambda_i)} h\gamma \tag{4.124}$$

For the case when there is a single repeated root $\lambda_{n-1} = \lambda_n = 0$, one should use (2.98), (4.47), (2.65) and (2.66). The final result can be put in the form

$$P^* = \sum_{i=1}^{n-2} e^{\lambda_i T} k_i \varkappa_i + (k_{n-1}\varkappa_{n-1} + k_n \varkappa_n) + T k_n \varkappa_{n-1}$$

$$+ \left\{ \sum_{i=1}^{n-2} \frac{e^{\lambda_i T_2}\left(e^{\lambda_i T_1} - 1\right)}{\lambda_i} k_i \varkappa_i + T_1 (k_{n-1}\varkappa_{n-1} + k_n \varkappa_n) \right.$$

$$\left. + \left(T_1 T_2 + \frac{T_1^2}{2}\right) k_n \varkappa_{n-1} \right\} h\gamma \tag{4.125}$$

or

$$P^* = \sum_{i=1}^{n-2} e^{\lambda_i T} \frac{F(\lambda_i)}{\Delta'(\lambda_i)} + \left[\frac{F(0)}{\Delta_{n-1}(0)}\right]^{(1)} + T \frac{F(0)}{\Delta_{n-1}(0)}$$

$$+ \left\{ \sum_{i=1}^{n-2} \frac{e^{\lambda_i T_2}\left(e^{\lambda_i T_1} - 1\right)}{\lambda_i} \frac{F(\lambda_i)}{\Delta'(\lambda_i)} + T_1 \left[\frac{F(0)}{\Delta_{n-1}(0)}\right]^{(1)} \right.$$

$$\left. + \left(T_1 T_2 + \frac{T_1^2}{2}\right) \frac{F(0)}{\Delta_{n-1}(0)} \right\} h\gamma \tag{4.126}$$

We will now construct the adjoint matrix $F^*(\lambda)$. In accordance with (2.55), we have

$$\left.\begin{aligned} \frac{F^*(\lambda)}{\Delta^*(\lambda)} &= (\lambda E - P^*)^{-1} \\ \Delta^*(\lambda) &= \det(\lambda E - P^*) \end{aligned}\right\} \tag{4.127}$$

With account of expressions (4.20)-(4.25), we can write the characteristic matrix in the form

$$\lambda E - P^* = K(\lambda E - M(T) - u'\beta) K^{-1}$$

$$= K[\lambda E - M(T)]\left\{E - [\lambda E - M(T)]^{-1} u'\beta\right\} K^{-1} \tag{4.128}$$

Taking (4.128) into account, we put (4.127) in the form (see the rules for inverting the product of several matrices, given in

Chapter 2):

$$\left.\begin{aligned}\frac{F^*(\lambda)}{\Delta^*(\lambda)} &= K\{E-[\lambda E-M(T)]^{-1}u'\beta\}^{-1}[\lambda E-M(T)]^{-1}K^{-1},\\ \Delta^*(\lambda) &= \det[\lambda E-M(T)]\det\{E-[\lambda E-M(t)]^{-1}u'\beta\}.\end{aligned}\right\} \quad (4.129)$$

The main difficulty now has to do with inverting the matrix inside the braces of expressions (4.129). We introduce the new notation

$$u'' = [\lambda E - M(t)]^{-1}u' \quad (4.130)$$

and rewrite this matrix in the form

$$E-[\lambda E-M(T)]^{-1}u'\beta = E-u''\beta \quad (4.131)$$

Let us transpose matrix (4.131) and write it in expanded form. We then have

$$(E-u''\beta)^T = \left\| \begin{matrix} 1-u_1''\beta_1 & -u_2''\beta_1 & \dots & -u_n''\beta_1 \\ -u_1''\beta_2 & 1-u_2''\beta_2 & \dots & -u_n''\beta_2 \\ \dots & \dots & \dots & \dots \\ -u_1''\beta_n & -u_2''\beta_n & \dots & 1-u_n''\beta_n \end{matrix} \right\| \quad (4.132)$$

Next we must find the cofactors for the elements of matrix (4.132) and divide them by the determinant of the matrix. The determinant and minors of (4.132) are given by (4.29). In the given case $z_i = 1$ for any index i and the c_{ij} are the elements of the matrix $-u''\beta$, which has rank unity. We therefore have

$$\det(E-u''\beta)^T = \det(E-u''\beta) = 1-\sum_{i=1}^{n} u_i''\beta_i = 1-\beta u'' \quad (4.133)$$

The minor obtained by cancelling the ith row and ith column has the same structure as (4.132) except that the elements u_i'' and β_i are absent. Consequently, the cofactor corresponding to the diagonal element can be written in the form

$$1-\sum_{j=1}^{n}{}' u_j''\beta_j = 1-\sum_{i=1}^{n} u_j''\beta_j + u_i''\beta_i \quad (4.134)$$

Here, the prime indicates that elements with index i should be omitted from the sum.

In determining the cofactor for the element $(-\beta_i u_j'')$ it is necessary to cancel the ith row and jth column and rearrange the rows or the columns in such a way that the remaining $(n-2)$ diagonal elements of (4.132) are again on the main diagonal of the minor obtained. The position on the main diagonal left is then occupied by the element $(-\beta_j u_i'')$. One next establishes that, for any element $(-\beta_i u_j)$ with $i \neq j$, the sum $(i+j)$ and the number of rearrangements of rows and columns that it was necessary to make is an odd number.

Thus, we establish that the cofactor D_{ij} for the element $(-\beta_i u''_j)$ has the form

$$D_{ij} = u''_i \beta_j \tag{4.135}$$

In accordance with (4.134) and (4.135), the matrix of the cofactors, i.e. the adjoint matrix for $E - u''\beta$, can be written as the sum of the two matrices

$$\left(1 - \sum_{i=1}^{n} u''_i \beta_i\right) \begin{Vmatrix} 1 & 0 & \dots & 0 \\ 0 & 1 & \dots & 0 \\ \cdot & \cdot & \cdot & \cdot \\ 0 & 0 & \dots & 1 \end{Vmatrix} + \begin{Vmatrix} u''_1\beta_1 & u''_1\beta_2 & \dots & u''_1\beta_n \\ u''_2\beta_1 & u''_2\beta_2 & \dots & u''_2\beta_n \\ \cdot & \cdot & \cdot & \cdot \\ u''_n\beta_1 & u''_n\beta_2 & \dots & u''_n\beta_n \end{Vmatrix} \tag{4.136}$$

We now divide the sum of the matrices (4.136) by the determinant (4.133). The result is the inverse matrix for $E - u''\beta$ which we will write in the compact form

$$(E - u''\beta)^{-1} = E + \frac{u''\beta}{1 - \beta u''} \tag{4.137}$$

If, in (4.133) and (4.137), we substitute the column u'' from (4.130), we obtain matrix expressions in the form

$$\left.\begin{aligned} &\{E - [\lambda E - M(T)]^{-1} u'\beta\}^{-1} = \\ &\qquad = E + \frac{[\lambda E - M(T)]^{-1} u'\beta}{1 - \beta [\lambda E - M(T)]^{-1} u'} \\ &\det\{E - [\lambda E - M(T)]^{-1} u'\beta\} = \\ &\qquad = 1 - \beta [\lambda E - M(T)]^{-1} u' \end{aligned}\right\} \tag{4.138}$$

Using (4.138), we rewrite (4.129) in the form

$$\frac{F^*(\lambda)}{\Delta^*(\lambda)} = K\left\{E + \frac{[\lambda E - M(T)]^{-1} u'\beta}{1 - \beta [\lambda E - M(T)]^{-1} u'}\right\}[\lambda E - M(T)]^{-1} K^{-1} \tag{4.139}$$

and

$$\Delta^*(\lambda) = \{\det [\lambda E - M(t)]\}\{1 - \beta [\lambda E - M(T)]^{-1} u'\} \tag{4.140}$$

The scalar expression (4.140) determines the characteristic polynomial of the operator matrix P^* (4.15). This expression is valid for any characteristic values λ_i of matrix P. In particular, one can obtain from this expression the characteristic values of the equation in the form (4.30) and (4.40). In all cases $M(T)$ is a triangular matrix whose diagonal elements are $e^{\lambda_i T}$. Consequently the inequality

$$\det [\lambda E - M(T)] = \prod_{i=1}^{n} (\lambda - e^{\lambda_i T}) = \Delta_p^*(\lambda) \tag{4.141}$$

always holds [see (4.64)].

In the matrix expressions (4.139) and (4.140), the matrix inverse to $\lambda E - M(T)$ appears. For those cases when the characteristic values λ_i of matrix P are simple, and when they include a double zero root $\lambda_{n-1} = \lambda_n = 0$, this matrix is written, respectively, in the forms

$$[\lambda E - M(T)]^{-1} = \left\| \begin{matrix} \frac{1}{\lambda - e^{\lambda_i T}} & 0 & \dots & 0 \\ 0 & \frac{1}{\lambda - e^{\lambda_i T}} & \dots & 0 \\ \dots & \dots & \dots & \dots \\ 0 & 0 & \dots & \frac{1}{\lambda - e^{\lambda_n T}} \end{matrix} \right\| \tag{4.142}$$

and

$$[\lambda E - M(T)]^{-1} = \left\| \begin{matrix} \frac{1}{\lambda - e^{\lambda_i T}} & \dots & 0 & 0 & 0 \\ \dots & \dots & \dots & \dots & \dots \\ 0 & \dots & \frac{1}{\lambda - e^{\lambda_{n-2} T}} & 0 & 0 \\ 0 & \dots & 0 & \frac{1}{\lambda - 1} & 0 \\ 0 & \dots & 0 & \frac{T}{(\lambda - 1)^2} & \frac{1}{\lambda - 1} \end{matrix} \right\| \tag{4.143}$$

These formulae are obtained quite easily using (2.94) and (2.98). Multiplying both sides of (4.139) by $\Delta^*(\lambda)$, and taking account of (4.140) and (4.141), we obtain the matrix relation

$$F^*(\lambda) = K \{\Delta^*(\lambda) [\lambda E - M(T)]^{-1}$$

$$+ \Delta_p^*(\lambda) [\lambda E - M(T)]^{-1} u' \beta [\lambda E - M(T)]^{-1}\} K^{-1} \tag{4.144}$$

which determines the adjoint matrix $F^*(\lambda)$ that we are seeking. This relation is valid for all the cases discussed in this book. Although this formula is of value just as it stands, it can also lead to a form which is more convenient for practical application.

Let us substitute into (4.144) u' and β from (4.21) and (4.25). We obtain

$$F^*(\lambda) = \Delta^*(\lambda) K [\lambda E - M(T)]^{-1} K^{-1} + \Delta_p^*(\lambda) K [\lambda E$$

$$-M(T)]^{-1}M(T_2)N(T_1)K^{-1}h\gamma K\times[\lambda E-M(T)]^{-1}K^{-1} \qquad (4.145)$$

We now make use of the expansion formula (2.24). For the case of simple roots we have the expansions

$$\left.\begin{aligned} &K[\lambda E-M(T)]^{-1}K^{-1}=\sum_{i=1}^{n}\frac{1}{\lambda-e^{\lambda_i T}}k_i\varkappa_i \\ &K[\lambda E-M(T)]^{-1}M(T_2)N(T_1)K^{-1} \\ &\qquad\qquad=\sum_{i=1}^{n}\frac{e^{\lambda_i T_2}\left(e^{\lambda_i T_1}-1\right)}{\lambda_i\left(\lambda-e^{\lambda_i T}\right)}k_i\varkappa_i \end{aligned}\right\} \qquad (4.146)$$

since $[\lambda E-M(T)]^{-1}$ is given by (4.142) and the matrix $M(T_2)\ N(T_1)$ is in this case diagonal with elements $\dfrac{e^{\lambda_i T_2}\left(e^{\lambda_i T_1}-1\right)}{\lambda_i}$. Next we use the matrix identities (2.64). Matrix (4.145) can then be put in the form

$$F^*(\lambda)=\left\{\Delta^*(\lambda)E+\Delta_p^*(\lambda)\sum_{i=1}^{n}\frac{e^{\lambda_i T_2}\left(e^{\lambda_i T_1}-1\right)}{\lambda_i\left(\lambda-e^{\lambda_i T}\right)}\frac{F(\lambda_i)}{\Delta'(\lambda_i)}h\gamma\right\}$$
$$\times\left\{\sum_{j=1}^{n}\frac{1}{\lambda-e^{\lambda_j T}}\frac{F(\lambda_j)}{\Delta'(\lambda_j)}\right\} \qquad (4.147)$$

Suppose the roots λ_k^* of the characteristic equation $\Delta^*(\lambda)=0$ are simple (for a stable discontinuous system the case of simple roots λ_k^* is very important). Then according to (2.80) with $g^*(m)\equiv 0$ and with the values of $F^*(\lambda_k^*)$ yielded by (4.147), we have

$$x(mT)=\sum_{k=1}^{n}\lambda_k^{*m}\frac{\Delta_p^*(\lambda_k^*)}{\Delta^{*\prime}(\lambda_k^*)}\left(\sum_{i=1}^{n}\frac{e^{\lambda_i T_2}\left(e^{\lambda_i T_1}-1\right)}{\lambda_i\left(\lambda_k^*-e^{\lambda_i T}\right)}\frac{F(\lambda_i)}{\Delta'(\lambda_i)}h\right)$$
$$\times\left(\gamma\sum_{j=1}^{n}\frac{1}{\lambda_k^*-e^{\lambda_j T}}\frac{F(\lambda_j)}{\Delta'(\lambda_j)}\right)x(0) \qquad (4.148)$$

Formula (4.148) determines the discrete free-motion of a discontinuous system expressed in terms of the assumed parameters.

To find the continuous motion it is necessary first to express the matrices (4.8) and (4.12) in the form of expansions [see, for example, (4.123) and (4.124)]

$$Q'(t)=\sum_{i=1}^{n}e^{\lambda_i t}\frac{F(\lambda_i)}{\Delta'(\lambda_i)}+\sum_{i=1}^{n}\frac{e^{\lambda_i t}-1}{\lambda_i}\frac{F(\lambda_i)}{\Delta'(\lambda_i)}h\gamma$$

$$Q''(t) = \sum_{i=1}^{n} e^{\lambda_i (t+T_1)} \frac{F(\lambda_i)}{\Delta'(\lambda_i)} + \sum_{i=1}^{n} \frac{e^{\lambda_i t}\left(e^{\lambda_i T_1} - 1\right)}{\lambda_i} \frac{F(\lambda_i)}{\Delta'(\lambda_i)} h\gamma \qquad (4.149)$$

Using (4.148) and (4.149), and the matrices (2.73), we write (4.7) and (4.11) in the form

$x[(m-1)T + \tau] =$

$$\sum_{k=1}^{n} \lambda_k^{*m-1} \frac{\Delta_p^*(\lambda_k^*)}{\Delta^{*'}(\lambda_k^*)} \left(\sum_{i=1}^{n} \frac{e^{\lambda_i \tau} e^{\lambda_i T_2}\left(e^{\lambda_i T_1} - 1\right)}{\lambda_i \left(\lambda_k^* - e^{\lambda_i T}\right)} \frac{F(\lambda_i)}{\Delta'(\lambda_i)} h + \sum_{i=1}^{n} \frac{e^{\lambda_i \tau} - 1}{\lambda_i} \frac{F(\lambda_i)}{\Delta'(\lambda_i)} h \right) \left(\gamma \sum_{j=1}^{n} \frac{1}{\lambda_k^* - e^{\lambda_j T}} \frac{F(\lambda_j)}{\Delta'(\lambda_j)} \right) x(0) \qquad (4.150)$$

$$0 \leqslant \tau \leqslant T_1$$

and

$x[(m-1)T + T_1 + \tau] =$

$$= \sum_{k=1}^{n} \lambda_k^{*m} \frac{\Delta_p^*(\lambda_k^*)}{\Delta^{*'}(\lambda_k^*)} \left(\sum_{i=1}^{n} \frac{e^{\lambda_i \tau}\left(e^{\lambda_i T_1} - 1\right)}{\lambda_i \left(\lambda_k^* - e^{\lambda_i T}\right)} \frac{F(\lambda_i)}{\Delta'(\lambda_i)} h \right) \times \left(\gamma \sum_{j=1}^{n} \frac{1}{\lambda_k^* - e^{\lambda_j T}} \frac{F(\lambda_j)}{\Delta'(\lambda_j)} \right) x(0), \quad 0 \leqslant \tau \leqslant T_2 \qquad (4.151)$$

In deriving these formulae use was made of the equality

$$1 = \sum_{i=1}^{n} \frac{e^{\lambda_i T_2}\left(e^{\lambda_i T_1} - 1\right)}{\lambda_i \left(\lambda_k^* - e^{\lambda_i T}\right)} \gamma \frac{F(\lambda_i)}{\Delta'(\lambda_i)} h \qquad (4.152)$$

which is obtained from (4.35) by putting $\lambda = \lambda_k^*$. For the boundary time values, $t = (m-1)T$ and $t = mT$, (4.150) and (4.151) naturally give the same values for $x[(m-1)T]$ and $x(mT)$, as does (4.148).

For the case of a double zero root $\lambda_{n-1} = \lambda_n = 0$, we have

$$K[\lambda E - M(T)]^{-1} K^{-1} = \sum_{i=1}^{n-2} \frac{1}{\lambda - e^{\lambda_i T}} k_i \varkappa_i + \frac{1}{\lambda - 1} (k_{n-1}\varkappa_{n-1} + k_n \varkappa_n) + \frac{T}{(\lambda - 1)^2} k_n \varkappa_{n-1}$$

$$K[\lambda E - M(T)]^{-1} M(T_2) N(T_1) K^{-1}$$

$$
\left.\begin{aligned}
&=\sum_{i=1}^{n-2}\frac{e^{\lambda_i T_2}\left(e^{\lambda_i T_1}-1\right)}{\lambda_i\left(\lambda-e^{\lambda_i T}\right)}k_i\varkappa_i \\
&+\frac{1}{\lambda-1}\left(k_{n-1}\varkappa_{n-1}+k_n\varkappa_n\right) \\
&\quad+\left(T_2+\frac{T_1}{2}+\frac{T}{\lambda-1}\right)\frac{T_1}{\lambda-1}k_n\varkappa_{n-1}
\end{aligned}\right\} \qquad (4.153)
$$

These expressions can be obtained by making use of the expansion formula (2.24) and expressions (4.143) for the matrix $[\lambda E - M(T)]^{-1}$ and (4.47) for the matrix $M(T_2)N(T_1)$.

Furthermore, remembering the matrices (2.64) and (2.66), we can write (4.145) in the form

$$
\begin{aligned}
F^*(\lambda) = \{\Delta^*(\lambda)E + \Delta_p^*(\lambda) &\left(\sum_{i=1}^{n-2}\frac{e^{\lambda_i T_2}\left(e^{\lambda_i T_1}-1\right)}{\lambda_i\left(\lambda-e^{\lambda_i T}\right)}\frac{F(\lambda_i)}{\Delta'(\lambda_i)}\right. \\
+\frac{T_1}{\lambda-1}\left[\frac{F(0)}{\Delta_{n-1}(0)}\right]^{(1)} &+\left.\left(T_2+\frac{T_1}{2}+\frac{T}{\lambda-1}\right)\frac{T_1}{\lambda-1}\frac{F(0)}{\Delta_{n-1}(0)}\right)n\gamma \\
\times\left(\sum_{j=1}^{n-2}\frac{1}{\lambda-e^{\lambda_j T}}\frac{F(\lambda_j)}{\Delta'(\lambda_j)}\right. &+\frac{1}{\lambda-1}\left[\frac{F(0)}{\Delta_{n-1}(0)}\right]^{(1)} \\
&\left.+\frac{T}{(\lambda-1)^2}\frac{F(0)}{\Delta_{n-1}(0)}\right)
\end{aligned} \qquad (4.154)
$$

In the case we are considering, the discrete motion is obtained from (2.80) with $g^*(m)\equiv 0$ if $F^*(\lambda_k^*)$ is found from (4.154). The corresponding formulae for a continuous motion can also be obtained by using the same arguments as were used for the case of simple roots λ_i.

a) Response of the system to a desired response signal

In accordance with (1.19), the argument of the control σ can be written in matrix form

$$
\sigma = \gamma(x - x_3) \qquad (4.155)
$$

where x_3 is the column of scalar function x_{3i} which determine the desired response. The equations which determine the response of the system to a desired response signal can be obtained very simply from the formulae of Section 4.1 if $\gamma(x - x_3)$ is inserted in the appropriate expressions in place of γx.

Thus, for example, by virtue of (4.7), (4.8), (4.11), (4.12), (4.15) and (4.16), and making the substitution just mentioned, we obtain the following relations:

$$x[(m-1)T+\tau] = Q'(\tau)\,x[(m-1)T]$$

$$-KN(\tau)K^{-1}h\gamma x_3[(m-1)T] \qquad 0 \leqslant \tau \leqslant T_1 \tag{4.156}$$

$$x[(m-1)T+T_1+\tau] = Q''(\tau)\,x[(m-1)T]$$

$$-KM(\tau)N(T_1)K^{-1}h\gamma x_3[(m-1)T] \quad 0 \leqslant \tau \leqslant T_2 \tag{4.157}$$

$$x(mT) = P^*x[(m-1)T] - KM(T_2)N(T_1)K^{-1}h\gamma x_3[(m-1)T] \tag{4.158}$$

Equation (4.158) is an inhomogeneous difference equation whose solution for zero initial conditions determines the response to a desired response signal of a system which is in discrete motion. After the solution of the difference equation (4.158) has been found, (4.156) and (4.157) permit one to determine the matrix coordinate x for any instant of time lying within any recurrence interval.

We will write the inhomogeneous difference equation (4.158) in the form

$$x(mT) = P^*x[(m-1)T] + g^*[(m-1)T] \tag{4.159}$$

where

$$g^*(mT) = -KM(T_2)N(T_1)h\gamma x_3(mT) \tag{4.160}$$

Clearly, $g^*(mT)$ is a column matrix. The difference equation (4.159) is the same as (2.76) with an accuracy up to that of the chosen step, so that the response of a pulsed system in discrete motion to a desired response signal will be determined by (2.78)-(2.83) if we put $x(0)=0$ in these formulae and take T as the new step of the discrete time. In future we shall make use of the formulae

$$x(mT) = \sum_{m'=0}^{m-1} P^{*m-m'-1} g^*(m'T) \tag{4.161}$$

and

$$x(mT) = \sum_{i=1}^{n} \frac{F^*(\lambda_i^*)}{\Delta^{*\prime}(\lambda_i^*)} \sum_{m'=0}^{m-1} \lambda_i^{*m-m'-1} g^*(m'T) \tag{4.162}$$

Equality (4.161) determines the response of the system in all cases, and (4.162) only in the case when the roots λ_i^* of the characteristic equation $\Delta^*(\lambda)=0$ are simple.

Consider now the response of a stable system when the desired response signal can be described by certain simple functions of time.

b) Response to a unit action

In this case $\gamma x_3(mT)=1$. Let us introduce the notation

$$-KM(T_2)N(T_1)K^{-1}h=g_0^* \tag{4.163}$$

Then, in accordance with (4.160), (4.161) can be rewritten in the form

$$x(mT)=\left(\sum_{m'=0}^{m-1} P^{*m-m'-1}\right)g_0^* \tag{4.164}$$

Applying the formula for finding the sum of a geometric progression, we obtain

$$x(mT)=\frac{E-P^{*m}}{E-P^*}g_0^* \tag{4.165}$$

Furthermore, from formulae of type (2.55) and (2.68), we can establish

$$x(mT)=-\left(\sum_{i=1}^{n}\lambda_i^{*m}\frac{F^*(\lambda_i^*)}{\Delta^{*\prime}(\lambda_i^*)}\right)\frac{F^*(1)}{\Delta^*(1)}g_0^*+\frac{F^*(1)}{\Delta^*(1)}g_0^* \tag{4.166}$$

c) Response to an exponential and to a harmonic action

Let $\gamma x_3(mT)=e^{\alpha_0 mT}$ Then, considering (4.160) and (4.163), according to (4.161) we have

$$x(mT)=\left(\sum_{m'=0}^{m-1} P^{*m-m'-1}e^{\alpha_0 m'T}\right)g_0^* \tag{4.167}$$

The sum on the right-hand side of this equality is a geometric progression whose denominator is equal to $P^*e^{-\alpha_0 T}$. Using the formula for the sum of a geometric progression and multiplying the numerator and denominator of the expression obtained by $e^{\alpha_0 T}$, we can write (4.167) in the form

$$x(mT)=\frac{e^{\alpha_0 mT}E-P^{*m}}{e^{\alpha_0 T}E-P^*}g_0^* \tag{4.168}$$

Applying formulae of type (2.55) and (2.68), we obtain

$$x(mT)=-\left(\sum_{i=1}^{n}\lambda_i^{*m}\frac{F^*(\lambda_i^*)}{\Delta^{*\prime}(\lambda_i^*)}\right)\frac{F^*(e^{\alpha_0 T})}{\Delta^*(e^{\alpha_0 T})}g_0^*$$

$$+\frac{F^*(e^{\alpha_0 T})}{\Delta^*(e^{\alpha_0 T})}g_0^*e^{\alpha_0 mT} \tag{4.169}$$

The formula obtained permits us to determine the response of a system in discrete motion to an external harmonic action. For example, in the case of a sinusoidal action, we have

$$\sin \omega mT = \frac{(e^{i\omega mT} - e^{-i\omega T})}{2i}$$

Making use of the property of superposition of linear systems, and using (4.169), we obtain

$$x(mT) = -\left(\sum_{j=1}^{n} \lambda_j^{*m} \frac{F^*(\lambda_j^*)}{\Delta^{*\prime}(\lambda_j^*)}\right) \mathrm{Im}\left\{\frac{F^*(e^{i\omega T})}{\Delta^*(e^{i\omega T})} g_0^*\right\}$$

$$+ \mathrm{Im}\left\{\frac{F^*(e^{i\omega T})}{\Delta^*(e^{i\omega T})} g_0^*\right\} \sin \omega mT - \mathrm{Re}\left\{\frac{F^*(e^{i\omega T})}{\Delta^*(e^{i\omega T})} g_0^m\right\} \cos \omega mT \qquad (4.170)$$

where Re and Im denote the real and imaginary parts of the corresponding complex number. With $\alpha_0 = 0$, (4.169) coincides with (4.166). Formula (4.169) loses its meaning if $e^{\alpha_0 T}$ is a root of the characteristic equation $\Delta^*(\lambda) = 0$.

d) Response to an action of the form

In this case we have $\gamma x_3(mT) = (mT)^k e^{\alpha_0 mT}$ or, in more compact form, $\gamma x_3(mT) = \left(\frac{d^k}{d\alpha^k} e^{\alpha mT}\right)_{\alpha=\alpha_0}$. Formula (4.161) can in the given case be written in the form

$$x(mT) = \left\{\frac{d^k}{d\alpha^k} \sum_{m'=0}^{m-1} P^{*m-m'-1} e^{\alpha m' T}\right\}_{\alpha=\alpha_0} g_0^* \qquad (4.171)$$

Identical sums of a geometric progression feature in (4.167) and (4.171). Consequently, on the basis of (4.169), (4.171) can be put in the form

$$x(mT) = -\left\{\sum_{i=1}^{n} \lambda_i^{*m} \frac{F^*(\lambda_i^*)}{\Delta^{*\prime}(\lambda_i^*)}\right\}\left\{\frac{d^k}{d\alpha^k} \frac{F^*(e^{\alpha T})}{\Delta^*(e^{\alpha T})}\right\}_{\alpha=\alpha_0} g_0^*$$

$$+ \left\{\frac{d^k}{d\alpha^k} \frac{F^*(e^{\alpha T})}{\Delta^*(e^{\alpha T})} e^{\alpha mT}\right\}_{\alpha=\alpha_0} g_0^* \qquad (4.172)$$

Formula (4.172) determines the response of a system in discrete motion to the action indicated above. In conjunction with the superposition principle this formula permits one to obtain quite easily the response of the system to a desired response signal of more general form when $\gamma x_3(t) = Q_k(t) e^{\alpha_0 T}$, where $Q_k(t)$ is a polynomial of degree k with respect to t.

In all the formulae derived above [(4.166), (4.169) and (4.172)], the first term on the right-hand side determines the discrete free-motion of a system excited by a desired response signal, and the second term determines the discrete forced motion. If a discontinuous system satisfies the stability conditions, then one can say that the first term determines the transient response of the discrete motion of the system towards the steady state or a state of equilibrium, whilst these states themselves are determined by the second term in the above formulae. In particular, it is interesting to note that, if the steady state of a system in discrete motion is in an equilibrium position, it will execute periodic oscillations when the system is in continuous motion.

If this equilibrium position happens to coincide with the point $x=0$, then this point also will be an equilibrium position of the system in continuous motion.

Formula (4.161) should be used to determine the responses of systems in discrete motion, since this formula enables the required results to be obtained in a very compact form. The same results can, of course, also be obtained by using the solution of the difference equation (4.159) in the form (4.162). Using the solution in the form (4.162), one can also determine the response of the system in those special cases when $e^{\alpha_0 T}$ is a root of the characteristic equation $\Delta^*(\lambda)=0$. Indeed, let us again set $\gamma x_3(mT) = e^{\alpha_0 m T}$. In that, taking (4.160) and (4.163) into account, we obtain from (4.162) the expression

$$x(mT)=\sum_{i=1}^{n}\frac{F^*(\lambda_i^*)}{\Delta^{*\prime}(\lambda_i^*)}\sum_{m'=0}^{m-1}\lambda_i^{*m-m'-1}e^{\alpha_0 m'T}g_0^* \tag{4.173}$$

Removing the second sum by means of the geometric progression formula, we obtain

$$x(mT)=\sum_{i=1}^{n}\frac{e^{\alpha_0 mT}-\lambda_i^{*m}}{e^{\alpha_0 T}-\lambda_i^*}\,\frac{F^*(\lambda_i^*)}{\Delta^{*\prime}(\lambda_i^*)}\,g_0^* \tag{4.174}$$

For the sake of definiteness, let $\lambda_n^*=e^{\alpha_0 T}$. Then, removing the indeterminacy for $i=n$, we put (4.174) in the form

$$x(mT)=-\sum_{i=1}^{n-1}\frac{\lambda_i^{*m}}{e^{\alpha_0 T}-\lambda_i^*}\,\frac{F^*(\lambda_i^*)}{\Delta^{*\prime}(\lambda_i^*)}\,g_0^*$$

$$+e^{\alpha_0 mT}\sum_{i=1}^{n-1}\frac{1}{e^{\alpha_0 T}-\lambda_i^*}\,\frac{F^*(\lambda_i^*)}{\Delta^{*\prime}(\lambda_i^*)}\,g_0^*+me^{\alpha_0(m-1)T}\,\frac{F^*(e^{\alpha_0 T})}{\Delta^{*\prime}(e^{\alpha_0 T})}\,g_0^* \tag{4.175}$$

Apart from the column g_0^*, all the elements of the formula just obtained were expressed earlier in terms of the initial parameters of the systems under consideration. The column matrix g_0^*, which is given by (4.163), enters as a factor into the second term of (4.15). Using (4.124) and (4.126), it is therefore not difficult to establish the following expansions for g_0^*

$$\left.\begin{aligned} g_0^* &= -\sum_{i=1}^{n} \frac{e^{\lambda_i T_2}\left(e^{\lambda_i T_1}-1\right)}{\lambda_i} \frac{F(\lambda_i)}{\Delta'(\lambda_i)} h \\ g_0^* &= -\sum_{i=1}^{n-2} \frac{e^{\lambda_i T_2}\left(e^{\lambda_i T_1}-1\right)}{\lambda_i} \frac{F(\lambda_i)}{\Delta'(\lambda_i)} h \\ &\quad - T_1\left[\frac{F(0)}{\Delta_{n-1}(0)}\right]^{(1)} h - \left(T_1 T_2 + \frac{T_1^2}{2}\right)\frac{F(0)}{\Delta_{n-1}(0)} h \end{aligned}\right\} \qquad (4.176)$$

which are valid respectively for the cases of simple roots λ_i and when, among the roots λ_i, there is a double zero root $\lambda_{n-1}=\lambda_n=0$.

According to the idealisation adopted in this book, the response of a discontinuous system does not depend on the values assumed by the desired response function at internal points of the recurrence intervals provided it is assumed that the beginning of each interval coincides with the start of the control signal.

The picture is somewhat altered when one is analysing systems acted upon by external perturbations. In this case, the response of the system depends substantially upon the detailed variation of the 'perturbing' function throughout the interval of variation of the time t.

e) Response of the system to an external perturbation

In this case, the motion of a discontinuous system is determined by a matrix differential equation (4.2) in which the column g is a known function of time defining the external perturbation. A solution to (4.2) can easily be constructed, starting from the following considerations. Suppose at time $t=(m-1)T$ (which is the start of the mth recurrence interval), the matrix coordinate $x=x[(m-1)T]$. Then, inside the mth interval, $\psi(\sigma)$ is a known function of time and so also is function g. Thus, inside the mth recurrence interval, (4.2) can be treated as an inhomogeneous linear differential equation into the right-hand side of which there enter two known functions of time which take the form of sums. One can therefore apply the principle of superposition. Let us first, for example, find the solution for $g\equiv 0$ satisfying the initial condition $x[(m-1)T]$ for $t_0=(m-1)T$ and then add to it the solution for $\psi(\sigma)\equiv 0$ satisfying the zero initial condition. The first solution we obtained earlier; it is given by (4.7) and (4.11).

The second solution is given by (2.128) for $x(t_0)=0$. Combining these two solutions, we obtain

$$\left.\begin{aligned} x[(m-1)T+\tau] &= Q'(\tau)\,x[(m-1)T] \\ &\quad+\int_0^{\tau} KM(\tau-\xi)K^{-1}g[(m-1)T+\xi]\,d\xi \\ &\qquad 0\leqslant\tau\leqslant T_1 \\ x[(m-1)T+T_1+\tau] &= Q''(\tau)\,x[(m-1)T] \\ &\quad+\int_0^{T_1+\tau} KM(T_1+\tau-\xi)K^{-1}g[(m-1)T+\xi]\,d\xi \\ &\qquad 0\leqslant\tau\leqslant T_2 \end{aligned}\right\} \tag{4.177}$$

where the matrices $Q'(\tau)$ and $Q''(\tau)$ are given by (4.8) and (4.12). from the second equality (4.177), for $\tau=T_2$ – see formula (4.14) – we obtain the expression

$$x(mT)=P^*x[(m-1)T]+\int_0^T KM(T-\xi)K^{-1}g[(m-1)T+\xi]\,d\xi \tag{4.178}$$

which can be written in the form of (4.159) if we put

$$g^*(mT)=\int_0^T KM(T-\xi)K^{-1}g(mT+\xi)\,d\xi \tag{4.179}$$

The inhomogeneous difference equation (4.178) determines the discrete motion of a discontinuous system when it is acted upon by an external perturbation. Once the discrete motion has been found, the continuous motion of the system is given by (4.177).

Thus, the discrete motion of the system due to either a desired response signal or to an external perturbation is given by one and the same inhomogeneous difference equation (4.159), in which the column $g^*(mT)$ is determined, respectively, by (4.160) and (4.179). Thus, the results obtained earlier will be valid in the given case also provided the necessary precautions are observed. In fact, the function of time enters into the column (4.160) as a scalar factor, so column (4.179) must first be put in the same form, for example by breaking it up into a sum by a suitable choice of columns.

Column (4.179) can be expressed in terms of the original parameters in the form of the expressions

$$
\left.
\begin{aligned}
g^*(mT) &= \sum_{i=1}^{n} \frac{F(\lambda_i)}{\Delta'(\lambda_i)} \int_0^T e^{\lambda_i (T-\xi)} g(mT+\xi)\, d\xi \\
g^*(mT) &= \sum_{i=1}^{n-2} \frac{F(\lambda_i)}{\Delta'(\lambda_i)} \int_0^T e^{\lambda_i (T-\xi)} g(mT+\xi)\, d\xi \\
&\quad + \left[\frac{F(0)}{\Delta_{n-1}(0)}\right]^{(1)} \int_0^T g(mT+\xi)\, d\xi \\
&\quad + \frac{F(0)}{\Delta_{n-1}(0)} \int_0^T (T-\xi)\, g(mT+\xi)\, d\xi
\end{aligned}
\right\} \qquad (4.180)
$$

which are valid, respectively, for simple roots λ_i and when a double zero root $\lambda_{n-1} = \lambda_n = 0$ is present.

4.7 LIMITING CASES OF THE DISCONTINUOUS CONTROL FUNCTION

The control function $\psi(\sigma)$ in the discontinuous systems so far considered has been represented by the curve given in Fig 1.7. For an arbitrary recurrence interval, the width of a pulse and the duration of the subsequent pause are to be regarded as finite and constant intervals of time T_1 and T_2.

One can consider the cases when T_1 or T_2 are vanishingly small quantities in relation to some characteristic time of the system. This characteristic time can conveniently be taken as the quantity $1/|\lambda_i|_{\max}$, where $|\lambda_i|_{\max}$ is the root with the greatest modulus belonging to the characteristic equation of the open system. With the assumptions we have made we are not interested in any change of coordinates at the internal points of the above intervals, so these changes can in the limit be put equal to zero.

In the first case the limiting transition must be made in a manner that preserves the basic physical characteristics of the control signal determined by the area of the pulses. Should the width of the pulses be decreased, their height must be proportionately increased so that their area will remain constant.

This kind of idealisation has been adopted in several other cases; in particular, in generating an impulse function (a Dirac δ-function) from a rectangular pulse [15]. Therefore, a control function consisting of instantaneous pulses will be referred to, for brevity, as an impulse control function. We will denote it by $\psi^*(\sigma)$.

Let us introduce the auxiliary control function $\overline{\psi}^*(\sigma)$ with a pulse width T_1^*, defined by the formula

$$\bar{\psi}^*(\sigma) = \frac{T_1}{T_1^*}\psi(\sigma) \tag{4.181}$$

For any value $T_1^* < T_1$, the area of the pulse $T_1^*\bar{\psi}^*(\sigma)$ will be equal to the constant quantity $T_1\psi(\sigma)$. In the limit, for $T_1^* = 0$, we obtain from $\bar{\psi}^*(\sigma)$ the impulse control function $\psi^*(\sigma)$. This function is represented in Fig 4.5 in the form of impulse functions equally spaced in time. The height of any δ-function pulse is conventionally taken to be equal to the magnitude of the argument of the control σ at the corresponding instants of time. This conventionally recognises the fact that the intensity of the instantaneous pulses is proportional to the corresponding values of σ.

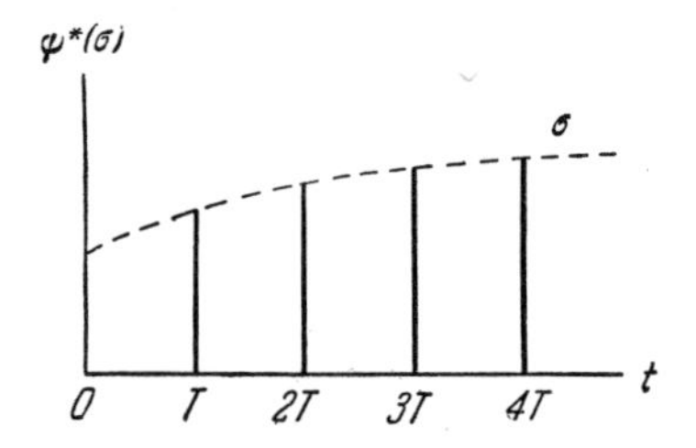

Fig 4.5 Graph of an impulse control function.

Let us consider the free-motion of a discontinuous system with a control function given by (4.181). In this case the value of the matrix coordinate x for $t = (m-1)T + T_1^*$ is given by the expression

$$x\left[(m-1)T + T_1^*\right] = Q'(T_1^*)\, x\,[(m-1)T] \tag{4.182}$$

where

$$Q'(T_1^*) = KM(T_1^*)K^{-1} + KN(T_1^*)K^{-1}\frac{T_1}{T_1^*}h\gamma \tag{4.183}$$

It is not difficult to obtain these equalities from (4.7) and (4.8) by setting $\tau = T_1^*$ and using (4.181). The matrices $M(t)$ and $N(t)$ are given by (2.94), (2.98) and (2.120)-(2.122). In accordance with these formulae, in the limit when $T_1^* = 0$, we obtain from (4.183) the expression

$$Q'(0) = E + T_1 h\gamma \tag{4.184}$$

By supposing that $T_1^* \to 0$ in (4.182) and assuming $Q'(0) \neq E$ (4.184), we obtain the result that, for $T_1^* = 0$, the values of the matrix coordinate x are not the same on the two sides of (4.182). In order to stress this fact, we will in the limiting equality write from the left $x[(m-1)T+0]$, and from the right $-x[(m-1)T-0]$. Taking this into account, and using the expression for $Q'(0)$ given in (4.184), in the limit when $T_1^*=0$ we can obtain from (4.182) a relation in the form

$$x[(m-1)T+0]=x[(m-1)T-0]+T_1 h\gamma x[(m-1)T-0] \quad (4.185)$$

The interval of time represented by the width of a pulse of the function $\psi(\sigma)$ has been contracted to a point. In the region close to this point we must distinguish a left- and right-hand side. The right-hand side corresponds to the passage of an instantaneous control pulse; its action on the system in this case reduces to a step-wise change of the matrix coordinate x in accordance with (4.185). In the interval of time $[(m-1)T+0]\leqslant t\leqslant(mT-0)$ of the mth recurrence interval that remains the motion of the system is described by the differential equation (4.9). The solution of this equation, which is valid for the interval indicated above, can be written in the form

$$x[(m-1)T+\tau]=KM(\tau)K^{-1}x[(m-1)T+0] \quad (4.186)$$

which is obtained from (4.10) for $T_1=0$, where account has been taken of the two-sidedness of the boundary points of the recurrence interval just mentioned. If we substitute in here $x[(m-1)T+0]$ from (4.185), we obtain the expression

$$x[(m-1)T+\tau]=Q''(\tau)x[(m-1)T-0] \quad (4.187)$$

where

$$Q''(\tau)=KM(\tau)K^{-1}+KM(\tau)K^{-1}T_1 h\gamma \quad (4.188)$$

With $\tau=T-0$, we will have

$$x(mT-0)=P^*x[(m-1)T-0] \quad (4.189)$$

where

$$P^*=Q''(T)=KM(T)K^{-1}+KM(T)K^{-1}T_1 h\gamma \quad (4.190)$$

The difference equation (4.189) determines the discrete motion of the system when the control function takes the form of instantaneous pulses. After the solution of this equation has been found, the value of the matrix coordinate x at intermediate points of the recurrence interval is determined using (4.187) and (4.188) for $0<\tau<T$.

The matrices (4.188) and (4.190) can be formally obtained from (4.12) and (4.15) if, in the latter expressions, we put $T_1=T_1^*$ replace h by $\frac{T_1}{T_1^*}h$ and go to the limit by taking $T_1^*=0$, assuming at the same time, of course, that $T_2=T$. Consequently, all the relations we obtained for the control function with a finite pulse width will be valid also for a system with instantaneous control pulses, provided that we carry out the above substitution and take the limit in the manner described.

In the second limiting case, when the duration of the pause is assumed to be equal to zero, the control function $\psi(\sigma)$ takes the form of a stepped function – as illustrated in Fig 4.6. In this case, all the required relations are obtained directly from the formulae derived for a control function $\psi(\sigma)$ with a pause of finite duration by going to the limit $T_2 \to 0$ preserving, as always, the equality $T_1 + T_2 = T$. In particular, from (4.15) we obtain the expression

$$P^* = KM(T)K^{-1} + KN(T)K^{-1}h\gamma \tag{4.191}$$

which determines the matrix operator of the discrete motion of a discontinuous system having a stepped control function $\psi(\sigma)$.

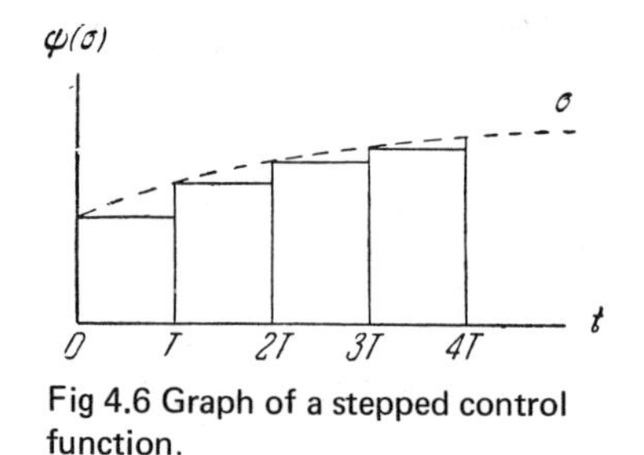

Fig 4.6 Graph of a stepped control function.

In conclusion, we note that a theory of linear discontinuous systems with impulse and stepped control functions can be constructed independently by following exactly the same procedure and going through the same arguments as those used for systems with control pulses of finite (non-zero) width and with pauses of finite duration. In developing such a theory, it is necessary to bear in mind that all expressions directly containing the column u', or columns determined through u', should be retained without change regardless of the form of the control function. Examples of such expressions are the characteristic equations in the form (4.30) and (4.40), the adjoint matrix in the form (4.144) and the transformed characteristic equations (4.56) and (4.58) etc. The column u' is expressed in terms of the elements of the column u by formulae which depend upon the form of the control function. These formulae are the following: relation (4.25) for a control function with a finite pulse width and length of pause; the expression

$$u' = T_1 M(T) u \tag{4.192}$$

for an impulse control function; and the equality

$$u' = N(T) u \tag{4.193a}$$

for a stepped control function. Formulae (4.25), (4.192) and (4.193) are represented in matrix form, which is valid for any form of the roots λ_i.

The reader can choose any of the above methods, or use them simultaneously, to provide a means of checking the results obtained.

As an exercise, let us take an impulse control function $\psi^*(\sigma)$ and reduce the transformed characteristic equation (4.58) to the form

$$\prod_{j=1}^{n-2}[\nu(1-e^{\lambda_j T})+(1+e^{\lambda_j T})]\Bigg\{1-T_1(\nu-1)$$

$$\times\left[\sum_{i=1}^{n-2}\frac{e^{\lambda_i T}}{\nu(1-e^{\lambda_i T})+(1+e^{\lambda_i T})}\frac{\gamma F(\lambda_i)h}{\Delta'(\lambda_i)}+\frac{1}{2}\gamma\left[\frac{F(0)}{\Delta_{n-1}(0)}\right]^{(1)}h\right.$$

$$\left.+\frac{T}{4}(\nu+1)\frac{\gamma F(0)h}{\Delta_{n-1}(0)}\right]\Bigg\}=0 \qquad (4.193b)$$

and when the condition $\gamma h=0$ is fulfilled, to the form

$$\prod_{j=1}^{n-2}[\nu(1-e^{\lambda_j T})+(1+e^{\lambda_j T})]$$

$$\times\Bigg\{(1-T_1(\nu+1)\left[\sum_{i=1}^{n-2}\frac{1}{\nu(1-e^{\lambda_i T})+(1+e^{\lambda_i T})}\frac{\gamma F(\lambda_i)h}{\Delta'(\lambda_i)}\right.$$

$$\left.+\frac{1}{2}\gamma\left[\frac{F(0)}{\Delta_{n-1}(0)}\right]^{(1)}h+\frac{T}{4}(\nu+1)\frac{\gamma F(0)h}{\Delta_{n-1}(0)}\right]\Bigg\}=0 \qquad (4.193c)$$

To derive equation (4.193c) it is suggested that the material given at the end of Section 2.3 (or an intermediate form of equation) be used, employing the column elements u and row elements β. In this case, in accordance with (4.21), we have $\beta u=\gamma h$.

4.8 NONLINEAR DISCONTINUOUS SYSTEMS

Let us now consider discontinuous systems with a control function generated in accordance with the pulse-width modulation principle. The control function in this case is given by Table 1.1 and is illustrated in Fig 1.9.

The basic philosophy of finding a solution to the initial differential equations which will be valid inside the mth recurrence interval, i. e. for $(m-1)T\leqslant t\leqslant mT$, remains unchanged for such a system. We therefore make use of the systems of relations obtained for linear discontinuous systems and endeavour to obtain the expressions required by introducing whatever changes are necessary.

Consider the free-motion of a nonlinear discontinuous system. Using (4.7)-(4.12) and with account of (1.11), we can write the solution in the mth recurrence interval in the form

$$\left.\begin{aligned} x[(m-1)T+\tau] &= KM(\tau)K^{-1}x[(m-1)T] \\ &\quad + KN(\tau)K^{-1}h\chi_{m-1} \qquad 0 \leqslant \tau \leqslant T_{1,m-1} \\ x[(m-1)T+T_{1,m-1}+\tau] & \\ = KM(T_{1,m-1}+\tau)K^{-1}&x[(m-1)T] \\ + KM(\tau)(T_{1,m-1})K^{-1}&h\chi_{m-1} \qquad 0 \leqslant \tau \leqslant T_{2,m-1} \end{aligned}\right\} \tag{4.194}$$

From the second equality (4.194), for $\tau = T_{2,m-1}$ we obtain

$$x(mT) = KM(T)K^{-1}x[(m-1)T]$$
$$+ KM(T_{2,m-1})N(T_{1,m-1})K^{-1}h\chi_{m-1} \tag{4.195}$$

remembering that $T_{1,m-1}$ is given by

$$T_{1,m-1} = T_1\chi_{m-1}\sigma[(m-1)T] \tag{4.196}$$

or

$$T_{1,m-1} = T_1\chi_{m-1}\gamma x[(m-1)T] \tag{4.197}$$

Considered together, (4.195) and (4.197) represent a nonlinear difference equation describing the discrete motion of a system with pulse-width modulation of the control signal. The nonlinear difference equation obtained is a recurrence relation. It permits one to determine step by step, in accordance with a given initial condition $x(0)$, the values x for any discrete instant of time mT. The values of the coordinate x at intermediate points in the recurrence interval are given by (4.194). In order to carry out these computations, it is necessary first to express all the elements of these formulae in terms of the input parameters.

It is possible to use the Lyapunov method to investigate the stability of the discrete motion of a nonlinear discontinuous system (see Chapter 3). For this purpose it is necessary to derive the nonlinear equation of the first approximation. According to Lyapunov terminology, (4.195) and (4.197) determine the nonlinear difference equation of the perturbed motion of the system; the equilibrium state is given by the identity $x \equiv 0$. With $x \equiv 0$, the left-hand side of (4.197) also vanishes and $T_{2,m-1}$ is changed to T, since $T_{1,m} + T_{2,m} = T$.

To construct the equation of the first approximation, we expand the right-hand side of (4.195) as a Taylor series, taking the centre of the expansion as $x=0$ and keeping only terms not greater than first order in x. Then, with account of the remarks made above, we obtain

$$x(mT)=KM(T)K^{-1}x[(m-1)T]+K\left[\frac{dM(\xi_2)}{d\xi_2}N(\xi_1)T_{2,m-1}\right.$$

$$\left.+M(\xi_2)\frac{dN(\xi_1)}{d\xi_1}T_{1,m-1}\right]_{\substack{\xi_1=0\\ \xi_2=T}}K^{-1}h\chi_{m-1} \quad (4.198)$$

Assuming that $M(0)=E$, $N(0)=0$, $\frac{dM}{dt}=JM(t)$, $\frac{dN(t)}{dt}=M(t)$ and, making use of (4.197), we obtain from (4.198) the expression

$$x(mT)=P^*x[(m-1)T] \quad (4.199)$$

where

$$P^*=KM(T)K^{-1}+KM(T)K^{-1}T_1h\gamma \quad (4.200)$$

which will be the equation of the first approximation for the nonlinear discontinuous system. The matrix operator P^* (4.199) of the equation of the first approximation so obtained coincides formally with matrix (4.190).

We can therefore conclude that the investigation of the stability of a nonlinear discontinuous system in the Lyapunov sense reduces to an analysis of the stability of a linear discontinuous system operating with an impulse control function $\psi^*(\sigma)$.

4.9 DISCONTINUOUS SYSTEMS WITH VANISHINGLY SMALL RECURRENCE PERIODS

The behaviour of discontinuous system with small recurrence periods T approaches that of the corresponding linear systems operating with continuous control. In fact, the 'invention' of the principle of discontinuous control with a forced rhythm of application of control pulses was actually an example of engineering intuition which suggested that such a method would, on average, achieve a proportional control. The average value $\psi_{av}^{(m)}(\sigma)$ of the control function in the mth recurrence interval for any of the discontinuous systems considered above is given by

$$\psi_{av}^{(m)}(\sigma)=\frac{T_1}{T}\gamma x[(m-1)T] \quad (4.201)$$

where, for a stepped control function, it is necessary to take $T_1=T$. It is therefore natural to compare the behaviour of discontinuous systems with small values of T with the behaviour of a linear continuous system whose motion is described by the differential equation

$$\dot{x}=Px+\frac{T_1}{T}h\gamma x \tag{4.202}$$

represented in matrix form. The solution of (4.202), which assumes a specified initial value $x(0)$ for $t=0$, can be written in the form [see (2.127) for $g(t)\equiv 0$)]

$$x(t)=e^{\left(P+\frac{T_1}{T}h\gamma\right)t}x(0) \tag{4.203}$$

From (4.203) one can, in particular, find the value of the matrix coordinate x at discrete instants of time mT, i.e.

$$x(mT)=\tilde{P}^{*m}x(0) \tag{4.204}$$

where the matrix $\tilde{P}^*$ is determined by the expression

$$\tilde{P}^*=e^{\left(P+\frac{T_1}{T}h\gamma\right)T} \tag{4.205}$$

If we expand the exponential on the right-hand side of (4.205) in the form of the series (2.85), retaining in this expansion terms of order not greater than the first with respect to T, we obtain the approximate equality

$$\tilde{P}^*=E+\left(P+\frac{T_1}{T}h\gamma\right)T \tag{4.206}$$

(since the inequality $T_1\leqslant T$ always holds).

The discrete free-motion of a linear discontinuous system is in all cases given by

$$x(mT)=P^{*m}x(0) \tag{4.207}$$

in which the matrix operator P^* is given by (4.15), (4.190) and (4.191), depending on the type of pulsed control function $\psi(\sigma)$.

Let us consider the matrix operator (4.15), since this operator has the most general form. To simplify later discussion we shall assume that the matrix P is non-singular, i.e. all the roots λ_i are non-zero. Considering (2.90), (2.91) and (2.107)-(2.110), we can then rewrite (4.15) in the form

$$P^*=e^{PT}+e^{PT_2}(e^{PT_1}-E)P^{-1}h\gamma \tag{4.208}$$

or

$$P^*=e^{PT}+(e^{PT}-e^{PT_2})P^{-1}h\gamma \tag{4.209}$$

Expanding the exponential in (4.209) as a series, we obtain the approximate equality

$$P^* \approx E + PT + T_1 h\gamma \tag{4.210}$$

which is the same as (4.206). Equality (4.210) remains valid for an impulse control function; for a stepped control function one must set $T_1 = T$.

Thus, if the recurrence period T is a sufficiently small quantity, the values of the matrix coordinate x for the linear continuous system (4.202) will be the same, to a first approximation, as those for the discontinuous systems we have been discussing. It is assumed here, of course, that identical initial conditions are taken in the two cases.

Let us consider now the characteristic equations of a discontinuous system with a matrix operator P^*, and of a continuous system described by (4.202). We have proved that, in first approximation, the matrices P^* and $\tilde{P}^*$ are the same [see (4.206) and (4.210)]. From Section 2.5, it follows that the characteristic values $\tilde{\lambda}_i^*$ of the matrix $\tilde{P}^*$ (4.205) are given by the expression $e^{\tilde{\lambda}_i T}$, where $\tilde{\lambda}_i$ are the characteristic values of the matrix $\left(P + \frac{T_1}{T} h\gamma\right)$ and consequently of (4.202) also. Therefore, in first approximation, the following equalities should hold:

$$\lambda_i^* = \tilde{\lambda}_i^* = e^{\tilde{\lambda}_i T} \tag{4.211}$$

Thus, if in the characteristic equation $\Delta^*(\lambda) = 0$ of matrix P^* which is given, for example, by (4.15) we replace λ by $e^{\tilde{\lambda} T}$ and take T to zero keeping the ratio T_1/T fixed, then, in the limit when $T = 0$, we obtain the characteristic equation for (4.202). We suggest that the reader carries out these calculations himself and for this purpose we recommend that the characteristic equation $\Delta^*(\lambda) = 0$ in the form (4.49) be used.

Let us consider further the transformation (4.54). Replacing λ in this transformation by $e^{\lambda T}$ and assuming that T is sufficiently small, we obtain the approximate relation

$$\nu \approx \frac{1}{\tilde{\lambda} \frac{T}{2}} \tag{4.212}$$

If we substitute (4.212) into the transformed characteristic equation (4.56) and (4.58) then, in the limit, we again obtain the characteristic equation for (4.202).

In (4.202) we can regard the positive coefficient T_1/T as the gain of the control signal. Therefore a change in the ratio of the duration of the control pulse to the recurrence period in the pulsed system is equivalent, as regards its action, to a corresponding change of gain in the linear system with continuous control.

One can also approach the results obtained above from the standpoint of an approximate integration of linear differential equations and, in particular, one can obtain an approximate representation of the matrix $\tilde{P}^*$ in the form (4.206) by appealing to the following simple arguments. For our approximate integration of (4.202) we will choose a small step T. In the interval of time $(m-1)T \leqslant t \leqslant mT$ we will assume that the right-hand side of (4.202) is constant and we will write the equation itself for this interval in the form

$$\dot{x}(t) = \left(P + \frac{T_1}{T} h\gamma\right) x[(m-1)T] \tag{4.213}$$

The solution of this equation, which reduces to $x[(m-1)T]$ when $t=(m-1)T$, is represented in the form

$$x[(m-1)T+\tau] = x[(m-1)T] + \left(P + \frac{T_1}{T} h\gamma\right) \tau x[(m-1)T]$$
$$0 \leqslant \tau \leqslant T \tag{4.214}$$

Whence, for $\tau = T$, we obtain the linear difference equation

$$x(mT) = \tilde{P}^* x[(m-1)T] \tag{4.215}$$

whose matrix operator $\tilde{P}^*$ is given by (4.206).

4.10 DISCONTINUOUS-ACTION AUTOMATIC PILOT

Let us consider now the free-motion of an automatic system for stabilising the course of an aircraft by means of a discontinuous-action automatic pilot. We will assume that the control we are applying is with respect to the coordinate and its first two derivatives, and that the conditions for idealising the control function, which will take the form of an impulse function $\psi^*(\sigma)$, are fulfilled. If, now, we suppose that in (1.6) $a=\infty$, $g_1'=0$, and we replace $\psi(\sigma)$ by $\psi^*(\sigma)$, we obtain the equations of motion in the form

$$\left.\begin{aligned} \ddot{\varphi} + M\dot{\varphi} &= -N\eta \\ \Theta\ddot{\eta} + \dot{\eta} &= h_2'\psi^*(\sigma) \\ \sigma &= \varphi + \alpha_1\dot{\varphi} + \alpha_2\ddot{\varphi} \end{aligned}\right\} \tag{4.216}$$

The impulse function $\psi^*(\sigma)$ is given by the graph in Fig 4.5. We will reduce the system of equations (4.216) to the normal form (1.15). For this purpose, in accordance with (1.14), we introduce the new notation

$$x_1 = \varphi \quad x_2 = \eta \quad x_3 = \dot{\varphi} \quad x_4 = \dot{\eta} \tag{4.217}$$

In the new variables x_i we can convert (4.216) to the form

$$\left.\begin{aligned} &\dot{x}_1 = x_3 \quad \dot{x}_2 = x_4 \quad \dot{x}_3 = -Nx_2 - Mx_3 \\ &\dot{x}_4 = -\frac{1}{\Theta} x_4 + \frac{h_2'}{\Theta} \psi^*(\sigma) \\ &\sigma = x_1 - N\alpha_2 x_2 + (\alpha_1 - M\alpha_2) x_3 \end{aligned}\right\} \tag{4.218}$$

In making this transformation, we must first replace $\ddot{\varphi}$ in the third equation of (4.216) by an expression obtained for this quantity from the first equation. We will write the normal form of equations (4.218) in matrix form

$$\left.\begin{aligned} &\begin{Vmatrix} \dot{x}_1 \\ \dot{x}_2 \\ \dot{x}_3 \\ \dot{x}_4 \end{Vmatrix} = \begin{Vmatrix} 0 & 0 & 1 & 0 \\ 0 & 0 & 0 & 1 \\ 0 & -N & -M & 0 \\ 0 & 0 & 0 & -\frac{1}{\Theta} \end{Vmatrix} \begin{Vmatrix} x_1 \\ x_2 \\ x_3 \\ x_4 \end{Vmatrix} + \begin{Vmatrix} 0 \\ 0 \\ 0 \\ h_2'/\Theta \end{Vmatrix} \psi^*(\sigma) \\ &\sigma = \begin{Vmatrix} 1 & -N\alpha_2 & (\alpha_1 - M\alpha_2) & 0 \end{Vmatrix} \begin{Vmatrix} x_1 \\ x_2 \\ x_3 \\ x_4 \end{Vmatrix} \end{aligned}\right\} \tag{4.219}$$

In our case the basic constant matrices have the form: matrix P of the open control system:

$$P = \begin{Vmatrix} 0 & 0 & 1 & 0 \\ 0 & 0 & 0 & 1 \\ 0 & -N & -M & 0 \\ 0 & 0 & 0 & -1/\Theta \end{Vmatrix} \tag{4.220}$$

the column matrix h which disseminates the actions of the control signal in the system:

$$h = \begin{Vmatrix} 0 \\ 0 \\ 0 \\ 1 \end{Vmatrix} \frac{h_2'}{\Theta} \tag{4.221}$$

the row matrix γ which determines the structure of the argument of the control σ:

$$\gamma = \begin{Vmatrix} 1 & -N\alpha_2 & (\alpha_1 - M\alpha_2) & 0 \end{Vmatrix} \tag{4.222}$$

Matrices (4.221) and (4.222) are such that the equality

$$\gamma h = 0 \tag{4.223}$$

holds. For the open loop system we can obtain the following matrix and scalar expressions (it is more usual to call the open system the continuous part of the discontinuous system):

the characteristic matrix

$$\lambda E - P = \left\| \begin{matrix} \lambda & 0 & -1 & 0 \\ 0 & \lambda & 0 & -1 \\ 0 & N & \lambda + M & 0 \\ 0 & 0 & 0 & \lambda + 1/\Theta \end{matrix} \right\| \tag{4.224}$$

the characteristic polynomial

$$\Delta(\lambda) = (\lambda + M)\left(\lambda + \frac{1}{\Theta}\right)\lambda^2 \tag{4.225}$$

the roots of the characteristic equation

$$\lambda_1 = -M \quad \lambda_2 = -\frac{1}{\Theta} \quad \lambda_3 = \lambda_4 = 0 \tag{4.226}$$

the transposed characteristic matrix

$$(\lambda E - P)^T = \left\| \begin{matrix} \lambda & 0 & 0 & 0 \\ 0 & \lambda & N & 0 \\ -1 & 0 & \lambda + M & 0 \\ 0 & -1 & 0 & \lambda + 1/\Theta \end{matrix} \right\| \tag{4.227}$$

the adjoint matrix

$$F(\lambda) = \left\| \begin{matrix} \lambda(\lambda + M)\left(\lambda + \frac{1}{\Theta}\right) & -N\left(\lambda + \frac{1}{\Theta}\right) & \lambda\left(\lambda + \frac{1}{\Theta}\right) & -N \\ 0 & \lambda(\lambda + M)\left(\lambda + \frac{1}{\Theta}\right) & 0 & \lambda(\lambda + M) \\ 0 & -N\lambda\left(\lambda + \frac{1}{\Theta}\right) & \lambda^2\left(\lambda + \frac{1}{\Theta}\right) & -\lambda N \\ 0 & 0 & 0 & \lambda^2(\lambda + M) \end{matrix} \right\| \tag{4.228}$$

the polynomial $\Delta_3(0)$ for the double zero root $\lambda_3 = \lambda_4 = 0$

$$\Delta_3(\lambda) = (\lambda + M)\left(\lambda + \frac{1}{\Theta}\right) \tag{4.229}$$

the derivative of the matrix $F(\lambda)/\Delta_3(\lambda)$

$$\frac{d}{d\lambda}\frac{F(\lambda)}{\Delta_3(\lambda)}=\frac{\begin{Vmatrix}\cdot & \cdot & \cdot & 0\\ \cdot & \cdot & \cdot & 2\lambda+M\\ \cdot & \cdot & \cdot & -N\\ \cdot & \cdot & \cdot & \cdot\end{Vmatrix}}{(\lambda+M)\left(\lambda+\frac{1}{\Theta}\right)}-\frac{\left(2\lambda+M+\frac{1}{\Theta}\right)\begin{Vmatrix}\cdot & \cdot & \cdot & -N\\ \cdot & \cdot & \cdot & \lambda(\lambda+M)\\ \cdot & \cdot & \cdot & -\lambda N\\ \cdot & \cdot & \cdot & \cdot\end{Vmatrix}}{(\lambda+M)^2\left(\lambda+\frac{1}{\Theta}\right)^2} \tag{4.230}$$

In our expression for the derivative (4.230) we have deliberately written out only the first three elements of the last column of the matrices, putting dots in all the other cells. In doing this we wish to draw attention to the fact that all the calculations can be made in an economical manner. One need only evaluate those quantities which enter into the final result.

Later, we shall investigate the stability of a discontinuous system. Scalar quantities enter into the characteristic equation which we obtain from (4.228) and (4.230) with $\lambda=\lambda_i$ by multiplying from the right by column h and from the left by row γ. Since the first three elements of column (4.221) and the last element of row (4.222) are equal to zero, it is obvious that these scalar coefficients depend only on the first three elements of the last column of (4.228) and (4.230). If, therefore, it is only stability that interests us, then, in the adjoint matrix (4.228), we should restrict ourselves to determining only the first three elements of the last column and put dots in all the other cells.

For a stability investigation we make use of the Hurwitz criterion, applying it to the transformed characteristic equation. In the present case there is a double zero root $\lambda_3=\lambda_4=0$ among the roots and (4.223) is fulfilled, so that the transformed characteristic equation can be taken in the form (4.193c).

From the preceding formulae it is not difficult to obtain expressions for the scalar coefficients of this equation in the form

$$\left.\begin{aligned}
\varepsilon_1&=\frac{\gamma F(\lambda_1)\,h}{\Delta'(\lambda_i)}=\frac{Nh_2'}{M\theta-1}\left(\frac{1}{M^2}-\frac{\alpha_1}{M}+\alpha_2\right)\\
\varepsilon_2&=\frac{\gamma F(\lambda_2)\,h}{\Delta'(\lambda_2)}=\frac{Nh_2'}{M\theta-1}(-\theta^2+\alpha_1\theta-\alpha_2)\\
\varepsilon_3&=\gamma\left[\frac{F(0)}{\Delta_3(0)}\right]^{(1)}h=Nh_2'\left(\frac{M\theta+1}{M^2}-\frac{\alpha_1}{M}\right)\\
\varepsilon_4&=\frac{\gamma F(0)\,h}{\Delta_3(0)}=-\frac{Nh_2'}{M}
\end{aligned}\right\} \tag{4.231}$$

It is easy to verify that ε_1, ε_2 and ε_3 satisfy the inequality

$$\varepsilon_1+\varepsilon_2+\varepsilon_3=0 \tag{4.232}$$

[it can be shown that (4.232) arises as a consequence of (4.223)]. Let us put the transformed equation in the form (4.59)

$$p_0'\nu^4+p_1'\nu^3+p_2'\nu^2+p_3'\nu+p_4'=0 \tag{4.233}$$

whereupon, according to (4.193c) for $n=4$, and with account of the new notation for ε_i, we have

$$\left.\begin{aligned}
p_0' &= -\frac{1}{4} T_1 T \varepsilon_4 (1 - e^{\lambda_1 T})(1 - e^{\lambda_2 T}) \\
p_1' &= -\frac{1}{2} T_1 \varepsilon_3 (1 - e^{\lambda_1 T})(1 - e^{\lambda_2 T}) \\
&\qquad -\frac{1}{2} T_1 T \varepsilon_4 [1 - e^{(\lambda_1+\lambda_2) T}] \\
p_2' &= (1 - e^{\lambda_1 T})(1 - e^{\lambda_2 T}) + T_1 \varepsilon e^{\lambda_2 T} + T_1 \varepsilon_2 e^{\lambda_1 T} \\
&\quad + T_1 \varepsilon_3 \left[e^{(\lambda_1+\lambda_2) T} - \frac{1}{2} (1 - e^{\lambda_1 T})(1 - e^{\lambda_2 T}) \right] \\
&\qquad - \frac{1}{2} T_1 T (e^{\lambda_1 T} + e^{\lambda_2 T}) \\
p_3' &= 2 [1 - e^{(\lambda_1+\lambda_2) T}] \\
&\quad - \frac{1}{2} T_1 \varepsilon_3 [-1 - e^{(\lambda_1+\lambda_2) T} + e^{\lambda_1 T} + e^{\lambda_2 T}] \\
&\qquad + \frac{1}{2} T_1 T \varepsilon_4 [1 - e^{(\lambda_1+\lambda_2) T}] \\
p_4' &= (1 + e^{\lambda_1 T})(1 + e^{\lambda_2 T}) - T_1 \varepsilon_1 (1 + e^{\lambda_2 T}) \\
&\quad - T_1 \varepsilon_2 (1 + e^{\lambda_1 T}) - \frac{1}{2} \varepsilon_3 (1 + e^{\lambda_1 T})(1 + e^{\lambda_2 T}) \\
&\qquad + \frac{1}{4} T_1 T \varepsilon_4 (1 + e^{\lambda_1 T})(1 + e^{\lambda_2 T})
\end{aligned}\right\} \qquad (4.234)$$

The coefficients p_i' were found with the help of some very simple transformations involving the use of (4.232). We will now expand the exponentials in the expressions of (4.234) as series in powers of the recurrence period T. For each coefficient we retain the first two terms of the expansion with respect to the small parameter T. For this purpose the ratio T_1/T will be regarded as a constant; it is a constant of the system which is independent of T. After some simple reductions, we obtain for the coefficients p_i' expressions in the form

$$\left.\begin{aligned}
p_0' &= -\frac{1}{4} \frac{T_1}{T} \varepsilon_4 \lambda_1 \lambda_2 \left[1 + \frac{\lambda_1+\lambda_2}{2} T \right] T^4 \\
p_1' &= \frac{1}{2} \frac{T_1}{T} [\varepsilon_3 \lambda_1 \lambda_2 + \varepsilon_4 (\lambda_1 + \lambda_2)] \left(1 + \frac{\lambda_1+\lambda_2}{2} T \right) T^3 \\
p_2' &= \left\{ \lambda_1 \lambda_2 + \frac{T_1}{T} [\varepsilon_1 \lambda_2 + \varepsilon_2 \lambda_1 + \varepsilon_3 (\lambda_1 + \lambda_2) - \varepsilon_4] \right\} \\
&\qquad \times \left(1 + \frac{\lambda_1+\lambda_2}{2} T \right) T^2 \\
p_3' &= -2 (\lambda_1 + \lambda_2) \left(1 + \frac{\lambda_1+\lambda_2}{2} T \right) T \\
p_4' &= 4 \left(1 + \frac{\lambda_1+\lambda_2}{2} T \right)
\end{aligned}\right\} \qquad (4.235)$$

The right-hand sides of all the equalities in (4.235) contain the same factor, which can be discarded since it can have no effect upon the stability of the system. (In the special case when this factor is equal to zero, it will be necessary to retain in the expansion terms of higher order with respect to T_1.) We therefore discard the common factor in the coefficients of (4.235) and form (4.233) by means of the substitution (4.212). Then, replacing the roots λ_i and the coefficients ε_i by their expressions given by (4.226) and (4.231), introducing the notation

$$c = \frac{T_1}{T} < 1 \tag{4.236}$$

and multiplying all the coefficients obtained by $\Theta/4$, we obtain an equation of the form

$$\Theta\lambda^4 + (1 + M\Theta)\lambda^3 + (M + Nch_2'\alpha_2)\lambda^2 + Nch_2'\alpha_1 + Nch_2' = 0 \tag{4.237}$$

This equation is the characteristic equation for a limiting linear continuous system which, in general form, is represented by the matrix equation (4.202).

The characteristic equation (4.237) can be obtained directly from (4.216) by first replacing $\psi(\sigma)$ in these equations by $c\sigma$ and introducing the coefficient c from (4.236). All the coefficients of the characteristic equation of the fourth power (4.237) are positive, so that for a stability analysis the only important inequality is

$$(1 + M\Theta)(M + Nch_2'\alpha_2)Nch_2'\alpha_1 - \Theta N^2c^2h_2'^2\alpha_1^2 - (1 + M\Theta)^2 Nch_2' > 0 \tag{4.238}$$

which establishes the positiveness of the penultimate Hurwitz determinant (4.60). Dividing through by $Nc > 0$, multiplying out the brackets and gathering together similar terms in c, (4.238) assumes the form

$$M(1 + M\Theta)\left(\alpha_1 - \frac{1}{M} - \Theta\right) + \alpha_1\Theta\left(-\alpha_1 + \alpha_2 M + \frac{\alpha_2}{\Theta}\right)Nch_2' > 0 \tag{4.239}$$

It is now not difficult to establish that, when the inequalities

$$\left.\begin{aligned} \alpha_1 &> \frac{1}{M} + \Theta, \\ \alpha_1 &< \alpha_2 M + \frac{\alpha_2}{\Theta} \end{aligned}\right\} \tag{4.240}$$

are fulfilled, the system will be stable for any positive value of the coefficient ch_2'. When these inequalities are reversed, i.e. when

$$\left.\begin{aligned} \alpha_1 &< \frac{1}{M}+\Theta \\ \alpha_2 &> \alpha_2 M+\frac{\alpha_2}{\Theta} \end{aligned}\right\} \qquad (4.241)$$

the system will be unstable for all values $ch_2' > 0$.

For the two remaining possibilities, i.e. when

$$\left.\begin{aligned} \alpha_1 &> \frac{1}{M}+\Theta \\ \alpha_1 &> \alpha_2 M+\frac{\alpha_2}{\Theta} \end{aligned}\right\} \qquad (4.242)$$

or, conversely, when

$$\left.\begin{aligned} \alpha_1 &< \frac{1}{M}+\Theta \\ \alpha_1 &< \alpha_2 M+\frac{\alpha_2}{\Theta} \end{aligned}\right\} \qquad (4.243)$$

there exists a positive value $ch_2'=c^*$ such that the left-hand side of (4.238) vanishes. When (4.242) holds, the system will be stable for values $ch_2'<c^*$, and when (4.243) holds it will, on the contrary, be stable when $ch_2'>c^*$. The value $ch_2'=c^*$ in these cases defines the boundary of oscillatory stability.

The results obtained can be illustrated geometrically with the help of a Vishnegradskii diagram. Consider a space of three parameters of the system: the coefficient of natural deformation of an aircraft M, the coefficient of artificial deformation α_1, and the gain ch_2'. On the M, α_1 plane we plot the coefficient M as an independent variable along the abscissa, and α_1, along the ordinate—obtaining the hyperbola

$$\alpha_1=\frac{1}{M}+\Theta \qquad (4.244)$$

and the straight line

$$\alpha_1=\alpha_2 M+\frac{\alpha_2}{\Theta} \qquad (4.245)$$

The hyperbola (4.244) and the straight line (4.245) divide the plane of the first quadrant into four regions, as shown in Fig 4.7. Inside regions I, II, III and IV, equalities (4.240), (4.241), (4.242) and (4.243), respectively, will hold Thus, in region I the system will be stable for any $ch_2'>0$, in region II the opposite is true; the system will be unstable for every value of the coefficient $ch_2'>0$.

In regions III and IV the system will be stable for $ch_2' < c^*$ and $ch_2' > c^*$, respectively. The positive coefficient c^* is found from the equality

$$c^* = \frac{M(1 + M\Theta)\left(\alpha_1 - \frac{1}{M} - \Theta\right)}{\alpha_1 \Theta N(\alpha_1 - \alpha_2 M - \alpha_2/\Theta)} \tag{4.246}$$

which is obtained by equating the left-hand side of (4.239) to zero.

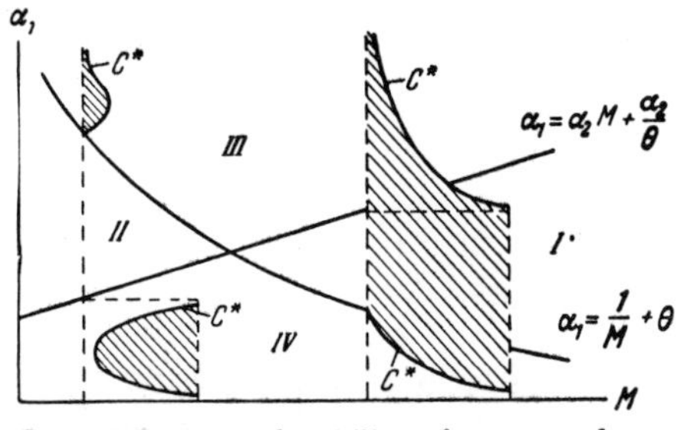

Fig 4.7 Regions of stability of a system for stabilising the course of an aircraft.

If, in (4.246), M and α_1 are assumed to be independent variables, then in the above space the parameters M, α_1 and ch_2' of (4.246) can be interpreted geometrically as a surface defining the boundary of stability of the system being considered.

Consider the two intersections of the bounding surface with the coordinate plane $M = \text{constant}$, situated to the left and right of the common point shared by the hyperbola (4.244) and the straight line (4.245). The cross-sections obtained are illustrated in Fig 4.7, where they are represented on the plane of the parameters M and α_1. In Fig 4.7 the shaded areas denote the regions of stability.

We investigated the stability using (4.237). This is the characteristic equation for the linear continuous control system obtained from the equations of motion (4.216) by the method indicated above. (It is interesting to note that it is as if we had effected a reduction of the over-all gain h_2 of the linear continuous system in the ratio $c = T_1/T < 1$.) However, this equation is also the characteristic equation for a linear discontinuous system provided we retain the first two terms of the expansion of the coefficients (4.234) as a series in powers of the small parameter T.

Consequently, if one wishes to analyse the stability of the system in greater detail, it is necessary to retain in the expansions terms of a higher order of smallness or, alternatively, one should carry out the investigation directly in accordance with the characteristic equation (4.233) without making any assumptions about the smallness of the recurrence period T.

In the last case we can be guided by rules established in engineering practice. In designing the regulator, the object of the control is usually regarded as specified. In addition, one can also assume that the time constants of some of the elements of which the regulator is composed are known, since they are associated parameters dependent upon the fact that the choice of such elements had to be made from a restricted range of devices with identical functions. If we accept this point of view, then in our problem we can consider the parameters M, N and Θ as known, or that at least we can consider their range of variation as specified. For the remaining parameters α_1, α_2 and ch'_2 characterising the control principle and the over-all amplification in the system, the analysis is then considerably simplified.

However, it is important to stress that, in the given case, a deeper meaning should be attached to the analysis of stability carried out in accordance with (4.237). This expression is not simply a characteristic equation for the limiting continuous system; it also in practice serves in first approximation (with an accuracy up to terms proportional to T^2) as a transformed characteristic equation for the discontinuous system. This property of the transformed equation is not an accident arising from the particular example taken; it is valid for any class of linear discontinuous control systems with an impulse control function characterised by (4.223).

As an exercise, let us deduce this property in a general form. For this purpose it is best to use the transformed characteristic equation (4.193b) or the analogous equation for simple roots λ_i. One can also use an intermediate form of this equation expressed in terms of the column elements u and the row elements β.

The problem is stated in the following way: find in general form an expression for the transformed characteristic equation in first approximation.

For the case of simple roots λ_i, the final result can be represented in the form

$$\prod_{j=1}^{n}(\tilde{\lambda}-\lambda_j)\left[1-\frac{T_1}{T}\left\{\sum_{i=1}^{n}\frac{1}{\tilde{\lambda}-\lambda_i}\,\frac{\gamma F(\lambda_i)\,h}{\Delta'(\lambda_i)}-\frac{T}{2}\sum_{i=1}^{n}\gamma_i h_i\right\}\right]=0 \qquad (4.247)$$

where $\tilde{\lambda}$ is given by (4.212).

Chapter 5

RELAY-OPERATED CONTROL SYSTEMS

There is now an extensive specialised literature on the theory of relay-operated control systems. The reader will find a short historical outline of the development of the theory of such systems in the monograph by Tsipkin [25], covering the period up to 1955. Here we shall mention the works of a few of these authors only. Andronov, with his students and collaborators, applied the method of point transformations of surfaces to investigate the structure in the large of the decomposition of phase space into trajectories for a number of classical control problems. They also applied these methods to the study of a system for stabilising the course of an aircraft equipped with an automatic pilot incorporating a constant speed servo-motor. The work of these authors has made an important contribution to the understanding of the processes taking place in relay-operated control systems.

Lur'e, in a series of papers summarised in a monograph [14], introduced into the discussion a special form of mathematical model which he called an indirect control system. This mathematical model is described in terms of differential equations in normal form of the same type as the system of equations (1.15). Applying the method of canonical variables as an intermediate means of investigation, Lur'e constructed in general form the symmetric periodic modes for this type of relay-operated control system.

In his monograph, Tsipkin has given a systematic exposition of the general theory of relay systems based on frequency representations. An interesting method of analysing relay-operated systems has been applied by Neimark [19]. It was this author in particular who first drew attention to the possibility of the existence of complex periodic modes of oscillation in relay-operated control systems.

In the last two sections of this book, the author discusses the free and forced oscillations of relay-operated systems, the self-oscillatory and sliding modes of behaviour using the methods of matrix calculus.

The author also applied matrix methods to the analysis of relay systems in some earlier works published at the beginning of the 1950s [30, 31].

5.1 GENERAL REMARKS

Relay-operated systems belong to a very complex class of nonlinear control systems. For nonlinear systems, no general solution exists for determining in closed form the motion of the system or its reaction to external disturbances. The superposition principle is not applicable to nonlinear systems and this complicates the study of the motion of the system when there are several actions present. As a result, approximate methods of investigation are widely used in the theory of nonlinear control systems. In many cases these methods permit one to obtain, to a first approximation, formulae which are applicable to actual engineering problems. However, owing to the special form of the nonlinearities that arise in the theory of relay systems, there are many important problems that yield exact solutions. This circumstance is important in itself, but it also provides a means of checking the accuracy of the results obtained by approximate methods.

To clarify the situation, consider the free-motion of a relay system which is described in matrix form by equations of the type

$$\dot{x} = Px + h\psi(\sigma) \qquad \sigma = \gamma x \tag{5.1}$$

In these equations the control function $\psi(\sigma)$ is represented geometrically in the form of relay characteristics, the most important types of which were given in Table 1.1. The relay control function $\psi(\sigma)$ belongs to the step-function class of functions and, in our case, it can assume two or three different values for the entire continua of values of the argument of the control σ. The changeover from one value of $\psi(\sigma)$ to another takes place in a series of sudden jumps for discrete values of σ, depending on the direction of the change. The passage of the argument of the control σ through the discrete values for which the control function flips from one level to another will be called switching, and the conditions that ensure that these jumps will happen will be called switching conditions. In the intervals of variation of σ inside which $\psi(\sigma)$ remains constant, (5.1) is an inhomogeneous linear differential equation with constant coefficients. Under these conditions, the right-hand side of (5.1) will have a very simple form: it will be equal to a column of constants, and will include the case when all the elements of this column are equal to zero. For each of the regions σ mentioned above it is possible to construct a general solution that depends on a corresponding number of arbitrary constants. If, then, at the instants when the system switches over from one region to another, we apply the method of fitting described at the beginning of the last chapter, we will be able to construct piece-by-piece the solution of the system for the specified initial conditions. It would seem however that, proceeding in this simple way, we might encounter situations in which the solution turns out to be noncontinuable when taken to the boundary of the region inside which $\psi(\sigma)$ remains constant.

On certain sections of the boundary of the regions in which $\psi(\sigma)$ remains constant, (5.1) would appear to be contradictory when $\psi(\sigma)$ has a relay characteristic. In these cases it is necessary to extend the definition of the system in a manner that will allow us to continue the solution indefinitely in time. In order to indicate the conditions under which the definitions of the system might reasonably be extended and to provide a clearer picture of the behaviour of the system during the switching process, it will be found convenient to change over to the geometrical representation in a state space.

For definiteness, consider the ideal relay characteristic illustrated in the first graph of Table 1.1. In this case there are two intervals of values of the argument of the control σ inside which the control function remains constant. For $\sigma > 0$, we have $\psi(\sigma) = 1$ and the motion of the system will be described by the matrix equation

$$\dot{x} = Px + h \tag{5.2}$$

for $\sigma < 0$ we have $\psi(\sigma)' = -1$ and correspondingly

$$\dot{x} = Px - h \tag{5.3}$$

The value $\sigma = 0$ represents the boundary between these intervals.

In expanded form this boundary can be written in the form of the equation

$$\sigma = \gamma x = \gamma_1 x_1 + \gamma_2 x_2 + \dots + \gamma_n x_n = 0 \tag{5.4}$$

In state space, (5.4) defines a hyperplane (in future we shall call it simply a plane) which passes through the origin of the coordinates. This plane divides the state space into two regions, or into two half-spaces for which $\sigma > 0$ and $\sigma < 0$.

In the half-spaces $\sigma > 0$ and $\sigma < 0$, the motion of the relay system is given by (5.2) and (5.3), respectively. Thus, in state space, the plane (5.4) is a switching plane.

An arbitrary initial value $x(t_0)$ determines a unique solution of (5.2) for $\sigma(t_0) = \gamma x(t_0) > 0$ and of (5.3) for $\sigma(t_0) = \gamma x(t_0) < 0$. In state space these solutions are represented by trajectories which, for $t = t_0$, pass through the point with the coordinate $x(t_0)$. The direction of a trajectory is determined by the velocity vector (column $\dot{x}$) directed along its tangent. State trajectories do not cross each other and they completely fill their half-spaces.

These two circumstances are a consequence of the fact that linear equations of the type (5.2) and (5.3) have a unique solution satisfying the specified initial condition $x(t_0)$* and a result of the fact that the initial value $x(t_0)$ can be specified arbitrarily inside the corresponding half-space (we note that two trajectories that have just a single common point must coincide entirely). Consequently, one and only one trajectory from each half-space $\sigma > 0$ and $\sigma < 0$ approaches each point in the switching plane.

The following possibilities are open to the trajectories in the two half-spaces in the immediate neighbourhood of the switching plane. They can have the same direction, or they can be directed towards or away from each other.

All three cases are represented in three-dimensional state space in Figs 5.1a and 5.1c. The first case corresponds to region I of the switching plane $\gamma x = 0$, the second and third cases to regions II and III of that plane. In the first case the trajectories directed towards the switching plane from different half-spaces behave as if they are natural continuations of each other. At the points where they cut the switching plane they join to form a single, continuous trajectory along which the image point freely travels from one half-plane to the other. In general, at the instant of crossing the plane (5.4), the components of the state velocity $\dot{x}$ and the velocity undergo a sudden but finite change of magnitude. When this occurs we shall say that the conditions of normal switching are fulfilled. At the same time, from an analytical point of view, any particular solution will be obtained alternately by the solutions of (5.2) and (5.3), and by joining them in accordance with the initial conditions.

When the image point falls on sections of the switching plane that correspond to the second case, it cannot immediately leave this plane and pass from one half-space to the other. It is forced to spend a certain time moving (sometimes called sliding) along the switching plane either towards the boundaries of that section or towards the coordinate origin which determines the controlled position of the system. This type of motion of a relay-operated system is called sliding. Physically, the existence of a sliding motion is due to the peculiar characteristic of a relay element: it begins to execute rapid oscillations from one extreme position to another while it injects into the system control signals whose maximum level alternates in sign and is of variable duration. For an ideal characteristic the switching frequency of the relay in the sliding mode must be theoretically considered as equal to zero.

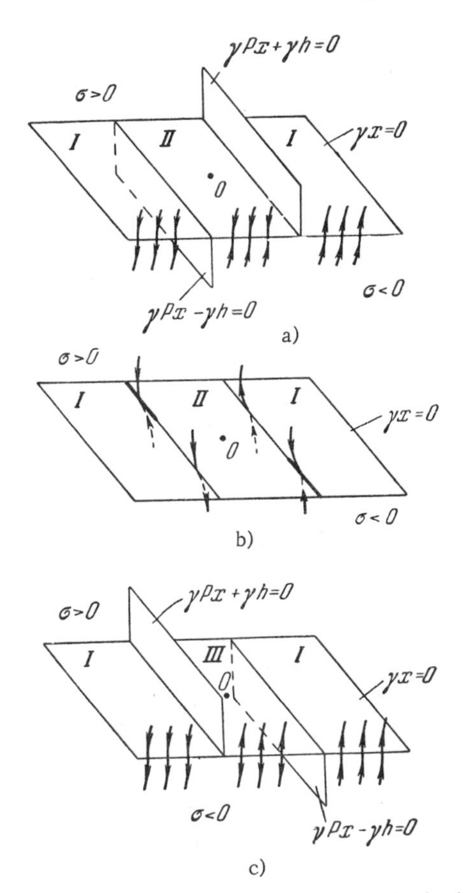

Fig 5.1 Direction of the state trajectories in the neighbourhood of the switching plane *a* and *b* — for $\gamma h < 0$; *c* — for $\gamma h > 0$.

It should be remembered, in this connection, that all the descriptions of systems given here are presented in a form applicable to ideal relay characteristics. If a hysteresis loop is taken into account, a more detailed description of the motion of a relay working in the sliding mode can be given, which is capable of yielding the finite frequency of switching (see Section 5.8).

The motion of a relay system working under sliding conditions cannot be described by alternating solutions of (5.2) and (5.3). It is necessary to introduce some new analytical relationships which will extend the definition of the initial equations (5.1). An analysis of the structure of state space in a band adjacent to the switching plane shows that, for any reasonable extension of the definitions, it is necessary to assume that in the sliding mode the image point must move along the switching plane. We shall assume that this is the case in Section 5.6, where we shall consider the sliding mode of motion of relay systems in greater detail.

In the third case, the behaviour of the relay system turns out to be indeterminate in that section of the switching plane where the trajectories emerge in opposite directions. When solving actual problems, this kind of section may be treated as continua of the unstable equilibrium of the system. Examples of this kind can be found not so much in control systems proper but in nonlinear systems of the valve oscillator type. These have a ʃ -shaped characteristic for which the operating conditions correspond to self-excited oscillations.

Let us turn now to the problem of finding the quantitative characteristics which determine the structure of the state space. These characteristics can be conveniently obtained from the following simple considerations. Each of the cases described is characterised by the fact that the projections of the velocity vectors to the perpendicular to the switching plane have either the same direction, are directed towards each other, or are directed away from each other. If the right-hand side of (5.4) is set equal to c and we consider c to be a real parameter, then we obtain a one-parameter family of mutually parallel planes. The vector with components $\gamma_1, \gamma_2, \ldots, \gamma_n$ (the row matrix γ) will be the gradient of this family. The vector gradient is perpendicular to the level surfaces of the family and indicates the direction of the steepest increase of the function in space. Thus, the vector γ is perpendicular to the switching plane and is directed towards the half-plane $\sigma > 0$.

The state velocity vector is given by the column matrix $\dot{x}$. The matrix product $\gamma\dot{x}$ can be regarded as the scalar product of the vectors γ and $\dot{x}$. Therefore, for $\gamma\dot{x} > 0$, the projection of the vector $\dot{x}$ on the perpendicular to the switching plane is directed along the vector γ and, conversely, for $\gamma\dot{x} < 0$, this projection has the opposite direction.

The velocity vector $\dot{x}$ is given by the right-hand sides of (5.2) and (5.3) for points in the half-spaces $\sigma > 0$ and $\sigma < 0$, respectively. Multiplying the matrices on the right-hand sides of these equations from the left by the row matrix γ, we obtain expressions for $\gamma\dot{x}$ which also determine $\dot{\sigma}$, the rate of change of the argument of the control σ. The conditions for normal switching which ensure that the image point moves from the half-space $\sigma < 0$ to the half-space $\sigma > 0$, or which ensure that the value of the function $\psi(\sigma)$ switches over from -1 to $+1$ are given by the relations

$$\gamma x(t_s) = 0, \quad \gamma P x(t_s) + \gamma h > 0 \quad \gamma P x(t_s) - \gamma h > 0 \tag{5.5}$$

For a movement of the image point in the opposite direction, i.e. for a switch in the value of $\psi(\sigma)$ from 1 to -1 the conditions are given by the expressions

$$\gamma x(t_s)=0 \quad \gamma P x(t_s)+\gamma h<0 \quad \gamma P x(t_s)-\gamma h<0 \tag{5.6}$$

In (5.5) and (5.6), t_s denotes the instant of time at which switching takes place. The left-hand sides (5.5) and (5.6) determine $\dot{\sigma}$ in the corresponding half-spaces. In crossing the switching plane, the value of $\dot{\sigma}$ jumps by $\pm 2\gamma h$. As a consequence of this we shall assume that, in a relay system, hard switching occurs when $\gamma h \neq 0$ and soft switching when $\gamma h=0$.

The sliding mode of behaviour is characterised by the relations

$$\gamma x(t)=0 \quad \gamma P x(t)+\gamma h<0 \quad \gamma P x(t)-\gamma h>0 \tag{5.7}$$

which must be fulfilled for a finite or infinite interval of time t.

Expressions (5.7) also determine the section of the switching plane inside which the sliding motion of the system takes place. Consider the equalities

$$\gamma P x+\gamma h=0 \quad \gamma P x-\gamma h=0 \tag{5.8a}$$

which in state space define two parallel planes symmetrically disposed with respect to the origin of the coordinates. These planes intersect the switching plane along two parallel straight lines which are also symmetrically situated with respect to the origin of the coordinates. The straight lines form the boundaries of the section in which the sliding motion occurs. These planes and straight lines are illustrated in Fig 5.1a, where region II is the zone of the sliding mode. The phase trajectories that pass out of the half-spaces $\sigma>0$ and $\sigma<0$ touch the switching plane if they abut the straight lines defined, respectively, by the first or second plane (5.8). This is made clear in Fig 5.1b.

Thus, the section where the sliding motion occurs takes the form of an infinite band on the switching plane bounded by the two parallel straight lines. The centre of this band is the origin of the coordinates, and its width depends on the scalar quantity γh. By changing those parameters of the elements h_i of column h which do not enter into the first terms of (5.8), one can directly affect the width of the sliding mode band. For $\gamma h=0$, the band contracts to a straight line.

Note

Each of the bounding straight lines of the sliding zones can be divided into two half-lines. The half-lines

$$\left.\begin{array}{ll} \gamma x(t) = 0 & \gamma x(t) = 0 \\ \gamma P x(t) + \gamma h = 0 & \gamma P x(t) - \gamma h = 0 \\ \gamma P^2 x(t) + \gamma P h < 0 & \gamma P^2 x(t) - \gamma P h > 0 \end{array}\right\} \quad (5.8\text{b})$$

relate to the sliding zone, and the half-lines

$$\left.\begin{array}{ll} \gamma x(t) = 0 & \gamma x(t) = 0 \\ \gamma P x(t) + \gamma h = 0 & \gamma P x(t) - \gamma h = 0 \\ \gamma P^2 x(t) + \gamma P h > 0 & \gamma P^2 x(t) - \gamma P h < 0 \end{array}\right\} \quad (5.8\text{c})$$

relate to the region of normal switching. Here, the inequalities appearing in (5.8b) and (5.8c) determine the signs of the projections of the vector $\ddot{x}$ of the state acceleration on to vector γ^*, which is perpendicular to the switching plane. That is, for $\dot{\sigma} = 0$, one can decide if switching will occur from the sign of the second derivative $\ddot{\sigma}$. For the trajectories of the half-spaces $\sigma > 0$ and $\sigma < 0$, the following two expressions hold, respectively

$$\left.\begin{array}{l} \gamma \ddot{x} = \gamma P^2 x + \gamma P h \\ \gamma \ddot{x} = \gamma P^2 x - \gamma P h \end{array}\right\} \quad (5.8\text{d})$$

Thus, when the left-hand and right-hand groups of conditions (5.8c) are fulfilled, the image point moves from the sliding zone to the half-spaces $\sigma > 0$ and $\sigma < 0$, respectively. In Fig5.1b the bounding straight lines belonging to the region of normal switching and to the sliding zone are represented by thin and thick lines, respectively. The single-type bounding half-lines are symmetric with respect to the origin of the coordinates.

In the third case the following relations hold

$$\gamma x(t) = 0 \quad \gamma P x(t) + \gamma h > 0 \quad \gamma P x(t) - \gamma h < 0 \qquad (5.9)$$

It is obvious that, in this case, (5.9) defines a band on the switching plane.

Relations (5.7) and (5.9) mutually contradict each other, so that the motions they represent cannot exist simultaneously in the same relay system. It is not difficult to establish that (5.7), which defines the sliding mode, can hold only when $\gamma h < 0$, so that (5.9) can hold only when $\gamma h > 0$. The direction of the state trajectories for the case $\gamma h > 0$ is illustrated in Fig 5.1c. Conditions (5.9) correspond to region III; in accordance with the direction of the state trajectories, region III is a sheet of unstable equilibrium positions. In true relay control systems, the condition $\gamma h < 0$ is always fulfilled or, more accurately, the inequality $\gamma h \leqslant 0$ always holds. This inequality has a rather deeper meaning. We recall that the fundamental problem of control consists of bringing the system to a stationary state, which in our case is represented by the equality $x=0$. In relay systems, the control action is introduced through the column matrix $\pm h$, through a suitable alternation of signs. The column h has dimensions of a velocity so that, in state space, it is a 'control velocity'. A relay system cannot solve the problem of control without switchings taking place in the relay control element since, from a mathematical point of view, there is no provision for such a solution in the structure of the solution of (5.2) and (5.3) – which describe the motion of the system in the half-spaces $\sigma > 0$ and $\sigma < 0$. Consequently, in these half-spaces the control velocity cannot be directed away from the switching plane. This condition is fulfilled if the scalar product of the vectors γ and h is not positive, i.e. if $\gamma h \leqslant 0$. For relay control systems, therefore, we shall consider only the first two cases: the normal switching mode and the sliding mode of motion along the switching plane. We should emphasise however that, since all the results obtained for the normal switching mode are independent of the sign of the scalar quantity γh, they will be equally valid for both control systems and nonlinear systems of the valve oscillator type mentioned above, having a step-function type of characteristic.

So far, we have been considering only relay elements with ideal characteristics. However, it is not difficult to generalise the results obtained above to other types of characteristic, and such generalisations will be made whenever it is necessary.

In the normal switching mode, (5.1) is nonlinear. Under certain conditions, this equation admits to the existence of a finite number (or a denumerable set) of periodic solutions which can exist throughout the region of variation of the system parameters. If the stationary mode of behaviour corresponding to the periodic system is stable, it will manifest itself in the form of oscillations of the system.

In many cases, a knowledge of the qualitative and quantitative characteristics of the stationary modes of oscillations described by the periodic solutions is a very important factor in making a general estimate of the behaviour of relay control systems.

We shall be concerned with the problem of determining the periodic solutions of (5.1) and, consequently, of the stationary modes of oscillations corresponding to these solutions. If it is undesirable to have such stationary modes of oscillation in the system, then we shall find that we have acquired en route the means of removing them. However, if our interest is not simply restricted to control systems but extends also to other systems in which oscillations represent the operating mode, then such an approach becomes attractive in its own right. We should note that even in control systems, small-amplitude high-frequency oscillations are sometimes permitted. This is done to suppress the effects of dry friction and backlash at the various places in the controller where there are parts in relative motion.

5.2 PERIODIC SOLUTIONS AND THE PERIODIC EQUATION

We shall seek the periodic solution of (5.1) for the case of normal switching, having represented the control function $\psi(\sigma)$ in the form of a relay characteristic with a hysteresis loop, as illustrated in the second and third graphs of Table 1.1. The conditions of normal switching (5.5) and (5.6) also retain their meaning in the case when σ_1^* and $-\sigma_1^*$ are inserted into the right-hand sides of the first inequalities appearing in (5.5) and (5.6), respectively. That is, for the transition form $\psi(\sigma)=-1$ to $\psi(\sigma)=1$ we will have the relations

$$\left.\begin{aligned} &\gamma x(t_s)=\sigma_1^* \\ &\gamma P x(t_s)-\gamma h>0 \\ &\gamma P x(t_s)+\gamma h>0 \end{aligned}\right\} \tag{5.10}$$

and, for the reverse transition from $\psi(\sigma)=1$ to $\psi(\sigma)=-1$ we have

$$\left.\begin{aligned} &\gamma x(t_s)=-\sigma_1^* \\ &\gamma P x(t_s)+\gamma h<0 \\ &\gamma P x(t_s)-\gamma h<0 \end{aligned}\right\} \tag{5.11}$$

(N. B. See Section 5.8 for greater detail. For a first reading it can be assumed that $\sigma_1^*=0$).

Suppose the conditions of normal switching (5.10) are fulfilled at a certain time $t=t_s$, which we shall take as our origin. Under these conditions, the motion of the system is given by (5.2). The solution of this equation, which satisfies the initial condition $x(0)$ is written in the form [see (2.129) for $t_0=0$]

$$x(\tau)=KM(\tau)K^{-1}x(0)+KN(\tau)K^{-1}h \tag{5.12}$$

The expression obtained will determine the motion of the system up to the time when the following inequality is valid:

$$\sigma(\tau)=\gamma x(\tau)=\gamma KM(\tau)K^{-1}x(0)+\gamma KN(\tau)K^{-1}h>-\sigma_1^* \tag{5.13}$$

Let us suppose that we have chosen the initial condition $x(0)=x^*(0)$ in such a way that, after a lapse of time T, we have

$$x(T)=-x^*(0) \tag{5.14}$$

and in the interval of time $0\leqslant\tau\leqslant T$ equality (5.13) is fulfilled.

For $\tau = T$ a new switching will occur since the matrix coordinate $x(T)$ will satisfy the condition of normal switching (5.11). This is a consequence of (5.14) and of the fact that, according to the condition, $x(0)$ satisfies (5.10). However, when $t > T$, the motion of the system will be determined by (5.3). Putting $t_0 = T$ in (3.129), changing the sign in front of h and with account of (5.14), we will write the solution of (5.3) in the form

$$x(T+\tau) = -KM(\tau)K^{-1}x(0) - KN(\tau)K^{-1}h \tag{5.15}$$

or

$$x(T+\tau) = -x(\tau) \tag{5.16}$$

where $x(\tau)$ is given by (5.12). This solution will be valid for $0 \leqslant \tau \leqslant T$, which corresponds to a time interval $T \leqslant t \leqslant 2T$ since, in this interval, the inequality

$$\sigma(T+\tau) = \gamma x(T+\tau) = -\gamma x(\tau) = -\sigma(\tau) < \sigma_1^* \tag{5.17}$$

will be fulfilled.

From (5.14) and (5.16) it follows that $x(2T) = x(0)$ – i.e., by the time $t = 2T$ has been reached, the autonomous system we are considering will have returned to its original state. Thus, we obtain a periodic solution with period $2T$ which is symmetric over the half-period T since it satisfies (5.16) in the interval $0 \leqslant \tau \leqslant T$.

For the actual construction of the periodic solution, one should seek a value $x^*(0)$ that will satisfy the conditions of normal switching (5.10) and for which (5.12) will satisfy (5.14) – the condition that the periodicity should be symmetric. Substituting $\tau = T$ into (5.12) and taking into account the periodicity condition (5.14), we obtain for $x^*(0)$ the matrix expression

$$-x^*(0) = KM(T)K^{-1}x^*(0) + KN(T)K^{-1}h \tag{5.18}$$

whence we shall have

$$-K[M(T)+E]K^{-1}x^*(0) = KN(T)K^{-1}h \tag{5.19}$$

Solving (5.19) with respect to $x^*(0)$, we obtain

$$x^*(0) = -K[M(T)+E]^{-1}N(T)K^{-1}h \tag{5.20}$$

We substitute $x^*(0)$ from here into the condition for normal switching (5.10). We then obtain

$$-\gamma K[M(T)+E]^{-1}N(T)K^{-1}h = \sigma_1^* \tag{5.21}$$

and

$$-\gamma PK[M(T)+E]^{-1}N(T)K^{-1}h > |\gamma h|$$

This inequality can be rewritten in the form

$$-\gamma KJ[M(T)+E]^{-1}N(T)K^{-1}h > |\gamma h| \tag{5.22}$$

if we substitute into it the matrix P expressed by a formula of type (2.42) in terms of the canonical matrix, and assume that $KK^{-1}=E$.

In (5.22) $|\gamma h|$ is the absolute value of the scalar quantity γh. Equality (5.21) is a transcendental equation with respect to T. This equation can have a finite or denumerable set of roots. Each positive root of (5.21) is a half-period T of the periodic solution if it satisfies (5.22) [since the roots of equation (5.21) are half-periods, we retain the notation T for them]. Equality (5.21) is known as the periodic equation. The oscillations taking place in the time interval $0 \leqslant t \leqslant T$ are described by the expression

$$x(t) = -KM(t)[M(T)+E]^{-1}N(T)K^{-1}h + KN(t)K^{-1}h \tag{5.23}$$

which is obtained from (5.12) by replacing τ by t and substituting $x(0)=x^*(0)$ from (5.20). The initial condition $x^*(0)$ can be regarded as the point where the closed trajectory corresponding to the periodic solution cuts the switching plane $\sigma=\sigma_1^*$. Relations (5.20)-(5.23) determine the periodic solution in matrix form.

The problem now is to put these relations in a form convenient for calculation. For this purpose we will express them in terms of the input parameters of the system in the form adopted in the preceding chapter.

The form of the matrix notation of the basic relations decides in essence the prescription to be used for the calculations. Let us consider first the case when the characteristic values λ_i of the matrix P are simple. We shall now go through the arguments for (5.20). The inverse matrix for $M(T)+E$ can be obtained from (4.142) if we put $\lambda=-1$ and change the sign. Matrix $N(T)$ is found from (2.120) with $t=T$. Thus the matrices $[M(T)+E]^{-1}$ and $N(T)$ are diagonal, with common elements equal to

$$\frac{1}{1+e^{\lambda_i T}} \quad \text{and} \quad \frac{e^{\lambda_i T}-1}{\lambda_i}$$

respectively. Therefore, we have

$$[M(T)+E]^{-1}N(T) = \begin{Vmatrix} \frac{1}{\lambda_1}\,\mathrm{th}\,\frac{\lambda_1 T}{2} & 0 & \dots & 0 \\ 0 & \frac{1}{\lambda_2}\,\mathrm{th}\,\frac{\lambda_2 T}{2} & \dots & 0 \\ \dots & \dots & \dots & \dots \\ 0 & 0 & \dots & \frac{1}{\lambda_n}\,\mathrm{th}\,\frac{\lambda_n T}{2} \end{Vmatrix} \tag{5.24}$$

since

$$\frac{e^{\lambda_i T}-1}{e^{\lambda_i T}+1}=\operatorname{th}\frac{\lambda_i T}{2}$$

Furthermore, using the expansion formula (2.24), we obtain

$$K[M(T)+E]^{-1}N(T)K^{-1}=\sum_{i=1}^{n}\frac{1}{\lambda_i}\operatorname{th}\frac{\lambda_i T}{2}k_i\varkappa_i \tag{5.25}$$

Substituting (5.25) in (5.20), we obtain a relation for finding the initial value $x^*(0)$ in the form

$$x^*(0)=-\sum_{i=1}^{n}\frac{1}{\lambda_i}\operatorname{th}\frac{\lambda_i T}{2}k_i\varkappa_i h \tag{5.26}$$

Substituting (5.25) in (5.21), we obtain the periodic equation in the form

$$-\sum_{i=1}^{n}\frac{1}{\lambda_i}\operatorname{th}\frac{\lambda_i T}{2}\gamma k_i\varkappa_i h=\sigma_1^* \tag{5.27}$$

To expand the left-hand side of (5.22), we proceed as follows. We multiply (5.24) from the left by the diagonal matrix $J=\Lambda$ given by (2.44). We shall then have

$$J[M(T)+E]^{-1}N(T)=\left\|\begin{matrix}\operatorname{th}\frac{\lambda_i T}{2} & 0 & \ldots & 0\\ 0 & \operatorname{th}\frac{\lambda_2 T}{2} & \ldots & 0\\ \cdot & \cdot & \cdot & \cdot\\ 0 & 0 & \ldots & \operatorname{th}\frac{\lambda_n T}{2}\end{matrix}\right\| \tag{5.28}$$

Applying the expansion formula (2.24) to this matrix, we obtain

$$KJ[M(T)+E]^{-1}N(T)K^{-1}=\sum_{i=1}^{n}\operatorname{th}\frac{\lambda_i T}{2}k_i\varkappa_i \tag{5.29}$$

Substituting (5.29) into (5.22), we write it in the form

$$-\sum_{i=1}^{n}\operatorname{th}\frac{\lambda_i T}{2}\gamma k_i\varkappa_i h>|\gamma h| \tag{5.30}$$

Finally, we go over to (5.23), which determines the periodic oscillation during the half-period T. Matrix $M(t)$ is given by (2.94); it is a diagonal matrix with common elements $e^{\lambda_i T}$, so that [see (5.24)].

$$M(t)[M(T)+E]^{-1}N(T)$$

$$=\begin{Vmatrix} \frac{e^{\lambda_1 t}}{\lambda_1}\operatorname{th}\frac{\lambda_1 T}{2} & 0 & 0 \ldots & 0 \\ 0 & \frac{e^{\lambda_2 t}}{\lambda_2}\operatorname{th}\frac{\lambda_2 T}{2} & 0 \ldots & 0 \\ \cdots & \cdots & \cdots & \cdots \\ 0 & 0 & 0 \ldots & \frac{e^{\lambda_n t}}{\lambda_n}\operatorname{th}\frac{\lambda_n T}{2} \end{Vmatrix} \quad (5.31)$$

Next, applying the expansion formula (2.24) to the diagonal matrices (2.120) and (5.31), we obtain

$$KM(t)[M(T)+E]^{-1}N(T)K^{-1}=\sum_{i=1}^{n}\frac{e^{\lambda_i t}}{\lambda_i}\operatorname{th}\frac{\lambda_i T}{2}k_i\varkappa_i \quad (5.32)$$

$$KN(t)K=\sum_{i=1}^{n}\frac{e^{\lambda_i t}-1}{\lambda_i}k_i\varkappa_i \quad (5.33)$$

Substituting these expressions into (5.23), we obtain a relation in the form

$$x(t)=\sum_{i=1}^{n}\frac{e^{\lambda_i t}}{\lambda_i}\operatorname{th}\frac{\lambda_i T}{2}k_i\varkappa_i h+\sum_{i=1}^{n}\frac{e^{\lambda_i t}-1}{\lambda_i}k_i\varkappa_i h \quad (5.34)$$

Finally, substituting into (5.27), (5.30) and (5.34) the $k_i\varkappa_i$ obtained from (2.64), we obtain the final form of the expressions for the periodic equation

$$-\sum_{i=1}^{n}\frac{1}{\lambda_i}\operatorname{th}\frac{\lambda_i T}{2}\frac{\gamma F(\lambda_i)h}{\Delta'(\lambda_i)}=\sigma_1^* \quad (5.35)$$

for the normal switching inequality [which is the name we shall give inequality (5.22), which combines the two inequalities expressing the conditions for normal switching (5.10)]:

$$-\sum_{i=1}^{n}\operatorname{th}\frac{\lambda_i T}{2}\frac{\gamma F(\lambda_i)h}{\Delta'(\lambda_i)}>|\gamma h| \quad (5.36)$$

and, for the periodic oscillation during the half-period T:

$$x(t)=-\sum_{i=1}^{n}\frac{e^{\lambda_i t}}{\lambda_i}\operatorname{th}\frac{\lambda_i T}{2}\frac{F(\lambda_i)h}{\Delta'(\lambda_i)}-\sum_{i=1}^{n}\frac{e^{\lambda_i t}-1}{\lambda_i}\frac{F(\lambda_i)h}{\Delta'(\lambda_i)} \quad (5.37)$$

$$0\leqslant t\leqslant T$$

It will be recalled that $t=0$ corresponds to the point at which the closed trajectory of the periodic solution cuts the switching plane $\gamma x=\sigma_1^*$. The matrix coordinate $x^*(0)$ of this point is obtained from (5.26) in the manner indicated above, or from (5.37) with $t=0$ in the form

$$x^*(0)=-\sum_{i=1}^{n}\frac{1}{\lambda_i}\operatorname{th}\frac{\lambda_i T}{2}\frac{F(\lambda_i)h}{\Delta'(\lambda_i)} \tag{5.38}$$

For the case when there is a single multiple root among the roots λ_i, namely the double zero root $\lambda_{n-1}=\lambda_n=0$, the procedure for multiplying out the matrix expressions (5.20)-(5.23) remains the same. Thus, using the method indicated above, we obtain the matrix $[M(T)+E]^{-1}$ from (4.143). This matrix, and the matrix $N(T)$ [formula (2.122) with $t=T$] is now quasidiagonal. If they are decomposed into blocks – as was done in expressions (4.41) and (4.42) – then only the lower blocks will be new. Multiplying out these cells, we obtain

$$\begin{Vmatrix}\frac{1}{2} & 0\\ -\frac{T}{4} & \frac{1}{2}\end{Vmatrix}\begin{Vmatrix}T & 0\\ \frac{T^2}{2} & T\end{Vmatrix}=\begin{Vmatrix}\frac{T}{2} & 0\\ 0 & \frac{T}{2}\end{Vmatrix} \tag{5.39}$$

Considering (5.24) and (5.39), we obtain

$$[M(T)+E]^{-1}N(T)$$

$$=\begin{Vmatrix}\frac{1}{\lambda_1}\operatorname{th}\frac{\lambda_1 T}{2} & \dots & 0 & 0 & 0\\ \dots & \dots & \dots & \dots & \dots\\ 0 & \dots & \frac{1}{\lambda_{n-2}}\operatorname{th}\frac{\lambda_{n-2}T}{2} & 0 & 0\\ 0 & \dots & 0 & \frac{T}{2} & 0\\ 0 & \dots & 0 & 0 & \frac{T}{2}\end{Vmatrix} \tag{5.40}$$

Then, according to the expansion formula (2.24), we shall have

$$K[M(T)+E]^{-1}N(T)K^{-1}$$

$$=\sum_{i=1}^{n-2}\frac{1}{\lambda_i}\operatorname{th}\frac{\lambda_i T}{2}k_i\varkappa_i+\frac{T}{2}(k_{n-1}\varkappa_{n-1}+k_n\varkappa_n) \tag{5.41}$$

Multiplying matrix (5.40) by matrix J given by (2.49b) with $\lambda_{n-1}=0$, we have

$$J[M(T)+E]^{-1}N(T)=\left\|\begin{matrix} \text{th}\,\frac{\lambda_1 T}{2} & \ldots & 0 & 0 & 0 \\ \cdot & \cdot & \cdot & \cdot & \cdot \\ 0 & \ldots & \text{th}\,\frac{\lambda_{n-2}T}{2} & 0 & 0 \\ 0 & \ldots & 0 & 0 & 0 \\ 0 & \ldots & 0 & \frac{T}{2} & 0 \end{matrix}\right\| \tag{5.42}$$

Applying the expansion formula (2.24) to this matrix, we obtain

$$KJ[M(T)+E]^{-1}N(T)K^{-1}=\sum_{i=1}^{n-2}\text{th}\,\frac{\lambda_i T}{2}\,k_i\varkappa_i+\frac{T}{2}\,k_n\varkappa_{n-1} \tag{5.43}$$

If we now multiply matrix (5.40) from the left by the quasi-diagonal matrix $M(t)$ (2.98), we obtain

$$M(t)[M(T)+E]^{-1}N(T)$$

$$=\left\|\begin{matrix} \frac{e^{\lambda_1 t}}{\lambda_1}\,\text{th}\,\frac{\lambda_1 T}{2} & \ldots & 0 & 0 & 0 \\ \cdot & \cdot & \cdot & \cdot & \cdot \\ 0 & \ldots & \frac{e^{\lambda_{n-2}t}}{\lambda_{n-2}}\,\text{th}\,\frac{\lambda_{n-2}T}{2} & 0 & 0 \\ 0 & \ldots & 0 & \frac{T}{2} & 0 \\ 0 & \ldots & 0 & \frac{tT}{2} & \frac{T}{2} \end{matrix}\right\| \tag{5.44}$$

Applying the expansion formula (2.24) to the quasidiagonal matrices $N(t)$ (2.122) and (5.44), we obtain

$$KN(t)K^{-1}=\sum_{i=1}^{n-2}\frac{e^{\lambda_i t}-1}{\lambda_i}\,k_i\varkappa_i+t\,(k_{n-1}\varkappa_{n-1}+k_n\varkappa_n)+\frac{t^2}{2}\,k_n\varkappa_{n-1} \tag{5.45}$$

$$KM(t)[M(T)+E]^{-1}N(T)K^{-1}=\sum_{i=1}^{n-2}\frac{e^{\lambda_i t}}{\lambda_i}\,\text{th}\,\frac{\lambda_i T}{2}\,k_i\varkappa_i$$
$$+\frac{T}{2}(k_{n-1}\varkappa_{n-1}+k_n\varkappa_n)+\frac{tT}{2}\,k_n\varkappa_{n-1} \tag{5.46}$$

Given (5.41), (5.43), (5.45) and (5.46), the subsequent calculations can be carried out without difficulty. In the final stage one should only use the matrix expansion (2.64) for simple roots λ_i and (2.66) for the double zero root $\lambda_{n-1}=\lambda_n=0$.

Thus, for the case of a double zero root $\lambda_{n-1}=\lambda_n=0.$, we obtain the final form of the calculated expressions for the perodic equation:

$$-\sum_{i=1}^{n-2}\frac{1}{\lambda_i}\text{th}\,\frac{\lambda_i T}{2}\,\frac{\gamma F(\lambda_i)\,h}{\Delta'(\lambda_i)}-\frac{T}{2}\gamma\left[\frac{F(0)}{\Delta_{n-1}(0)}\right]^{(1)}h=\sigma_1^* \tag{5.47}$$

for the normal switching inequality:

$$-\sum_{i=1}^{n-2}\text{th}\,\frac{\lambda_i T}{2}\,\frac{\gamma F(\lambda_i)\,h}{\Delta'(\lambda_i)}-\frac{T}{2}\,\frac{\gamma F(0)\,h}{\Delta_{n-1}(0)}>|\gamma h| \tag{5.48}$$

for the periodic oscillation during the half-period:

$$\begin{aligned}x(t)=&-\sum_{i=1}^{n-2}\frac{e^{\lambda_i t}}{\lambda_i}\text{th}\,\frac{\lambda_i T}{2}\,\frac{F(\lambda_i)\,h}{\Delta'(\lambda_i)}-\frac{T}{2}\left[\frac{F(0)}{\Delta_{n-1}(0)}\right]^{(1)}h\\&-\frac{tT}{2}\,\frac{F(0)\,h}{\Delta_{n-1}(0)}+\sum_{i=1}^{n-2}\frac{e^{\lambda_i t}-1}{\lambda_i}\,\frac{F(\lambda_i)\,h}{\Delta'(\lambda_i)}\\&+t\left[\frac{F(0)}{\Delta_{n-1}(0)}\right]^{(1)}h+\frac{t^2}{2}\,\frac{F(0)\,h}{\Delta_{n-1}(0)},\qquad 0\leqslant t\leqslant T\end{aligned} \tag{5.49}$$

and, finally, for the initial coordinate of the periodic oscillation:

$$x^*(0)=-\sum_{i=1}^{n-2}\frac{1}{\lambda_i}\text{th}\,\frac{\lambda_i T}{2}\,\frac{F(\lambda_i)\,h}{\Delta'(\lambda_i)}-\frac{T}{2}\left[\frac{F(0)}{\Delta_{n-1}(0)}\right]^{(1)}h \tag{5.50}$$

5.3 STABILITY OF THE PERIODIC OSCILLATIONS AND THE CHARACTERISTIC EQUATION

The stability of the periodic oscillations is determined by studying the character of the solutions of (5.1) that satisfy initial conditions close to those corresponding to the periodic solutions. In state space there is an isolated, closed state trajectory corresponding to the periodic solution (5.23). Since each point of the closed trajectory can serve as the initial point for a periodic solution, there exists an infinite set of periodic solutions corresponding to the closed solution, each solution having a different phase. This is a consequence of the autonomy of the system of equations (5.1); their solution does not depend on the choice of the origin in time.

Because of the above circumstances, the periodic solution can possess ordinary Lyapunov stability but it cannot be asymptotically stable. In fact, it will be possible to find an element of its arc in any neighbourhood, however small, about an arbitrary point of the closed trajectory. Each point of this arc may be chosen as the initial point for the perturbed motion. This perturbed motion is a periodic solution differing from the initial motion by a phase factor. The continuum of periodic solutions possesses a single 'amplitude' and a single 'frequency'.

These quantities are usually of interest when we are dealing with eigenproblems in automatic control. The 'custodian' of these quantities is the isolated, closed state trajectory, called the limit cycle. It is therefore of interest to consider the properties of such a cycle. The study of the stability of a limit cycle requires no additional information apart from that which has already been given in the preceding chapters. One can show that ordinary stability of the periodic solution follows from the asymptotic stability of the limit cycle.

We will present our discussion in accordance with the following plan. The initial value (5.20) determined the matrix coordinate $x^*(0)$ of the point where the closed trajectory intersects the switching plane when the image point passes from the half-space $\sigma < 0$ to the half-space $\sigma > 0$. This direction of movement will be called intersection from below to above. After a time equal to the period $2T$ has elapsed, the motion of the image point along the closed trajectory takes it once more to the point of intersection. In this sense, then, the point $x^*(0)$ is fixed on the switching plane. The other solutions of (5.1) corresponding to initial conditions close to $x^*(0)$ determine, in general, the unclosed trajectories. These trajectories form a time sequence of points of intersection from below to above the switching plane.

Since trajectories do not intersect each other, we can decide about the stability of the limit cycle from the character of the time sequence of points on the switching plane. If this sequence converges to the fixed point, then the limit cycle will be asymptotically stable. In this sense the problem reduces to one of constructing a point transformation of the switching plane in the neighbourhood of the point corresponding to the periodic solution. This point is fixed for the point transformation being sought. Analytically, the point transformation is given by a difference equation.

For the periodic solution found above, which is symmetric during the half-period T, it is possible to simplify the discussion. For this case it is sufficient to establish a dependence between successive points of the switching plane, but not between points which intersect the plane only from below to above. Let $x^*(0)$ be given by (5.20), let it satisfy the condition for normal switching (5.10) and let it be the initial condition for the symmetric periodic solution (5.23). Then, after a time T, the matrix coordinate will assume the opposite value $-x^*(0)$. We shall choose a matrix coordinate $x^*(0)+\delta x(0)$ close to $x^*(0)$ and require that it shall satisfy the condition for normal switching (5.10). We shall take the coordinate $x^*(0)+\delta x(0)$ for the new initial condition which will, in general, determine the nonperiodic solution of (5.1). For this solution, the next switching time does not occur after a time T has elapsed but after a time $T+\delta T_0$. The corresponding matrix will now differ slightly from $-x^*(0)$, so we shall denote it through $-[x^*(0)+\delta x(T)]$.

The matrix coordinate $-[x^*(0)+\delta x(T)]$ must satisfy the conditions for normal switching (5.11). We shall now insert in solution (5.12) the new initial condition $x^*(0)+\delta x(0)$ in place of $x(0)$. Assuming $\tau=T+\delta T_0$ and using the notation we have introduced, we obtain the expression

$$-[x^*(0)+\delta x(T)]$$

$$=KM(T+\delta T_0)K^{-1}[x^*(0)+\delta x(0)]+KN(T+\delta T_0)K^{-1}h \tag{5.51}$$

Whence the equality appearing in the conditions for normal switching (5.11) can be written in the form

$$\gamma KM(T+\delta T_0)K^{-1}[x^*(0)+\delta x(0)]$$

$$+\gamma KN(T+\delta T_0)K^{-1}h=-\sigma_1^* \tag{5.52}$$

The two last inequalities of (5.11) are fulfilled for $-[x^*(0)+\delta x(T)]$ for reasons of continuity, since they are fulfilled for $-x^*(0)$ and the elements of the column $\delta x(T)$ are small quantities.

We shall denote the coordinates and successive switching times through $x^*(0)+\delta x(2T)$, $-[x^*(0)+\delta x(3T)]$ etc, $T+\delta T_1$, $T+\delta T_2$ etc, respectively. It is then not difficult to show that expressions (5.50) and (5.51) establish a relation between

$$x^*(0)+\delta x[(m-1)T],\quad T+\delta T_{m-1}$$

and

$$-[x^*(0)+\delta x(mT)],\quad T+\delta T_m$$

corresponding to any two successive normal switchings. Taking this fact into account, and after some simple re-grouping, we can write (5.51) and (5.52) in the form

$$\left.\begin{aligned}\delta x(mT)=-x^*(0)-KM(T+\delta T_{m-1})K^{-1}\{x^*(0)\\+\delta x[(m-1)T]\}-KN(T+\delta T_{m-1})K^{-1}h\\ \gamma KM(T+\delta T_{m-1})K^{-1}\{x^*(0)+\delta x[(m-1)T]\}\\+\gamma KN(T+\delta T_{m-1})K^{-1}h=-\sigma_1^*\end{aligned}\right\} \tag{5.53}$$

Expressions (5.53) are nonlinear difference equations with respect to the so-called variations δx of the matrix coordinate x. The variations $\delta x(mT)$ may be regarded as the matrix coordinates of the time sequence of points indicated above, and they are measured from the point $x^*(0)$. The analysis of the stability of the periodic solutions reduces now to a question of finding the conditions for which $\delta x(mT) \to 0$ as $m \to \infty$. This problem can be solved within the framework of Lyapunov's stability theory, as explained in Chapter 3. Indeed, (5.53) may be regarded as difference equations of the perturbed motion of a discrete dynamical system, so it is not difficult to show that the identity $\delta x \equiv 0$ is a trivial solution for these equations.

In order to investigate the stability in the Lyapunov sense, we shall construct the equations of the first approximation. For this purpose we shall expand the functions entering into (5.53) as a Taylor series in powers of the variations $\delta x[(m-1)T]$ and δT_{m-1}. All terms not containing variations obviously cancel. Then, discarding all terms in the expansion of order above the first with respect to the variations $\delta x[(m-1)T]$ and δT_{m-1}, we obtain

$$\left.\begin{aligned} \delta x(mT) = &-KM(T)K^{-1}\delta x[(m-1)T] - \\ &-[K\dot{M}(T)K^{-1}x^*(0) + K\dot{N}(T)K^{-1}h]\delta T_{m-1} \\ \gamma KM(T)K^{-1}&\delta x[(m-1)T] + \\ &+\gamma[K\dot{M}(T)K^{-1}x^*(0) + K\dot{N}(T)K^{-1}h]\delta T_{m-1} = 0 \end{aligned}\right\} \tag{5.54}$$

where

$$\dot{M}(T) = \left[\frac{dM(t)}{dt}\right]_{t=T} \qquad \dot{N}(T) = \left[\frac{dN(t)}{dt}\right]_{t=T}$$

According to (2.102), (2.109), (2.110) and (5.20), we have

$$\left.\begin{aligned} \dot{M}(T) = M(T)J \quad \dot{N}(T) = M(T) \\ x^*(0) = -K[M(T)+E]^{-1}N(T)K^{-1}h \end{aligned}\right\} \tag{5.55}$$

With the help of expressions (5.55) we can reduce equations (5.54) to the form

$$\left.\begin{aligned} \gamma x(mT) = &-KM(T)K^{-1}\delta x[(m-1)T] - \\ &-KS(T)K^{-1}h\,\delta T_{m-1} \\ \gamma KM(T)K^{-1}&\delta x[(m-1)T] + \gamma KS(T)K^{-1}h\,\delta T_{m-1} = 0 \end{aligned}\right\} \tag{5.56}$$

where

$$S(T) = M(T)\{-J[M(T)+E]^{-1}N(T)+E\} \tag{5.57}$$

From the second scalar equation (5.56), we obtain

$$\delta T_{m-1} = -\frac{\gamma KM(T)K^{-1}\delta x[(m-1)T]}{\gamma KS(T)K^{-1}h} \tag{5.58}$$

Substituting the value of δT_{m-1} obtained in the first equation in (5.56), we transform it into

$$\delta x\,(mT) = K\left[-M\,(T) + \frac{S\,(T)\,K^{-1}h\gamma KM\,(T)}{\gamma KS\,(T)\,K^{-1}h}\right]K^{-1}\,\delta x\,[(m-1)\,T] \qquad (5.59)$$

Thus, we have obtained in matrix form the equation of the first approximation which takes the form of a linear difference equation with a constant matrix operator

$$P^* = K\left[-M\,(T) + \frac{S\,(T)\,K^{-1}h\gamma KM\,(T)}{\gamma KS\,(T)\,K^{-1}h}\right]K^{-1} \qquad (5.60)$$

Consequently, the periodic solution that has been found will be asymptotically stable in the Lyapunov sense with an accuracy up to the phase if all the roots λ_i of the characteristic equation of the matrix operator (5.60) which can be written in the form

$$\det\left[\lambda E + M(T) - \frac{S\,(T)\,K^{-1}h\gamma KM\,(T)}{\gamma KS\,(T)\,K^{-1}h}\right] = 0 \qquad (5.61)$$

have moduli less than unity; and it will be unstable if the modulus of at least one of the roots is greater than unity. We now reduce (5.61) to a form which is more amenable for calculation. We use (4.21) and take the matrix $\lambda E + M(T)$ outside the square brackets. Equation (5.61) is then written in the form (see Section 2.1)

$$\det\,[\lambda E + M(T)]\,\det\left\{E - \frac{[\lambda E + M\,(T)]^{-1}\,S\,(T)\,u\beta M\,(T)}{\beta S\,(T)\,u}\right\} = 0 \qquad (5.62)$$

The second term in the braces is the product of column

$$\tilde{u} = [\lambda E + M\,(T)]^{-1}\,S\,(T)\,u \qquad (5.63)$$

and row

$$\tilde{\beta} = \frac{\beta M\,(T)}{\beta S\,(T)\,u} \qquad (5.64)$$

since $\beta S(T)\,u$ is a scalar quantity. Using (4.133), we put (5.62) in the form

$$\{\det\,[\lambda E + M(T)]\}\left\{1 - \frac{\beta M\,(T)\,[\lambda E + M\,(T)]^{-1}\,S\,(T)\,u}{\beta S\,(T)\,u}\right\} = 0 \qquad (5.65)$$

If we make use of the identity

$$M\,(T)\,[\lambda E + M\,(T)]^{-1} = [M\,(T) + \lambda E - \lambda E]\,[\lambda E + M\,(T)]^{-1}$$

$$= E - \lambda\,[\lambda E + M\,(T)]^{-1} \qquad (5.66)$$

then, after some simple reductions and after discarding the scalar quantity $\beta S(T) u$, the characteristic equation (5.65) can be written in the form

$$\{\det [\lambda E + M(T)]\} \lambda\beta [\lambda E + M(T)]^{-1} S(T) u = 0 \tag{5.67}$$

From this it follows that one root of the characteristic equation is always equal to zero. This fact is of the greatest importance, since it shows that, in an autonomous relay system, asymptotic stability of the periodic solution depends on the $(n-1)$ characteristic values of the linear difference equation of the first approximation (5.59) with an accuracy up to the phase (that is, up to the limit cycle).

Thus, we shall take as our characteristic equation for solving the stability problem an equation of the $(n-1)$th degree in the form

$$\{\det [\lambda E + M(T)]\} \beta [\lambda E + M(T)]^{-1} S(T) u = 0 \tag{5.68}$$

Before we can apply the various stability criteria discussed in Section 4.3, we must introduce into the discussion a rational function of

$$W^*(\lambda) = -\frac{\beta M(T) [\lambda E + M(T)]^{-1} S(T) u}{\beta S(T) u} \tag{5.69a}$$

which appears in the characteristic equation in the form (5.65).

The form of writing the matrix expressions (5.68) and (5.69a) decides the order and character of the preliminary operations which must be carried out in order to obtain the final computing formulae.

We shall expand first the matrix expressions (5.68) and (5.69a) for the case of simple roots λ_i. In this case all the square matrices entering into (5.68) and (5.69a) are diagonal. Thus, from (4.142), by changing the sign of $M(T)$, we obtain

$$[\lambda E + M(T)]^{-1} = \left\| \begin{matrix} \frac{1}{\lambda + e^{\lambda_1 T}} & 0 & \dots & 0 \\ 0 & \frac{1}{\lambda + e^{\lambda_2 T}} & \dots & 0 \\ \cdot & \cdot & \cdot & \cdot \\ 0 & 0 & \dots & \frac{1}{\lambda + e^{\lambda_n T}} \end{matrix} \right\| \tag{5.69}$$

Taking into account the form of the matrix (5.28) and the identity

$$e^{\lambda_i T}\left(1 - \operatorname{th}\frac{\lambda_i T}{2}\right) = 1 + \operatorname{th}\frac{\lambda_i T}{2} \tag{5.70}$$

we obtain for matrix (5.57) an expression in the form

$$S(T)=\left\|\begin{matrix} 1+\operatorname{th}\frac{\lambda_i T}{2} & 0 & \dots & 0 \\ 0 & 1+\operatorname{th}\frac{\lambda_i T}{2} & \dots & 0 \\ \dots & \dots & \dots & \dots \\ 0 & 0 & \dots & 1+\operatorname{th}\frac{\lambda_n T}{2} \end{matrix}\right\| \tag{5.71}$$

Multiplying out matrices (5.69) and (5.71), we obtain

$$[\lambda E+M(T)]^{-1}S(T)$$

$$=\left\|\begin{matrix} \frac{1+\operatorname{th}\frac{\lambda_1 T}{2}}{\lambda+e^{\lambda_1 T}} & 0 & \dots & 0 \\ 0 & \frac{1+\operatorname{th}\frac{\lambda_2 T}{2}}{\lambda+e^{\lambda_2 T}} & \dots & 0 \\ \dots & \dots & \dots & \dots \\ 0 & 0 & \dots & \frac{1+\operatorname{th}\frac{\lambda_n T}{2}}{1+e^{\lambda_n T}} \end{matrix}\right\| \tag{5.72}$$

Multiplying matrix (5.72) from the left by the diagonal matrix $M(T)$ with common elements $e^{\lambda_i T}$, we obtain

$$M(T)\,[\lambda E+M(T)]^{-1}S(T)$$

$$=\left\|\begin{matrix} \frac{e^{\lambda_1 T}\left(1+\operatorname{th}\frac{\lambda_1 T}{2}\right)}{\lambda+e^{\lambda_1 T}} & \dots & 0 \\ \dots & \dots & \dots \\ 0 & \dots & \frac{e^{\lambda_n T}\left(1+\operatorname{th}\frac{\lambda_n T}{2}\right)}{\lambda+e^{\lambda_n T}} \end{matrix}\right\| \tag{5.73}$$

For any roots λ_i the inequality

$$\det[\lambda E+M(T)]=\prod_{i=1}^{n}(\lambda+e^{\lambda_i T}) \tag{5.74}$$

holds where each root is counted a number of times equal to its multiplicity. Multiplying (5.71) from the left by the row β and from the right by the column u we obtain by direct calculation

$$\alpha=\beta S(T)\,u=\sum_{i=1}^{n}\left(1+\operatorname{th}\frac{\lambda_i T}{2}\right)u_i\beta_i \tag{5.75}$$

where, for brevity, the quantity $\beta S(T)\,u$ has been denoted by α.

Carrying out the same operation on the matrices (5.72) and (5.73), and taking account of (5.74) and (5.75), we obtain an expression for the characteristic equation (5.68) in the form

$$\prod_{j=1}^{n}(\lambda+e^{\lambda_j T})\sum_{i=1}^{n}\frac{1+\operatorname{th}\frac{\lambda_i T}{2}}{\lambda+e^{\lambda_i T}}u_i\beta_i=0 \qquad (5.76)$$

and an expression for the rational function (5.69) in the form

$$W^*(\lambda)=-\frac{1}{\alpha}\sum_{i=1}^{n}\frac{e^{\lambda_i T}\left(1+\operatorname{th}\frac{\lambda_i T}{2}\right)}{\lambda+e^{\lambda_i T}}u_i\beta_i \qquad (5.77)$$

For the case when among the roots λ_i there is one multiple root $\lambda_{n-1}=\lambda_n=0$, the general procedure of calculation remains the same, of course, although the expressions obtained are somewhat more complicated. Thus, from the matrix (4.143), by replacing λ by $-\lambda$ and by changing the signs in front of all the terms, we obtain

$$[\lambda E+M(T)]^{-1}=\left\|\begin{matrix}\frac{1}{\lambda-e^{\lambda_1 T}} & \cdots & 0 & 0 & 0\\ \cdots & \cdots & \cdots & \cdots & \cdots\\ 0 & \cdots & \frac{1}{\lambda-e^{\lambda_{n-2}T}} & 0 & 0\\ 0 & \cdots & 0 & \frac{1}{\lambda+1} & 0\\ 0 & \cdots & 0 & -\frac{T}{(\lambda+1)^2} & \frac{1}{\lambda+1}\end{matrix}\right\| \qquad (5.78)$$

With account of the form of matrix (5.42), we first form the matrix

$$-J[M(T)+E]^{-1}N(T)+E$$

$$=\left\|\begin{matrix}1-\operatorname{th}\frac{\lambda_1 T}{2} & \cdots & 0 & 0 & 0\\ \cdots & \cdots & \cdots & \cdots & \cdots\\ 0 & \cdots & 1-\operatorname{th}\frac{\lambda_{n-2}T}{2} & 0 & 0\\ 0 & \cdots & 0 & 1 & 0\\ 0 & \cdots & 0 & -\frac{T}{2} & 1\end{matrix}\right\| \qquad (5.79)$$

Multiplying matrix (5.79) by the quasidiagonal matrix $M(T)$ [(2.98) for $t=T$], and taking account of the identity (5.75), we obtain for the matrix (5.57), the expression

$$S(T)=\left\|\begin{matrix} 1+\operatorname{th}\frac{\lambda_1 T}{2} & \dots & 0 & 0 & 0 \\ \cdot & \cdot & \cdot & \cdot & \cdot \\ 0 & \dots & 1+\operatorname{th}\frac{\lambda_{n-2}T}{2} & 0 & 0 \\ 0 & \dots & 0 & 1 & 0 \\ 0 & \dots & 0 & \frac{T}{2} & 1 \end{matrix}\right\| \tag{5.80}$$

Multiplying out the matrices (5.78) and (5.80), we obtain

$$[\lambda E+M(T)]^{-1}S(T)$$

$$=\left\|\begin{matrix} \frac{1+\operatorname{th}\frac{\lambda_1 T}{2}}{1+e^{\lambda_1 T}} & \dots & 0 & 0 & 0 \\ \cdot & \cdot & \cdot & \cdot & \cdot \\ 0 & \dots & \frac{1+\operatorname{th}\frac{\lambda_{n-2}T}{2}}{\lambda+e^{\lambda_{n-2}T}} & 0 & 0 \\ 0 & \dots & 0 & \frac{1}{\lambda+1} & 0 \\ 0 & \dots & 0 & \frac{T}{2}\frac{\lambda-1}{(\lambda+1)^2} & \frac{1}{\lambda+1} \end{matrix}\right\| \tag{5.81}$$

Multiplying (5.81) by the quasidiagonal matrix $M(T)$ [(2.98) for $t=T$)], we obtain

$$M(T)[\lambda E+M(T)]^{-1}S(T)$$

$$=\left\|\begin{matrix} \frac{e^{\lambda_1 T}\left(1+\operatorname{th}\frac{\lambda_1 T}{2}\right)}{\lambda+e^{\lambda_1 T}} & \dots & 0 & 0 & 0 \\ \cdot & \cdot & \cdot & \cdot & \cdot \\ 0 & \dots & \frac{e^{\lambda_{n-2}T}\left(1+\operatorname{th}\frac{\lambda_{n-2}T}{2}\right)}{\lambda+e^{\lambda_{n-2}T}} & 0 & 0 \\ 0 & \dots & 0 & \frac{1}{\lambda+1} & 0 \\ 0 & \dots & 0 & \frac{T}{2}\frac{2\lambda+1}{(\lambda+1)^2} & \frac{1}{\lambda+1} \end{matrix}\right\| \tag{5.82}$$

Multiplying (5.8) from the left by row β and from the right by column u, we obtain by direct calculation

$$\alpha=\beta S(T)u=\sum_{i=1}^{n}\left(1+\operatorname{th}\frac{\lambda_i T}{2}\right)u_i\beta_i+u_{n-1}\beta_{n-1}+u_n\beta_n+\frac{T}{2}u_{n-1}\beta_n \tag{5.83}$$

Carrying out the same operation on matrices (5.81) and (5.82), and with account of (5.74) and (5.83), we obtain an expression for the characteristic equation (5.68) in the form

$$(\lambda+1)^2\prod_{j=1}^{n-2}\left(\lambda+e^{\lambda_j T}\right)\left\{\sum_{i=1}^{n-2}\frac{1+\operatorname{th}\frac{\lambda_i T}{2}}{\lambda+e^{\lambda_i T}}u_i\beta_i\right.$$

$$\left.+\frac{1}{\lambda+1}(u_{n-1}\beta_{n-1}+u_n\beta_n)+\frac{T}{2}\frac{\lambda-1}{(\lambda+1)^2}u_{n-1}\beta_n\right\}=0 \tag{5.84}$$

and an expression for the rational function (5.69) in the form

$$W^*(\lambda)=-\frac{1}{\alpha}\left\{\sum_{i=1}^{n-2}\frac{e^{\lambda_i T}\left(1+\operatorname{th}\frac{\lambda_i T}{2}\right)}{\lambda+e^{\lambda_i T}}u_i\beta_i\right.$$

$$\left.+\frac{1}{\lambda+1}(u_{n-1}\beta_{n-1}+u_n\beta_n)+\frac{T}{2}\frac{2\lambda+1}{(\lambda+1)^2}u_{n-1}\beta_n\right\} \tag{5.85}$$

The rational functions (5.77) and (5.85) enable us to carry out the stability analysis in accordance with the argument principle discussed in Section 4.3, since these functions play the same role as the functions (4.63).

In order to put these functions into their final form, we substitute into (5.75), (5.83) and (5.85) the quantities u_i, β_i, found from (4.51); for simple roots λ_i we should take the first of formulae (4.51), and for a double zero root we should take last two expressions. Then, for the rational functions (5.77) and (5.85) we obtain, respectively, expressions in the form

$$\left.\begin{aligned} W^*(\lambda)&=-\frac{1}{\alpha}\sum_{i=1}^{n}\frac{e^{\lambda_i T}\left(1+\operatorname{th}\frac{\lambda_i T}{2}\right)}{\lambda+e^{\lambda_i T}}\frac{\gamma F(\lambda_i)h}{\Delta'(\lambda_i)}\\ \alpha&=\sum_{i=1}^{n}\left(1+\operatorname{th}\frac{\lambda_i T}{2}\right)\frac{\gamma F(\lambda_i)h}{\Delta'(\lambda_i)}\end{aligned}\right\} \tag{5.86}$$

and

$$\left.\begin{aligned} W^*(\lambda)&=-\frac{1}{\alpha}\left\{\sum_{i=1}^{n-2}\frac{e^{\lambda_i T}\left(1+\operatorname{th}\frac{\lambda_i T}{2}\right)}{\lambda+e^{\lambda_i T}}\frac{\gamma F(\lambda_i)h}{\Delta'(\lambda_i)}\right.\\ &\quad\left.+\frac{1}{\lambda+1}\gamma\left[\frac{F(0)}{\Delta_{n-1}(0)}\right]^{(1)}h+\frac{T}{2}\frac{2\lambda+1}{(\lambda+1)^2}\frac{\gamma F(0)h}{\Delta_{n-1}(0)}\right\}\\ \alpha&=\sum_{i=1}^{n-2}\left(1+\operatorname{th}\frac{\lambda_i T}{2}\right)\frac{\gamma F(\lambda_i)h}{\Delta'(\lambda_i)}+\gamma\left[\frac{F(0)}{\Delta_{n-1}(0)}\right]^{(1)}h\\ &\quad+\frac{T}{2}\frac{\gamma F(0)h}{\Delta_{n-1}(0)}\end{aligned}\right\} \tag{5.87}$$

By making the substitution (4.53) and using (4.51), we transform the characteristic equations (5.76) and (5.84), respectively, to the form

$$\prod_{j=1}^{n}\left(\nu-\text{th}\,\frac{\lambda_j T}{2}\right)\sum_{i=1}^{n}\frac{1-\text{th}^2\,\dfrac{\lambda_i T}{2}}{\nu-\text{th}\,\dfrac{\lambda_i T}{2}}\,\frac{\gamma F(\lambda_i)\,h}{\Delta'(\lambda_i)}=0 \tag{5.88}$$

and

$$\nu^2\prod_{j=1}^{n-2}\left(\nu-\text{th}\,\frac{\lambda_j T}{2}\right)\left\{\sum_{i=1}^{n-2}\frac{1-\text{th}^2\,\dfrac{\lambda_i T}{2}}{\nu-\text{th}\,\dfrac{\lambda_i T}{2}}\,\frac{\gamma F(\lambda_i)\,h}{\Delta'(\lambda_i)}\right.$$
$$\left.+\frac{1}{\nu}\gamma\left[\frac{F(0)}{\Delta_{n-1}(0)}\right]^{(1)}h+\frac{T}{2\nu^2}\,\frac{\gamma F(0)\,h}{\Delta_{n-1}(0)}\right\}=0 \tag{5.89}$$

The transformed characteristic equations of the $(n-1)$th degree (5.88) and (5.89) are convenient in the respect that they enable us to solve the stability problem with the help of the Hurwitz criterion.

Note

From the asymptotic stability of the limit cycle there follows the ordinary stability of the original periodic solution. This assertion can easily be proved by showing that, with a suitable choice of $\delta x(0)$, the sum of all the variations δT_m can be made smaller than any preassigned number. From (5.58), (5.59) and (5.60) it is easily established that

$$\sum_{m=1}^{m}\delta T_m=-\frac{\gamma KM(T)K^{-1}}{\gamma KS(T)K^{-1}h}\sum_{m=0}^{\infty}P^{*m}\,\delta x(0)$$

From the asymptotic stability of the limit cycle it follows that $P^{*m}\to 0$ as $m\to\infty$ and the matrix P^*-E is non-singular. Consequently,

$$\sum_{m=1}^{\infty}\delta T_m=\frac{\gamma KM(T)K^{-1}[P^*-E]^{-1}\,\delta x(0)}{\gamma KS(T)K^{-1}h}$$

5.4 INTERCHANGE OF STABILITY

When seeking the periodic solutions it is necessary to determine the positive roots of the equation of periods. The equation of periods is a transcendental equation and it is convenient to solve it by graphical methods. For the purpose of making a graphical construction we shall consider the left-hand side of the equation of periods as a function of a positive parameter τ and denote it by $\sigma^*(\tau)$. Then, for example, in the case of simple roots λ_i we can write, in accordance with (5.35),

$$\sigma^*(\tau) = -\sum_{i=1}^{n} \frac{1}{\lambda_i} \operatorname{th} \frac{\lambda_i \tau}{2} \frac{\gamma F(\lambda_i) h}{\Delta'(\lambda_i)} \tag{5.90}$$

We shall call the graph of function $\sigma^*(\tau)$ the curve of periods. Along with the curve of periods we shall consider also a straight line parallel to the abscissa, lying at a distance σ_1^* away from it in the positive direction; for brevity, we shall refer to this line as the straight line σ_1^*. The abscissae of the points of intersection of the curve of periods $\sigma^*(\tau)$ with the straight line σ_1^* which satisfy the normal switching inequality, are the half-periods T. For the curve of periods (5.90), the normal switching inequality is given by (5.36).

We shall differentiate $\sigma^*(\tau)$ with respect to τ. When $\tau = T$ we shall have

$$\left[\frac{d\sigma^*(\tau)}{d\tau}\right]_{\tau=T} = \dot{\sigma}^*(T) = -\frac{1}{2} \sum_{i=1}^{n} \frac{1}{\operatorname{ch}^2 \frac{\lambda_i T}{2}} \frac{\gamma F(\lambda_i) h}{\Delta'(\lambda_i)} \tag{5.91}$$

Geometrically, $\dot{\sigma}^*(T)$ determines the slope of the tangent at those points of the curve of periods whose abscissae can be half-periods. With the help of identity

$$\frac{1}{\operatorname{ch}^2 y} = 1 - \operatorname{th}^2 y \tag{5.92}$$

(5.91) can be converted to the form

$$\dot{\sigma}^*(T) = -\frac{1}{2} \sum_{i=1}^{n} \left(1 - \operatorname{th}^2 \frac{\lambda_i T}{2}\right) \frac{\gamma F(\lambda_i) h}{\Delta'(\lambda_i)} \tag{5.93}$$

We now conceptually expand the transformed characteristic equation (5.88) and represent it in the form (4.59). The coefficient of ν of the highest power is obtained from (5.88) by setting the second term in the expressions $\nu - \operatorname{th} \frac{\lambda_i T}{2}$ equal to zero. Thus, it is not difficult to obtain the expression

$$p_0' = \sum_{i=1}^{n} \left(1 - \operatorname{th}^2 \frac{\lambda_i T}{2}\right) \frac{\gamma F(\lambda_i) h}{\Delta'(\lambda_i)} \tag{5.94}$$

Comparing (5.93) and (5.94), we obtain the relation

$$\dot{\sigma}^*(T) = -\frac{1}{2} p_0' \tag{5.95}$$

This relation is characteristic for the system being considered and it does not depend on the type of roots λ_i. Thus, half of the highest coefficients of the transformed characteristic equation coincide with the slope of the tangent to the curve of periods at the points where it intersects the straight line σ_1^* if the sign of one of these quantities is reversed. In particular, from (5.95) it follows that the highest coefficient of p_0' will be negative or, conversely, positive according to whether the curve of periods intersects the straight line σ_1^* from below to above, or from above to below, respectively. If a unique and continuous curve of periods intersects the straight line σ_1^* at several points, then obviously at these points the signs of the slopes of the tangents to the curve of periods will alternate. However, in that case, the signs of the highest coefficient p_0' in the transformed characteristic equation will also alternate.

The Hurwitz conditions have different forms, depending on the sign of the highest coefficient p_0'. For $p_0' > 0$ the Hurwitz conditions take the form of (4.60) while, for $p_0' < 0$, they are given by (4.61). If for example, an oscillatory mode of behaviour corresponds to a point of intersection then, in the majority of cases one meets in practice, one can expect that the next (or the preceding) point of intersection will correspond to an unstable periodic solution since, to preserve stability, it is necessary to change the signs of all the Hurwitz determinants with odd indices as well as the sign of the coefficient p_0'. Relation (5.95) is of great practical importance because it helps one to formulate the stability conditions for the periodic solutions correctly in the form of (4.60) and (4.61).

5.5 OSCILLATIONS IN A SYSTEM FOR STABILISING THE COURSE OF AN AIRCRAFT

The free-motion of the stabilisation system will be described by equations of the type (4.216) in the form

$$\left.\begin{aligned} \ddot{\varphi} + M\dot{\varphi} &= -N\eta \\ \Theta\ddot{\eta} + \dot{\eta} &= h_2'\psi(\sigma) \\ \sigma &= \varphi + \alpha_1\dot{\varphi} + \alpha_2\ddot{\varphi} \end{aligned}\right\} \qquad (5.96)$$

The control function $\psi(\sigma)$ will be represented in the form of an ideal relay characteristic or in the form of a hysteresis characteristic of the backlash type. The graphs of these characteristics are shown in the first two diagrams of Table 1.1.

Before we can use the general computational formulae for finding the periodic solutions, it is necessary to determine the characteristics of the open control system. These characteristics are – the characteristic polynomial $\Delta(\lambda)$, its roots λ_i, the adjoint matrix $F(\lambda)$ and its row elements γ, and column elements h. However, all these quantities have already been obtained in Section 4.10 since, for $\psi(\sigma) \equiv 0$, (4.216) and (5.96) are identical.

The system we are considering belongs to the class of relay systems with soft switching, since (4.223) holds, i.e. $\gamma h = 0$. Consequently, the entire switching plane – with the exception of the straight line or its segment – is a normal switching zone.

Among the roots λ_i determined by (4.226) there is one multiple root, namely $\lambda_3 = \lambda_4 = 0$. The periodic solutions will therefore be given by (5.47)-(5.50) and their stability will be given by the transformed characteristic equation (5.89). Bearing in mind the form of the roots (4.226) and the coefficients (4.231), we will write the equation of periods (5.47) in the form

$$\frac{Nh_2'}{M\Theta - 1}\left[\left(-\frac{1}{M^3} + \frac{\alpha_1}{M^2} - \frac{\alpha_2}{M}\right) \operatorname{th}\frac{MT}{2} + (\Theta^3 - \alpha_1\Theta^2 + \alpha_2\Theta) \operatorname{th}\frac{T}{2\Theta}\right] + \frac{Nh_2'}{M}\left(\alpha_1 - \frac{1}{M} - \Theta\right)\frac{T}{2} = \sigma_1^* \tag{5.97}$$

We represent (5.97) in the more symmetric form

$$-\frac{Nh_2'}{M\Theta - 1}\left[\left(-\frac{1}{M^3} + \frac{\alpha_1}{M^2} - \frac{\alpha_2}{M}\right)\left(\frac{MT}{2} - \operatorname{th}\frac{MT}{2}\right) + (\Theta^3 - \alpha_1\Theta^2 + \alpha_2\Theta)\left(\frac{T}{2\Theta} - \operatorname{th}\frac{T}{2\Theta}\right)\right] = \sigma^* \tag{5.98}$$

After replacing T by τ, the left-hand side of (5.98) determines the curve of periods $\sigma^*(\tau)$ (see preceding section). We shall expand the function $\sigma^*(\tau)$ as a power series in τ. Taking account of the formula

$$\operatorname{th} y = y - \frac{y^3}{3} + \frac{2y^5}{3 \cdot 5} - \cdots \tag{5.99}$$

we obtain for small values of τ the expression

$$\sigma^*(\tau) \approx \frac{Nh_2'}{\Theta}\left(-\alpha_1 + \alpha_2 M + \frac{\alpha_2}{\Theta}\right)\frac{\tau^3}{24} \tag{5.100}$$

For large values τ one may neglect the hyperbolic tangents with respect to τ. We then have

$$\sigma^*(\tau) \approx \frac{Nh_2'}{M}\left(\alpha_1 - \frac{1}{M} - \Theta\right)\frac{\tau}{2} \tag{5.101}$$

Thus, the behaviour of the curve of periods depends on the sign of the expressions inside the brackets in (5.100) and (5.101). These expressions are the same as those inside the brackets of (4.239). Therefore, on the plane of the parameters α_1, M, we can again construct the four regions illustrated in Fig 4.7. Inside these regions the combination of signs of the brackets mentioned above remains unchanged. In region I both brackets are positive; in region II they are both negative. In regions III and IV the brackets have opposite signs. In region III the bracket in (5.100) is negative, while in region IV it is positive. The function $\sigma^*(\tau)$ is the sum of the two monotonic functions.

Taking all these factors into account it is easy to establish the character of the behaviour of the curve of periods in all four regions shown in Fig 4.7. The shape of the curves of periods in these regions is shown in Fig 5.2. In addition to the curves, the straight line σ_1^* is shown. The periodic solutions correspond to the points where this straight line intersects the curve of periods. If we are considering an ideal relay characteristic, then we should assume that $\sigma_1^*=0$; the periodic solutions will then correspond to the points of intersection of the curve of periods with the abscissa axis.

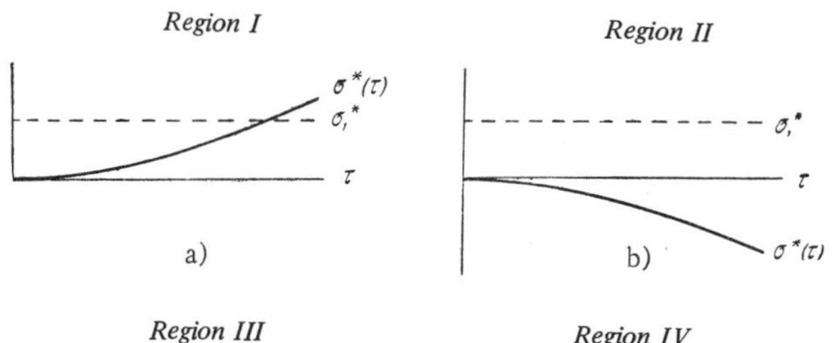

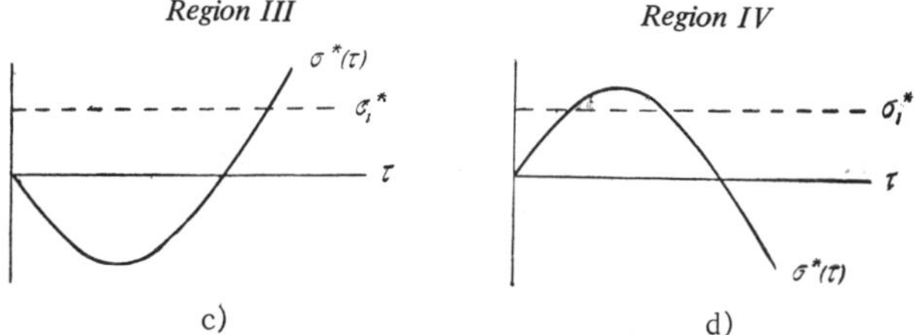

Fig 5.2 Curves of periods of self-oscillatory modes.

It is necessary to check that the normal switching inequality, which in this case is obtained from (5.48), is fulfilled at the points of intersection. With account of the parameters (4.223), (4.226) and (4.231), inequality (5.48) is written in the form

$$\frac{Nh_2'}{M\Theta-1}\left[\left(\frac{1}{M^2}-\frac{\alpha_1}{M}+\alpha_2\right)\operatorname{th}\frac{MT}{2}\right.$$

$$\left.+(-\Theta^2+\alpha_1\Theta-\alpha_2)\operatorname{th}\frac{T}{2\Theta}\right]+\frac{Nh_2'}{M}\frac{T}{2}>0 \qquad (5.102)$$

For sufficiently large values of T this inequality is fulfilled. It is fulfilled also for small values of T since, with an accuracy up to terms proportional to T^3, it can in this case be written in the form

$$\frac{Nh_2'\alpha_2}{\Theta}\frac{T}{2} > 0$$

From this we can deduce that (5.102) holds for all T, since its left-hand side is a linear combination of monotonically increasing functions. Consequently, for the case being considered, all the roots of the equation of periods (5.98) will determine the half-periods T of the symmetric periodic solutions.

Thus, for an ideal relay characteristic we have one periodic solution in each of the regions III and IV (see Fig 5.2). In regions I and II periodic solutions do not exist. The presence of a loop in the relay characteristic leads to the appearance of a periodic solution in region I and, for small values of σ_l^*, to the appearance of a further periodic solution in region IV.

To determine the stability of the periodic solutions we have found, we must make use of the transformed characteristic equation (5.98). In the given case it is an equation of the third degree, of the form

$$p_0'\nu^3 + p_1'\nu^2 + p_2'\nu + p_3' = 0 \tag{5.103}$$

whose coefficients p_i' are given by the expressions

$$\left.\begin{aligned}
p_0' &= \frac{Nh_2'}{M\Theta - 1}\left[\left(-\frac{1}{M^2} + \frac{\alpha_1}{M} - \alpha_2\right)\text{th}^2\frac{MT}{2}\right.\\
&\qquad\left. + (\Theta^2 - \alpha_1\Theta + \alpha_2)\,\text{th}^2\frac{T}{2\Theta}\right]\\
p_1' &= \frac{Nh_2'\left(1 + \text{th}\frac{MT}{2}\,\text{th}\frac{T}{2\Theta}\right)}{M\Theta - 1}\\
&\qquad\times\left[\left(-\frac{1}{M^2} + \frac{\alpha_1}{M} - \alpha_2\right)\text{th}\frac{MT}{2}\right.\\
&\qquad\left. + (\Theta^2 - \alpha_1\Theta + \alpha_2)\,\text{th}\frac{T}{2T}\right] - \frac{Nh_2'T}{2M}\\
p_2' &= Nh_2'\left[\left(\frac{M\Theta + 1}{M^2} - \frac{\alpha_1}{M}\right)\text{th}\frac{MT}{2}\,\text{th}\frac{T}{2\Theta} - \frac{T}{2M}\right.\\
&\qquad\left.\times\left(\text{th}\frac{MT}{2} + \text{th}\frac{T}{2\Theta}\right)\right]\\
p_3' &= -\frac{Nh_2'T}{2M}\,\text{th}\frac{MT}{2}\,\text{th}\frac{T}{2\Theta}
\end{aligned}\right\} \tag{5.104}$$

It is well known that the necessary condition for stability is that all the coefficients p_i' should simultaneously be either positive or negative quantities. From (5.104) we can establish that $p_3' < 0$.

However, at the points where the curve of periods cuts the straight line σ_1^* from above to below, we have $p_0' > 0$ (see preceding section). In this case, therefore, the necessary condition for stability is not fulfilled, so the corresponding periodic solution will be unstable. This situation obtains in region IV for the second point of intersection. In all the other cases represented in Fig 5.2, the curve of periods cuts the straight line σ_1^* from below to above, so for these points $p_0' < 0$. In accordance with (4.61), the Hurwitz inequalities should in this case be written in the form

$$p_0' < 0, \ \Delta_1 = p_1' < 0 \ \ \Delta_2 = p_1'p_2' - p_0'p_3' > 0, \ p_3' < 0 \tag{5.105}$$

In some later calculations we shall find it more convenient to make use of the stability condition in the form of the inequalities

$$p_3' < 0, \ p_2' < 0 \ \ \Delta_2' = p_2'p_1' - p_3'p_0' > 0, \ p_0' < 0 \tag{5.106}$$

which are equivalent to (5.105). Further, it is easy to establish by a direct test that the coefficient p_2' can be written in the form

$$p_2' = Nh_2'\left[-\frac{1}{M^2}\,\text{th}\,\frac{T}{2\Theta}\left(\frac{MT}{2} - \text{th}\,\frac{MT}{2}\right) - \frac{\Theta}{M}\,\text{th}\,\frac{MT}{2} \right.$$
$$\left. \times\left(\frac{T}{2\Theta} - \text{th}\,\frac{T}{2\Theta}\right) - \frac{\alpha_1}{M}\,\text{th}\,\frac{MT}{2}\,\text{th}\,\frac{T}{2\Theta}\right] \tag{5.107}$$

Since $y - \text{th}\, y > 0$ when $y > 0$, then from (5.107) we can establish that the coefficient $p_2' < 0$.

Let us transform the expression which determines the coefficient p_1'. For this purpose we shall multiply and divide the expression in the square brackets in the second equality (5.104) by $\left(\text{th}\,\frac{MT}{2} + \text{th}\,\frac{T}{2\Theta}\right)$. Then, with account of the remaining equalities (5.104), we obtain an expression for the coefficient p_1' in the form

$$p_1' = \frac{1 + \text{th}\,\frac{MT}{2}\,\text{th}\,\frac{T}{2\Theta}}{\text{th}\,\frac{MT}{2} + \text{th}\,\frac{T}{2\Theta}}\,(p_0' + p_2') - p_3' \tag{5.108}$$

Using this expression, we represent the penultimate Hurwitz determinant in the form

$$\Delta_2' = (p_0' + p_2')\left[\frac{1 + \text{th}\,\frac{MT}{2}\,\text{th}\,\frac{T}{2\Theta}}{\text{th}\,\frac{MT}{2} + \text{th}\,\frac{T}{2\Theta}}\,p_2' - p_3'\right] \tag{5.109}$$

Finally, taking into account the last two equalities of (5.104), we reduce this expression to the form

$$\Delta_2' = \frac{Nh_0'(p_0'+p_2')}{M}\left[\frac{\operatorname{th}\frac{MT}{2}\operatorname{th}\frac{T}{2\Theta}\left(1+\operatorname{th}\frac{MT}{2}\operatorname{th}\frac{T}{2\Theta}\right)}{\operatorname{th}\frac{MT}{2}+\operatorname{th}\frac{T}{2\Theta}}\right.$$

$$\left.\times\left(-\alpha_1+\frac{1}{M}+\Theta\right)-\frac{T}{2}\right] \tag{5.110}$$

Since $p_0'<0$ and $p_2'<0$, then, when the inequality

$$\alpha_1 > \frac{1}{M}+\Theta \tag{5.111}$$

is fulfilled, we will have $\Delta_2>0$. From Fig 4.7 it is evident that (5.111) is fulfilled in regions I and III. The conditions for stability are fulfilled in regions I and III and the system has self-oscillating modes corresponding to the periodic solutions there. The form of (5.110) does not allow us to reach a conclusion about the stability of the periodic solution in region IV which corresponds to an intersection from below to above. However, this solution derives from the loop of the relay characteristic. The period T of this solution is small and so also is the quantity σ_1^*. If we now expand the expression in (5.110) in the square brackets as a series in powers of T, then, with an accuracy up to the order T, we obtain the relation

$$\Delta_2' = \frac{-Nh_0'(p_0'+p_2')}{M\Theta+1}\,\frac{T}{2} \tag{5.112}$$

Thus, in this case $\Delta_2'>0$, and consequently the periodic solution also is stable.

We can use (5.49) to construct the oscillations during the half-period T provided we specify the actual numerical values of the system parameters. However we can, with the aid of (5.50), arrive at the general conclusion that, for $T=0$, $x^*(0)=0$. Whence, for reasons of continuity, it follows that for small periods T there correspond small values of the coordinates $x_1^*(0)$, $x_2^*(0), \ldots, x_n^*(0)$ which are elements of the column $x^*(0)$. From this we can conclude that the amplitudes of the corresponding periodic oscillations will also be small quantities.

It follows from these considerations and from the analysis given above that the presence of a loop in the relay characteristic gives rise to self-oscillations in the system in regions I and III. If the width of the loop is small, the self-oscillations have low amplitude and high frequency.

The self-oscillations in region III and the unstable periodic solution in region IV are a consequence of the particular control principle chosen.

5.6 SLIDING MOTION AND THE STABILITY OF THE EQUILIBRIUM POSITION

In relay systems with hard switching, there exist sliding motions. Motions of this type were described in the first section of this chapter for an ideal relay characteristic. During a sliding motion, the image point moves along the switching plane and the relay control element oscillates between its extreme positions with a theoretically infinite frequency. This type of motion continues for a certain interval of time, which may be finite or infinite, and goes on until the image point crosses the boundary of the sliding zone or arrives at the equilibrium position $x=0$. The sliding motion is not given by (5.1). In order to describe this motion it is necessary to redefine the control function $\psi(\sigma)$.

We will redefine $\psi(\sigma)$ in accordance with the following considerations (Note: sliding motions and their conditions of existence have been discussed by Dolgolenko [8] for systems very similar to those investigated in this book, where the author maintains symmetry for the coordinates of the system in the case of sliding motions also. This approach simplifies the solution of the problem and enables one to express the final results in terms of the original parameters in a very simple form. Although at first glance this may seem strange, this approach actually came to the author from a consideration of indeterminate Lagrange factors). During a sliding motion the first equality in (5.7) must be satisfied identically, i.e. in accordance with the second equation (5.1) the argument of the control σ should be equated identically to zero and, consequently, its time derivative $\dot{\sigma}$ also. We shall differentiate the second equation in (5.1), insert in the expression obtained the column $\dot{x}$ found from the first equation in (5.1), and we shall equate the result identically to zero. We then obtain

$$\dot{\sigma}=\gamma\dot{x}=\gamma Px+\gamma h\psi(\sigma)\equiv 0 \tag{5.113}$$

Since in hard switching $\gamma h\neq 0$, solving (5.113) with respect to $\psi(\sigma)$, we obtain the expression

$$\psi(\sigma)\equiv-\frac{\gamma Px}{\gamma h} \tag{5.114}$$

with the help of which we shall redefine the control function in the sliding mode. Substituting $\psi(\sigma)$ from (5.114) into the first equation in (5.1), we obtain the matrix equation

$$\dot{x}=Px-\frac{1}{\gamma h}h\gamma Px \tag{5.115}$$

which is a compact way of writing a system of linear differential equations with constant coefficients.

The motion of a relay system in the sliding mode will be defined by (5.115). If we multiply both sides of (5.115) from the left by the row γ, we obtain the identity $\gamma\dot{x}\equiv 0$. In Section 5.1 it was shown that, in state space, row γ is a vector perpendicular to the family of planes parallel to the switching plane $\sigma=\gamma x=0$. The identity $\gamma\dot{x}\equiv 0$ shows that, by virtue of (5.115), the vector $\dot{x}$ of the state velocity is always perpendicular to the vector γ. However, in that case the image point will always move along a plane parallel to the switching plane which passes through the initial point $x(0)$. In particular, for the initial condition $\sigma(0)=\gamma x(0)=0$, the image point will always move along the switching plane $\sigma=\gamma x=0$.

Thus, (5.115) defines the plane-parallel motion of the image point throughout state space. The sliding motion is determined by the partial integral of this equation which is dependent upon the initial value $x(0)$ satisfying the condition $\gamma x(0)=0$.

Equation (5.115) was obtained by redefining the function $\psi(\sigma)$ in accordance with (5.114). We show below that the equation obtained defines the limiting motion of the image point in the case when the ideal relay characteristic assumes as a limit a loop-shaped characteristic of the backlash type where the width of the loop tends to zero.

We shall turn now to the problem of determining the structure of the solution of (5.115) satisfying for $t=0$ the condition $\gamma x(0)=0$. It will be shown below that the characteristic equation corresponding to this problem automatically has a zero root. Anticipating this result, we will denote the characteristic polynomial through $\lambda\tilde{\Delta}(\lambda)$ and the adjoint matrix through $\tilde{F}(\lambda)$. We shall assume that $\tilde{\lambda}_1, \tilde{\lambda}_2, \ldots, \tilde{\lambda}_{n-1}$ are simple roots of the polynomial $\tilde{\Delta}(\lambda)$ and that none of them are equal to zero. Then, in accordance with (2.126) and (2.131) for $g(t)=h=0$ we can write the solution of (5.115) in the form

$$x(t)=\frac{\tilde{F}(0)}{\tilde{\Delta}(0)}x(0)+\sum_{i=1}^{n-1}e^{\tilde{\lambda}_i t}\frac{F(\tilde{\lambda}_i)}{\tilde{\lambda}_i\tilde{\Delta}'(\tilde{\lambda}_i)}x(0) \tag{5.116}$$

where

$$\left.\begin{aligned}\frac{\tilde{F}(\lambda)}{\lambda\tilde{\Delta}(\lambda)}&=\left(\lambda E-P+\frac{1}{\gamma h}h\gamma P\right)^{-1}\\ \lambda\tilde{\Delta}(\lambda)&=\det\left(\lambda E-P+\frac{1}{\gamma h}h\gamma P\right)\end{aligned}\right\} \tag{5.117}$$

We shall use the canonical representation of P, i.e. we will put $P=KJK^{-1}$ and express γ and h in terms of β and u given by (4.21). Then (5.117) becomes:

$$\left.\begin{aligned} \frac{\tilde{F}(\lambda)}{\lambda\tilde{\Delta}(\lambda)} &= K\left(\lambda E - J + \frac{1}{\beta u}\, u\beta J\right)^{-1} K^{-1} \\ \lambda\tilde{\Delta}(\lambda) &= \det\left(\lambda E - J + \frac{1}{\beta u}\, u\beta J\right) \end{aligned}\right\} \quad (5.118)$$

In the expressions obtained we take the matrix $\lambda E - J$ outside the brackets. Then, obviously, we obtain

$$\left.\begin{aligned} \frac{\tilde{F}(\lambda)}{\lambda\tilde{\Delta}(\lambda)} &= K\left[E + \frac{1}{\beta u}(\lambda E - J)^{-1} u\beta J\right]^{-1} (\lambda E - J)^{-1} K^{-1} \\ \lambda\tilde{\Delta}(\lambda) &= \det(\lambda E - J)\det\left[E + \frac{1}{\beta u}(\lambda E - J)^{-1} u\beta J\right] \end{aligned}\right\} \quad (5.119)$$

Introducing the notation

$$\left.\begin{aligned} u' &= -\frac{1}{\beta u}(\lambda E - J)^{-1} u \\ \beta' &= \beta J \end{aligned}\right\} \quad (5.120)$$

it is not difficult to see that, in the new notation, the term inside the square brackets in (5.119) reduces to the form of the matrix on the right-hand side of (4.131). If now we make use of (4.133) and (4.137) and return to the original notation, we can put (5.119) in the form

$$\left.\begin{aligned} \frac{\tilde{F}(\lambda)}{\lambda\tilde{\Delta}(\lambda)} &= K\left[E - \frac{(\lambda E - J)^{-1} u\beta J}{\beta u\left[1 + \frac{\beta J(\lambda E - J)^{-1} u}{\beta u}\right]}\right](\lambda E - J)^{-1} K^{-1} \\ \lambda\tilde{\Delta}(\lambda) &= [\det(\lambda E - J)]\left[1 + \frac{\beta J(\lambda E - J)^{-1} u}{\beta u}\right] \end{aligned}\right\} \quad (5.121)$$

Taking account of the matrix identity

$$J(\lambda E - J)^{-1} = -E + \lambda(\lambda E - J)^{-1} \quad (5.122)$$

and multiplying out the brackets and cancelling, we can express (5.121) in the form

$$\left.\begin{aligned} \frac{\tilde{F}(\lambda)}{\lambda\tilde{\Delta}(\lambda)} &= K(\lambda E - J)^{-1} K^{-1} + \frac{K(\lambda E - J)^{-1} u\beta K^{-1}}{\lambda\beta(\lambda E - J)^{-1} u} \\ &\quad - \frac{K(\lambda E - J)^{-1} u\beta(\lambda E - J)^{-1} K^{-1}}{\beta(\lambda E - J)^{-1} u} \\ \tilde{\Delta}(\lambda) &= \frac{\Delta(\lambda)\,\beta(\lambda E - J)^{-1} u}{\beta u} = \frac{\tilde{\Delta}_*(\lambda)}{\beta u} \end{aligned}\right\} \quad (5.123)$$

where $\Delta(\lambda)$ denotes a polynomial of the nth degree.

$$\Delta(\lambda) = \det(\lambda E - J) = \prod_{i=1}^{n} (\lambda - \lambda_i) \qquad (5.124)$$

i.e. the characteristic polynomial of the matrix P. A factor of λ has been dropped in the writing of the second equality of (5.123). Consequently the characteristic polynomial for this problem actually has the form $\lambda\tilde{\Delta}(\lambda)$ and one root is automatically equal to zero. The polynomial $\tilde{\Delta}(\lambda)$ is strictly of degree $n-1$ so that the coefficient of λ^{n-1} that it contains is equal to unity. (It is not difficult to see that the leading coefficient of the polynomial $\tilde{\Delta}_*(\lambda)$ is equal to $\beta u = \gamma h \neq 0$.)

Let us form now the columns $\tilde{F}(\tilde{\lambda}_i)\, x(0)$ appearing in the solution of (5.116). Recalling that $\tilde{\Delta}(\tilde{\lambda}_i) = 0$ and that for the sliding mode $\beta K^{-1} x(0) = \gamma x(0) = 0$, from (5.123) it is easy to obtain the expression

$$\tilde{F}(\tilde{\lambda}_i)\, x(0) = -\frac{\tilde{\lambda}_i \Delta(\tilde{\lambda}_i)}{\beta u} K(\tilde{\lambda}_i E - J)^{-1} u\beta (\tilde{\lambda}_i E - J)^{-1} K^{-1} x(0) \qquad (5.125)$$

and, in particular

$$\tilde{F}(0)\, x(0) = 0 \qquad (5.126)$$

From (5.126) it follows that the zero root of the characteristic polynomial does not influence the motion of the system in the sliding mode since the term corresponding to this root is missing from the solution (5.116). Thus, the sliding motion depends on the $n-1$ roots $\tilde{\lambda}_i$ of the polynomial $\tilde{\Delta}(\lambda)$ [see the second formula in (5.123)] and it is given by the expression

$$x(t) = -\sum_{i=1}^{n-1} e^{\tilde{\lambda}_i t} \frac{\Delta(\tilde{\lambda}_i)}{\tilde{\Delta}'_*(\tilde{\lambda}_i)} K(\tilde{\lambda}_i E - J)^{-1} u\beta (\tilde{\lambda}_i E - J)^{-1} K^{-1} x(0) \qquad (5.127)$$

which is obtained from (5.116) with account of (5.125) and (5.126).

Solution (5.127) remains valid up to the time when the condition for the sliding mode (5.7) is fulfilled. The first equality of (5.7) is fulfilled identically. This can be checked once again if we multiply both sides of (5.127) from the left by the row γ, take the second formula of (5.123) into account and assume $\gamma K = \beta$.

The last two inequalities of (5.7) can be written in the form

$$\gamma h < \gamma P x(t) < -\gamma h \qquad (5.128)$$

Inequalities (5.128) can be represented in terms of the matrices βu and J in the form

$$\beta u < \beta J K^{-1} x < -\beta u \qquad (5.129)$$

Let us substitute into (5.129) x from (5.127) and make use of (5.122) with $\lambda=\tilde{\lambda}_i$. Then, allowing for the fact that the $\tilde{\lambda}_i$ are roots of the polynomial $\tilde{\Delta}(\lambda)$ (5.123), we obtain

$$-1<\sum_{i=1}^{n-1} e^{\tilde{\lambda}_i t}\frac{\Delta(\tilde{\lambda}_i)}{\tilde{\Delta}'_*(\lambda_i)}\beta(\tilde{\lambda}_i E-J)^{-1}K^{-1}x(0)<1 \qquad (5.130)$$

In obtaining the conditions (5.130) we cancelled the scalar quantity $\beta u=\gamma h$ and changed the signs of the inequalities since, according to Section 5.1, the sliding modes can exist only when $\gamma h<0$. To obtain formulae suitable for computation it is necessary to express not only the characteristic equation but also the solution and the condition for the sliding mode in terms of the initial parameters.

After substituting into (5.123), (5.127) and (5.130) $\beta=\gamma K$ and $u=K^{-1}h$, we obtain the matrix

$$K(\lambda E-J)^{-1}K^{-1}=(\lambda E-P)^{-1}=\frac{F(\lambda)}{\Delta(\lambda)} \qquad (5.131)$$

Using this expression we obtain the following computational formulae:

for the characteristic polynomial

$$\tilde{\Delta}_*(\lambda)=\gamma F(\lambda)h \qquad (5.132)$$

for the sliding motion

$$x(t)=-\sum_{i=1}^{n-1} e^{\tilde{\lambda}_i t}\frac{F(\tilde{\lambda}_i)}{\Delta(\tilde{\lambda}_i)}h\gamma\frac{F(\tilde{\lambda}_i)}{\tilde{\Delta}'_*(\tilde{\lambda}_i)}x(0) \qquad (5.133)$$

for the sliding mode condition

$$-1<\sum_{i=1}^{n-1} e^{\tilde{\lambda}_i t}\frac{\gamma F(\tilde{\lambda}_i)x(0)}{\tilde{\Delta}'_*(\tilde{\lambda}_i)}<1 \qquad (5.134\text{a})$$

where the $\tilde{\lambda}_i$ are the roots of the polynomial (5.132).

The behaviour of the sliding state trajectories at the boundary of the sliding mode can be deduced as follows. The row γP can be regarded as a vector gradient of the family of parallel planes $\gamma Px=c$, where c is a real parameter. The plane (5.8) belongs to this family. The sign of the expression

$$\gamma P\dot{x}=\gamma P^2x-\frac{1}{\gamma h}\gamma Ph\gamma Px \qquad (5.134\text{b})$$

obtained by multiplying both sides of (5.115) from the left by the row γP determines the direction of the projection of the state velocity vector $\dot{x}$ of the 'sliding' trajectory on to the gradient.

On the bounding straight lines given by the intersection of the switching plane with the planes (5.8), we have

$$\left.\begin{aligned} \gamma P\dot{x} &= \gamma P^2 x + \gamma P h \\ \gamma P\dot{x} &= \gamma P^2 x - \gamma P h \end{aligned}\right\} \qquad (5.134c)$$

The signs of these expressions with respect to the half-lines (5.8b) and (5.8c) are such that, in the first case, the sliding state trajectories are directed inwards towards the sliding zone and, in the second case, outwards. Thus the directions of the three pieces of phase trajectory abutting on the half-lines (5.8b) and (5.8c) from the direction of the sliding zone and half-spaces $\sigma > 0$ and $\sigma < 0$ turn out to be consistent in the sense that they uniquely determine the motion of the image point with respect to the natural direction of time. In the opposite direction the uniqueness of the motion is destroyed. A similar situation obviously holds also for all internal points of the sliding zone.

In conclusion we should note that the image point can leave the sliding zone and move into the half-spaces $\sigma > 0$ or $\sigma < 0$ only by way of the half-lines (5.8c). The left-hand of (5.8c) gives the exit point into the half-space $\sigma > 0$, and the right-hand the exit point into the half-space $\sigma < 0$.

To illustrate these remarks, the interlinking of the pieces of the state trajectories abutting on the sliding zone and its boundaries is shown in Fig 5.3 in three-dimensional state space.

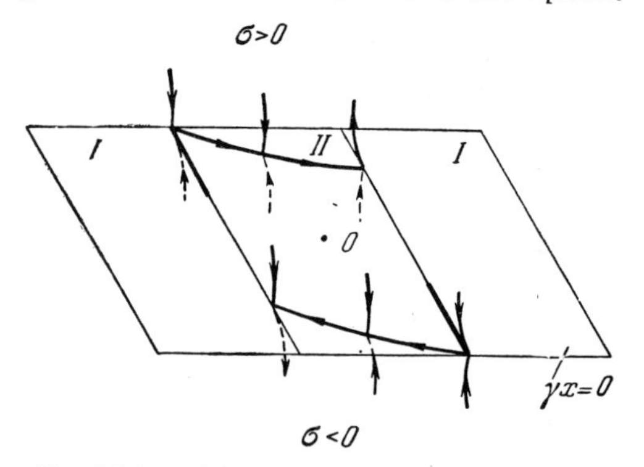

Fig 5.3 Interlinking of state trajectories in the sliding zone and on its boundaries.

Let us now try to determine some of the properties of the sliding motion. Consider a linear system of differential equations in matrix form

$$\dot{x} = Px + ch\gamma x \qquad (c > 0) \qquad (5.135)$$

This equation is obtained from (5.1) if we put $\psi(\sigma)=c\sigma$ and assume $c>0$. We shall determine the structure of the solution (5.135). For this equation, we have

$$\left.\begin{aligned}\frac{\tilde{\tilde{F}}(\lambda)}{\tilde{\tilde{\Delta}}(\lambda)} &= (\lambda E - P - ch\gamma)^{-1}\\ \tilde{\tilde{\Delta}}(\lambda) &= \det(\lambda E - P - ch\gamma)\end{aligned}\right\} \tag{5.136}$$

If, further, we go through the same arguments as before, we can reduce (5.136) to the form

$$\left.\begin{aligned}\frac{\tilde{\tilde{F}}(\lambda)}{\tilde{\tilde{\Delta}}(\lambda)} &= K\left[E+\frac{c(\lambda E-J)^{-1}u\beta}{1-c\beta(\lambda E-J)^{-1}u}\right](\lambda E-J)^{-1}K^{-1}\\ \tilde{\tilde{\Delta}}(\lambda) &= [\det(\lambda E-J)]\left[1-c\beta(\lambda E-J)^{-1}u\right]\end{aligned}\right\} \tag{5.137}$$

Taking the second formula of (5.123), and (5.124) into account we can write $\tilde{\tilde{\Delta}}(\lambda)$ in the form

$$\tilde{\tilde{\Delta}}(\lambda) = c\tilde{\tilde{\Delta}}_*(\lambda) = c\left[\frac{\Delta(\lambda)}{c} - \tilde{\Delta}_*(\lambda)\right] \tag{5.138}$$

where the polynomial $\tilde{\Delta}_*(\lambda)$ is given by (5.123) and (5.132). Then the solution of (5.135) for the case of simple roots $\tilde{\tilde{\lambda}}_i$ of the characteristic polynomial $\tilde{\tilde{\Delta}}(\lambda)$ and consequently, the solution of the polynomial $\tilde{\tilde{\Delta}}_*(\lambda)$ also, can be written in the form

$$x(t) = \sum_{i=1}^{n} e^{\tilde{\tilde{\lambda}}_i t}\frac{\Delta(\tilde{\tilde{\lambda}}_i)}{\tilde{\tilde{\Delta}}'_*(\tilde{\tilde{\lambda}}_i)}K(\tilde{\tilde{\lambda}}_i E - J)^{-1}u\beta(\tilde{\tilde{\lambda}}_i E - J)^{-1}K^{-1}x(0) \tag{5.139}$$

As shown above, in the polynomial $\tilde{\Delta}_*(\lambda)$ the coefficient of λ^{n-1} is equal to $\beta u = \gamma h < 0$. Taking this into account, we obtain in accordance with (5.138) the following limiting equalities:

$$\left.\begin{aligned}&\lim_{c=\infty}\tilde{\tilde{\Delta}}_*(\lambda) = -\tilde{\Delta}_*(\lambda)\\ &\lim_{c=\infty}\tilde{\tilde{\lambda}}_i = \tilde{\lambda}_i, \qquad i=1\ \ 2, \ldots, n-1\\ &\lim_{c=\infty}\tilde{\tilde{\lambda}}_n = -\infty\end{aligned}\right\} \tag{5.140}$$

It is now easy to establish that (5.127) represents the limit of (5.139) as $c\to\infty$. This fact can be used for the direct construction of the solution of (5.127). However, in addition, we can deduce from this the useful result that the sliding mode of a relay system coincides with the motion of the corresponding linear system with an infinitely large gain $(c\to\infty)$. This therefore gives rise to the interesting technical possibility of building control systems analogous to high-gain linear systems.

5.6.1 Stability of the equilibrium position

Investigation of the sliding mode has a direct bearing upon the problem of determining the stability of the equilibrium position of a relay system.

From the form of the solution (5.133) of (5.115) with the constraints imposed by (5.134) it follows that a sliding motion will be asymptotically stable if the real parts of all the roots $\tilde{\lambda}_1, \tilde{\lambda}_2, \ldots, \tilde{\lambda}_{n-1}$ of the polynomial $\tilde{\Delta}_*(\lambda)$ (5.132) are less than zero. From the point of view of Lyapunov theory and the geometry of state space, this fact can be described in the following way. Suppose on the switching plane inside the sliding zone we choose an arbitrary neighbourhood (ε) of the point $x=0$ which may be made as small as we please. Then, on this same plane, it is always possible to find another neighbourhood (δ) such that a phase trajectory of the sliding motion which starts from the neighbourhood (δ) cannot pass outside the limits of the first neighbourhood, and such that an image point sliding along the switching plane will asymptotically approach the equilibrium position, i.e. it will tend towards the point $x=0$.

It will be recalled that, for autonomous relay systems of the kind we are considering, motion along phase trajectories butting against points of the sliding zone from the direction of the half-spaces $\sigma>0$ and $\sigma<0$ are directed towards each other and that the trajectories themselves are independent of time.

It is then obvious that in the state space of the system one can find a neighbourhood of the point $x=0$ (that is, a neighbourhood that will include the points of the half-spaces $\sigma>0$ and $\sigma<0$) such that a state trajectory which begins in this neighbourhood will necessarily pass inside the neighbourhood (δ) of the switching plane and, furthermore, the image point will execute a sliding motion that will cause it to approach asymptotically the equilibrium position.

Thus, the asymptotic stability of the equilibrium position of a relay system follows from the asymptotic stability of the sliding motion. The problem of the stability of the equilibrium position of a relay system possessing a sliding mode reduces to the Hurwitz problem for the polynomial $\tilde{\Delta}_*(\lambda)$ (5.132). A sliding mode exists when $\gamma h<0$ and the polynomial $-\tilde{\Delta}_*(\lambda)$(5.132) is limiting for the polynomial $\tilde{\tilde{\Delta}}_*(\lambda)$ (5.138) as $c\to\infty$.

It follows therefore that the equilibrium position of a relay system with hard switching will be asymptotically stable if $\gamma h<0$ and the limiting characteristic polynomial $\tilde{\Delta}_*(\lambda)$ of the corresponding linear system will be Hurwitzian (that is, all its roots $\tilde{\lambda}_i$ will lie to the left of the imaginary axis).

On the other hand, for $\gamma h < 0$ and for a Hurwitzian polynomial $\Delta_*(\lambda)$, the corresponding linear system will be stable for large gains $c > 0$ – see (5.140) – and for $\gamma h > 0$ it is not difficult to show that it will be unstable under the same conditions.

The stability condition can also be formulated as follows: the equilibrium position of a relay system will be stable if and only if the corresponding linear system obtained by replacing the relay characteristic $\psi(\sigma)$ by the line $c\sigma$ is stable for $c > c^*$, where c^* is an arbitrarily small positive number.

In this formulation the stability condition is universal. It is valid for relay systems with soft switching as well, since such systems may be regarded as a limiting case of a hard switching relay system with $\gamma h \to 0$.

For systems with soft switching one can consider also the problem of finding the conditions which, taken in conjunction with the requirement that the limiting polynomial $\tilde{\Delta}_*(\lambda)$ shall be Hurwitzian, will ensure the stability of a high-gain linear system $(c > c^*)$. This problem was solved in a general form by Tsipkin [25] who investigated the stability of relay systems by treating the relay characteristic as a limiting piece-wise linear characteristic having an infinitely large slope over the linear section.

5.7 STABILITY AND SLIDING MODES OF AIRCRAFT STABILISING SYSTEMS

In this section we shall apply the general results obtained in the previous section to particular problems. As an example let us consider the various kinds of relay systems for stabilising the course of an aircraft. The schematic diagrams of such systems were discussed in Section 2.2.

a) Automatic pilot controlled by angle, angular velocity and angular acceleration signals allowing for the inertia of the servo-motor

In Section 5.5 we investigated the possibility that oscillations might arise in an aircraft stabilising system of this type, and we derived the corresponding equations of motion in the form of (5.96). The open loop control system for this case is given by the continuous part of (5.96) and it coincides with the corresponding part of the discontinuous stabilisation system considered in Section 4.10. Since (4.223) holds, we are dealing with a relay system with soft switching, and in a system with soft switching the sliding zone degenerates into a straight line. The stability of the equilibrium position is the same as the stability of the corresponding linear systems ($\psi(\sigma) = c\sigma$) with an infinitely large gain $c > 0$. The characteristic equation for such a linear system is given by (4.237) where the restriction (4.236) imposed on the coefficient c is now removed.

The question of whether or not the system is stable for the different values of c can be decided by inspecting the diagram given in Fig 4.7. It is clear that, in regions I and IV, the corresponding linear system will be stable for sufficiently large $c > 0$; in regions II and III, on the other hand, it will be unstable under equivalent conditions. We can therefore deduce that the equilibrium position of a relay stabilisation system will be asymptotically stable in the Lyapunov sense in regions I and IV lying below the straight line (4.245). In regions II and III lying above this line, the equilibrium position will be unstable. Stability in this sense is sometimes called stability in the small.

To determine stability in the large it is necessary to find in state space a limiting region such that any motion starting out from this region converges towards the origin of the coordinates defining the equilibrium position of the system. Andronov and Bautin [2], investigating the decomposition of the whole of state space into trajectories for the case of a particular automatic pilot problem, showed that the problem of stability is closely connected with the existence of periodic modes of behaviour. An unstable periodic oscillation mode exists in region IV (Fig 4.7). It is to be expected that the separatrix surface which divides the phase space into two parts will pass through the closed state trajectory corresponding to the periodic mode. The part of the state space which contains the origin of the coordinates will define the region of stability in the large.

In region I periodic modes do not exist, so that here stability in the large for all initial deviations should follow from stability of the equilibrium position in the Lyapunov sense.

In region III an oscillation mode does exist and the equilibrium position is unstable in the Lyapunov sense. Self-oscillations will be excited in the system and these represent the only stable mode of motion that is present.

In region I one expects the system to be unstable in the large for all initial deviations from the controlled equilibrium position.

To find an exact solution of the problem of stability in the large and of the behaviour of the system throughout state space is a very complex problem; the arguments given above are therefore of a rather heuristic nature.

However, the attempt to connect together the individual results and to draw up on that basis a general picture of the behaviour of the system is an important and very useful part of the over-all task of solving practical problems.

b) Automatic pilot controlled by angle, angular velocity and angular acceleration signals, with a fixed-speed servo-motor

This stabilising system possesses hard switching so that a sliding mode will exist. Let us investigate the sliding mode and the stability of the equilibrium position.

The equations of motion have the form

$$\left.\begin{aligned} &\ddot{\varphi} + M\dot{\varphi} = -N\eta \\ &\dot{\eta} = h_2'\psi(\sigma) \\ &\sigma = \varphi + \alpha_1\dot{\varphi} + \alpha_2\ddot{\varphi} \end{aligned}\right\} \qquad (5.141)$$

They are obtained from (5.96) by putting $\Theta = 0$. In normal form these equations can be rewritten

$$\left.\begin{aligned} &\dot{x}_1 = x_3 \quad \dot{x}_2 = h_2'\psi(\sigma) \quad \dot{x}_3 = -Nx_2 - Mx_3 \\ &\sigma = x_1 - N\alpha_2 x_2 + (\alpha_1 - M\alpha_2)\, x_3 \end{aligned}\right\} \qquad (5.142)$$

where x_1, x_2 and x_3 are given by (4.217) and the argument of the control σ obviously has the same form as it has in (4.218).

For (5.142), we can write

$$P = \begin{Vmatrix} 0 & 0 & 1 \\ 0 & 0 & 0 \\ 0 & -N & -M \end{Vmatrix} \qquad (5.143)$$

$$h = \begin{Vmatrix} 0 \\ 1 \\ 0 \end{Vmatrix} h_2' \qquad (5.144)$$

$$\gamma = \| 1 \quad -N\alpha_2 \quad (\alpha_1 - M\alpha_2) \| \qquad (5.145)$$

Multiplying column (5.144) from the left by row (5.145), we obtain

$$\gamma h = -Nh_2'\alpha_2 \qquad (5.146)$$

Since all the letter coefficients of (5.141) determine positive quantities then, in accordance with (5.146), $\gamma h < 0$. According to the general theory, a sliding mode exists for $\gamma h < 0$. The motion in the sliding zone is given by (5.132)-(5.134). For the matrix P (5.143), we have

$$\lambda E - P = \begin{Vmatrix} \lambda & 0 & -1 \\ 0 & \lambda & 0 \\ 0 & N & \lambda + M \end{Vmatrix} \tag{5.147}$$

$$(\lambda E - P)^T = \begin{Vmatrix} \lambda & 0 & 0 \\ 0 & \lambda & N \\ -1 & 0 & \lambda + M \end{Vmatrix} \tag{5.148}$$

$$F(\lambda) = \begin{Vmatrix} \lambda(\lambda + M) & -N & \lambda \\ 0 & \lambda(\lambda + M) & 0 \\ 0 & -\lambda N & \lambda^2 \end{Vmatrix} \tag{5.149}$$

$$\left.\begin{aligned} \Delta(\lambda) &= \det(\lambda E - P) = \lambda^2(\lambda + M) \\ \lambda_1 &= -M, \quad \lambda_2 = \lambda_3 = 0 \end{aligned}\right\} \tag{5.150}$$

Further, from (5.144), (5.145) and (5.149) we form the expressions

$$F(\lambda) h = \begin{Vmatrix} -N \\ \lambda(\lambda + M) \\ -\lambda N \end{Vmatrix} h_2' \tag{5.151}$$

$$\gamma F(\lambda) = \| \lambda(\lambda + M) \quad N(1 + \lambda\alpha_1 + \lambda^2\alpha_2) \quad \lambda(1 + \lambda\alpha_1 - \lambda M\alpha_2) \| \tag{5.152}$$

and finally [see (5.132)]

$$\gamma F(\lambda) h = \tilde{\Delta}_*(\lambda) = -(\alpha_2\lambda^2 + \alpha_1\lambda + 1) N h_2' \tag{5.153}$$

Differentiating (5.153) with respect to λ, we obtain

$$\tilde{\Delta}_*'(\lambda) = -(2\alpha_2\lambda + \alpha_1) N h_2' \tag{5.154}$$

The roots $\tilde{\lambda}_1$ and $\tilde{\lambda}_2$ of the polynomial $\tilde{\Delta}_*(\lambda)$ (5.153) are given by

$$\tilde{\lambda}_{1,2} = \frac{-\alpha_1 \pm \sqrt{\alpha_1^2 - 4\alpha_2}}{2\alpha_2} \tag{5.155}$$

For $\alpha_1^2 > 4\alpha_2$ we have two negative roots, while for $\alpha_1^2 < 4\alpha_2$ we have two complex conjugate roots with negative real parts. Thus, the sliding motion together with the equilibrium position of the system will be asymptotically stable.

Using (5.150)-(5.154) we can represent (5.133) and (5.134), which define the sliding mode and its condition for existence, in the respective forms

$$\left\| \begin{matrix} x_1(t) \\ x_2(t) \\ x_3(t) \end{matrix} \right\| = \sum_{i=1}^{2} e^{\tilde{\lambda}_i t} \frac{\tilde{\lambda}_i(\tilde{\lambda}_i + M)\, x_1(0) + \tilde{\lambda}_i\,(1 + \tilde{\lambda}_i \alpha_1 - \tilde{\lambda}_i M \alpha_2)\, x_3(0)}{\tilde{\lambda}_i^2 N\,(2\tilde{\lambda}_i \alpha_2 + \alpha_1)(\tilde{\lambda}_i + M)} \times \left\| \begin{matrix} -N \\ \tilde{\lambda}_i\,(\tilde{\lambda}_i + M) \\ -\tilde{\lambda}_i N \end{matrix} \right\| \qquad (5.156)$$

and

$$-h_2' < \sum_{i=1}^{2} e^{\tilde{\lambda}_i t} \frac{\tilde{\lambda}_i(\tilde{\lambda}_i + M)\, x_1(0) + \tilde{\lambda}_i\,(1 + \tilde{\lambda}_i \alpha_1 - \tilde{\lambda}_i M \alpha_2)\, x_3(0)}{N\,(2\tilde{\lambda}_i \alpha_2 + \alpha_1)} < h_2' \qquad (5.157)$$

[By virtue of (5.153)-(5.155) the coefficient of $x_2(0)$ vanishes and does not therefore appear in (5.156) and (5.157).] The coefficient h_2' specifies the constant velocity of the servo-motor. The roots $\tilde{\lambda}_1$ and $\tilde{\lambda}_2$ are given by (5.155); they do not depend on h_2'. Their values are determined entirely by the coefficients α_1 and α_2, which determine the structure of the argument of the control σ.

The sliding mode does not depend on the speed of the servo-motor. However, (5.157) becomes less restrictive as the gain h_2' is increased. Inequalities (5.157) are linear expressions with respect to the initial values of the coordinates $x_i(0)$, so they can be written in the form

$$-h_2' < \sum_{j=1}^{3} a_j(t)\, x_j(0) < h_2' \qquad (5.158)$$

or

$$\left(\sum_{j=1}^{3} a_j(t)\, x_j(0) \right)^2 < h_2'^2 \qquad (5.159)$$

where the functions $a_j(t)$ in turn depend linearly on the exponentials $e^{\tilde{\lambda}_1 t}$ and $e^{\tilde{\lambda}_2 t}$. With the help of the Schwartz-Bunyakovskii inequality (see, for example [11]) we will rewrite (5.159) in the form [i.e. we will make inequality (5.159) more stringent]

$$\left(\sum_{j=1}^{3} a_j(t)\, x(0) \right)^2 \leqslant \left(\sum_{j=1}^{3} a_j^2(t) \right) \left(\sum_{j=1}^{3} x_j^2(0) \right) \leqslant R^2 \sum_{j=1}^{3} x_j^2(0) < h_2'^2 \qquad (5.160)$$

where R^2 is a positive number equal to the maximum of the function

$\left(\sum_{j=1}^{3} a_j(t)\right)^2$. We know that this maximum exists since, for a stable system, the above function is bounded and continuous in the interval $0 \leqslant t < \infty$; when $t \to \infty$ it tends to zero. The number R^2 does not depend on the coefficient h_2'. The point with the coordinates $x_i(0)$ is chosen on the switching plane inside the sliding zone.

On the switching plane, consider a circle of radius h_2'/R with the coordinate origin as centre. For the problem we are considering the coordinates $x_j(0)$ of the points of this circle must satisfy the conditions

$$\left.\begin{aligned} & x_1^2(0) + x_2^2(0) + x_3^2(0) < \left(\frac{h_2'}{R}\right)^2 \\ & x_1(0) - N\alpha_2 x_2(0) + (\alpha_1 - M\alpha_2)\, x_3(0) = 0 \end{aligned}\right\} \tag{5.161}$$

As a consequence of the inequalities (5.158)-(5.160), the conditions in (5.161) are sufficient to ensure that the image point does not pass outside the limits of the sliding zone. With an increase of h_2' the radius of the circle (5.16) increases and consequently the region of initial values $x_j(0)$ for which the motion possesses the above property becomes larger.

Consider now the following qualitative arguments. As h_2' increases, the number of state trajectories abutting on the circle (5.161) from the direction of the half-spaces $\sigma > 0$ and $\sigma < 0$ also increases. As a consequence, in the state space of this system, the neighbourhood of the zero point from which the motion will be drawn into the sliding mode is enlarged. Motions which start out from a section of such a neighbourhood will be drawn into the sliding mode during the first approach of the image point to the switching plane. The sliding motion becomes in effect the operating mode of the system if, as a result of energy, physical or some other restriction, the initial deviations are unable to take the system beyond the limits of such a neighbourhood. In this case we say that the relay system has been rendered linear through the agency of the sliding mode.

c) Automatic pilot with a fixed-speed servo-motor, controlled by angle and angular velocity signals

The equations of motion can be obtained from (5.141) by putting $\alpha_2 = 0$. In accordance with (5.146), with $\alpha_2 = 0$ we have $\gamma h = 0$, showing that this stabilisation system is a relay system with soft switching. The characteristic equation of the corresponding linear system can be obtained directly from (5.151) with $\alpha_2 = 0$ and $\psi(\sigma) = c\sigma$, or from (4.237) by putting $\Theta = \alpha_2 = 0$. We therefore have

$$\lambda^3 + M\lambda^2 + Nh_2'c\alpha_1\lambda + Nh_2'c = 0 \quad (5.162)$$

The Hurwitz conditions reduce to

$$Nh_2'cM\left(\alpha_1 - \frac{1}{M}\right) > 0 \quad (5.163)$$

Figure 5.4 is the stability diagram, the shaded area being the region of stability. The boundary of the region of stability is the hyperbola

$$\alpha_1 = \frac{1}{M} \quad (5.164a)$$

For points lying above and below hyperbola (5.164a), a linear system will be, respectively, stable and unstable for all values of the gain $c > 0$, so that the controlled position of a relay stabilising system will be stable in the shaded region only of Fig 5.4.

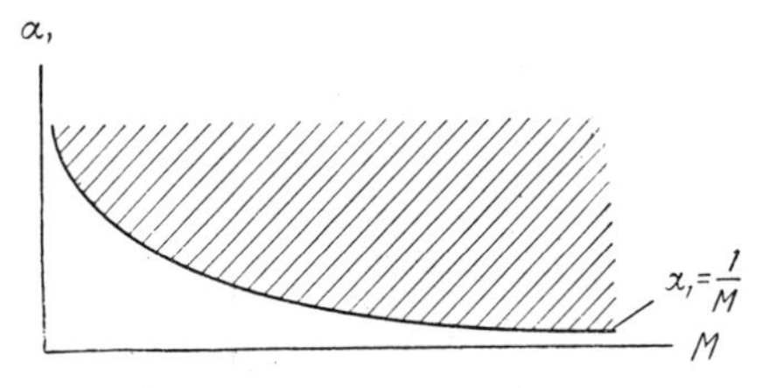

Fig 5.4 Diagram illustrating the stability of the periodic modes.

When $\Theta = \alpha_2 = 0$, the equation of periods (5.97) reduces to the form

$$\frac{Nh_2'}{M^2}\left(\alpha_1 - \frac{1}{M}\right)\left(\frac{MT}{2} - \text{th}\,\frac{MT}{2}\right) = \sigma_1^* \quad (5.164b)$$

Since, when $y > 0$ the inequality $y - \text{th}\,y > 0$ holds, the equation of periods (5.164b) has no positive roots for $\sigma_1^* = 0$. Thus, for an ideal relay characteristic, symmetric periodic modes do not exist in the stabilising system we are considering. One can therefore expect that a relay stabilising system operating in the shaded region will be stable under all initial deviations.

As has already been demonstrated, the stabilising system we are discussing now is a relay system with soft switching, so that no sliding zone can exist in it. Such a zone can be obtained if the control element is encompassed by a negative feedback relay loop via an inertial element as, for example, in the automatic pilot illustrated by the schematic diagram of Fig 1.6.

d) Automatic pilot incorporating relay negative feedback, controlled by angle and angular velocity signals

The equations of motion of this type of system can be obtained from (1.9) by putting $\alpha_2 = a^{-1} = 0$. For the free-motion of the system we obtain an equation in the form

$$\left.\begin{aligned} \ddot{\varphi} + M\dot{\varphi} &= -N\eta \\ \dot{\eta} &= h_2'\varphi(\sigma) \\ T_c\dot{\mu} + \mu &= h_3'\psi(\sigma) \\ \sigma &= \varphi + \alpha_1\dot{\varphi} - b\mu \end{aligned}\right\} \tag{5.165}$$

Introducing the notation

$$x_1 = \varphi \quad x_2 = \eta \quad x_3 = \dot{\varphi} \quad x_4 = \mu \tag{5.166}$$

we can rewrite (5.165) in normal form

$$\left.\begin{aligned} \dot{x}_1 = x_3 \quad \dot{x}_2 = h_2'\psi(\sigma) \quad \dot{x}_3 &= -Nx_2 - Mx_3 \\ \dot{x}_4 = -\frac{1}{T_c}x_4 + \frac{h_3'}{T_c}\psi(\sigma) \\ \sigma = x_1 + \alpha_1 x_3 - bx_4 \end{aligned}\right\} \tag{5.167}$$

For (5.167), column h and row γ have the form

$$h = \begin{Vmatrix} 0 \\ h_2' \\ 0 \\ \dfrac{h_3'}{T_c} \end{Vmatrix} \qquad \gamma = \| 1 \ \ 0 \ \ \alpha_1 \ \ -b \| \tag{5.168}$$

With account of (5.168), we obtain

$$\gamma h = -\frac{bh_3'}{T_c} \tag{5.169}$$

Since all the letter coefficients of (5.165) specify positive quantities we have, in accordance with (5.169), $\gamma h < 0$. Thus, a sliding zone exists in this stabilising system which depends on the parameters of the relay negative feedback loop.

Column h (5.168) depends on the coefficients h_2' and h_3'. These coefficients contain a common factor representing the output voltage of the relay amplifier illustrated in Fig 1.5. However, the roots $\tilde{\lambda}_i$ of the polynomial $\tilde{\Delta}_*(\lambda)$ do not depend on the common factor of the elements of the column h which, obviously, enters as a scalar multiplier on the right-hand side of (5.132). This multiplier will be in the denominator of the expression appearing in (5.134). An increase in the magnitude of this multiplier means that (5.134) imposes a less rigid restriction on the choice of an initial value $x(0)$ that will not take the system outside the limits of the sliding zone. By suitably choosing the value of this multiplier, the sliding mode can be made the operating mode of the relay stabilising system within the meaning we gave to this concept when discussing the second example of the present section.

It is easy to demonstrate in general form that a sliding zone can always be obtained by suitably choosing the parameters for the relay negative feedback system. However, for a proper choice of negative feedback parameters it is necessary to construct and analyse the sliding motion in accordance with the general formulae.

5.8 RELAY CHARACTERISTICS WITH A HYSTERESIS LOOP

Relay characteristics with a hysteresis loop are illustrated in the second and third graphs of Table 1.1, where the hysteresis is presumed to arise from backlash or dry friction in the coupling of the drive mechanism. These characteristics enable us to trace out the values assumed by the control function $\psi(\sigma)$ as a function of the magnitude and direction of change of the argument of the control σ.

The case of a drive mechanism with backlash is included in the discussion because it leads to a better understanding of the nature of the step-wise transition or jump in the value of the function $\psi(\sigma)$ that occurs at points corresponding to the corners of the relay characteristic. For $\sigma > \sigma_1^*$ and $\sigma < -\sigma_1^*$ we have, respectively, $\psi(\sigma) = 1$ and $\psi(\sigma) = -1$. In the interval $-\sigma^* < \sigma < \sigma_1^*$ the values of the control function $\psi(\sigma)$ depend on the previous history of variation of the argument of the control σ. The control function maintains that value (1 or –1) for which the argument of the control falls inside the above interval. A sudden jump in the value of the function $\psi(\sigma)$ from -1 to 1, or vice versa, occurs at points σ_1^* and $-\sigma_1^*$, respectively. In contrast to an ideal relay characteristic, this transition will occur if the sign of the derivative $\dot{\sigma}$ at these points, as determined from the magnitude of $\psi(\sigma)$ for which the changeover occurs, is the same as the sign of the finite increment of the function during the process of this changeover. However, the character of the subsequent motion depends substantially upon the sign acquired by $\dot{\sigma}$ after the changeover in the value of $\psi(\sigma)$ has taken place.

Suppose that, during the changeover in the values of $\psi(\sigma)$, the sign of $\dot{\sigma}$ does not change. In that case normal switching occurs. If, during the changeover the sign of $\dot{\sigma}$ is reversed, then after the switching the argument of the control σ begins to change in the reverse direction. The argument of the control will obviously vary within the closed interval $-\sigma_1^* \leqslant \sigma \leqslant \sigma_1^*$, when the above situation will take place at a series of successive switching points. For example, in the case of the second diagram in Table 1.1, the argument of the control will execute oscillations inside the backlash of the drive mechanism.

For $\psi(\sigma)=1$ and $\psi(\sigma)=-1$, the first equation in (5.1) is in the forms (5.2) and (5.3), respectively. If these equations are used to determine $\dot{\sigma}$, and σ is expressed in accordance with (5.1), the switching conditions can be written in the following way:

$$\gamma x=\sigma_1^* \quad \gamma P x-\gamma h>0 \quad \gamma P x+\gamma h>0 \tag{5.170}$$

and

$$\gamma x=\sigma_1^* \quad \gamma P x-\gamma h>0 \quad \gamma P x+\gamma h<0 \tag{5.171}$$

for a transition from $\psi(\sigma)=-1$ to $\psi(\sigma)=1$ and

$$\gamma x=-\sigma_1^* \quad \gamma P x+\gamma h<0 \quad \gamma P x-\gamma h<0 \tag{5.172}$$

and

$$\gamma x=-\sigma_1^* \quad \gamma P x+\gamma h<0 \quad \gamma P x-\gamma h>0 \tag{5.173}$$

for a transition from $\psi(\sigma)=1$ to $\psi(\sigma)=-1$. Expressions (5.170) and (5.172) determine the conditions for normal switching. When conditions (5.171) and (5.173) are fulfilled, the argument of the control oscillates inside the loop of the relay characteristic.

It is convenient to give a geometrical description of the results obtained, and for this purpose we shall make use of the idea of a three-dimensional state space. Figure 5.5 illustrates in three-dimensional state space the half-planes $\gamma x=\sigma_1^*$ and $\gamma x=-\sigma_1^*$ bounded by the state (5.8). These half-planes divide the state space into regions in which $\psi(\sigma)=1$ or $\psi(\sigma)=-1$; inside the boxes formed by the intersection of the four planes the control function can assume the values $\psi(\sigma)=1$ and $\psi(\sigma)=-1$. The half-planes $\gamma x=\sigma_1^*$ and $\gamma x=-\sigma_1^*$ are half-switching planes. The state trajectories abutting on the half-planes $\gamma x=\sigma_1^*$ and $\gamma x=-\sigma_1^*$ from both sides to the right of the first plane and to the left of the second plane (5.8), respectively, have identical directions. The portions of the half-planes $\gamma x=\sigma_1^*$ and $\gamma x=-\sigma_1^*$ shown here determine the region of normal switching.

Certain trajectories corresponding to $\psi(\sigma)=1$ penetrate inside the boxes during their common motion from above to below and they end on the lower face of the box, since the trajectories that cut this face from above, which correspond to $\psi(\sigma)=-1$, have opposite directions at these points. For a similar reason,

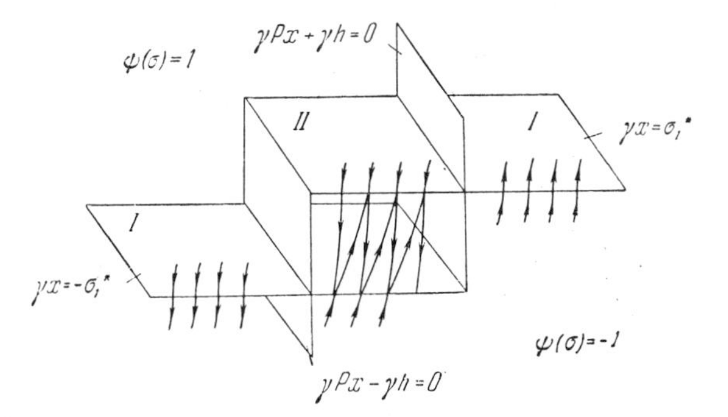

Fig 5.5 State space for a backlash type of relay characteristic. I – normal switching region; II – upper face of 'box'.

certain trajectories corresponding to $\psi(\sigma)=-1$ will end on the upper face. During its motion along these trajectories, the image point will execute oscillations between the lower and upper faces of the box. In this case the image point can pass outside the box only by going through a side face or through certain portions of its edges.

The motion we have just considered corresponds to a alternation of the switching conditions (5.171) and (5.173). We have already discussed in Section 2.2 the motion of a relay system with a hysteresis characteristic of the backlash type in the normal switching regions, since the switching conditions (5.10) and (5.11) which appear there are identical with expressions (5.170) and (5.172). We shall therefore consider now a motion in which successive switchings are determined by (5.171) and (5.173). For this purpose we shall suppose that σ_1^* is a sufficiently small quantity. The justification for this supposition is the fact that σ_1^* determines the half-width of the backlash or friction in the drive mechanism couplings which, in precision equipment, should be small quantities.

Suppose for $t=0$ the initial value of the matrix coordinate $x(0)$ satisfies the switching conditions (5.171). Then, in the interval of time $0 \leqslant t \leqslant T_1$ (up to the next switching), the motion is given by (5.2), whose solution is obtained from (2.129) with $t_0=0$. Assuming in addition that in (2.129) $\tau=T_1$, we obtain

$$x(T_1)=KM(T_1)K^{-1}x(0)+KN(T_1)K^{-1}h \tag{5.174}$$

Suppose the matrix coordinate satisfies switching conditions (5.173). In that case, applying similar arguments, putting $t_0=T_1$ and $\tau=T_2$ in (2.129) and replacing h by $-h$, we obtain

$$x(T_1+T_2)=KM(T_2)K^{-1}x(T_1)-KN(T_2)K^{-1}h \tag{5.175}$$

where the coordinate $x(T_1+T_2)$ satisfies the original switching conditions (5.171). In accordance with (4.198) for small values of the scalar quantity ξ, we will have

$$\left. \begin{aligned} M(\xi) &= M(0) + \xi \dot{M}(0) = E + \xi J \\ N(\xi) &= N(0) + \xi \dot{N}(0) = \xi E \end{aligned} \right\} \tag{5.176}$$

Putting successively $\xi = T_1$ and $\xi = T_2$, we substitute the matrices (5.176) into (5.174) and (5.175). Then, with account of the matrix relations $KK^{-1} = E$ and $KJK^{-1} = P$, we obtain

$$x(T_1) = (E + T_1 P)\, x(0) + T_1 h \tag{5.177}$$

and

$$x(T_1 + T_2) = (E + T_2 P)\, x(T_1) - T_2 h \tag{5.178}$$

We multiply both sides of (5.177) and (5.178) from the left by the row γ. Then, remembering that the columns $x(0)$ and $x(T_1 + T_2)$ satisfy (5.171), and that the column $x(T_1)$ satisfies (5.173), we obtain

$$T_1 = -\frac{2\sigma_1^*}{\gamma P x(0) + \gamma h} > 0 \quad T_2 = \frac{2\sigma_1^*}{\gamma P x(T_1) - \gamma h} > 0 \tag{5.179}$$

According to (5.179), the intervals of time T_1 and T_2 between two successive switchings on the faces of the 'box' are proportional to the hysteresis parameter σ_1^*. We now substitute into (5.178) the value $x(T_1)$ from (5.177) and, in multiplying out the expression obtained, we neglect terms containing the factor $T_1 T_2$. We then obtain

$$x(T_1 + T_2) = [E + (T_1 + T_2) P]\, x(0) + (T_1 - T_2)\, h \tag{5.180}$$

Let us multiply both sides of this equality from the left by the row γ. Taking into account the fact that the columns $x(0)$ and $x(T_1 + T_2)$ satisfy the first equality in (5.171), we obtain

$$\frac{T_1 - T_2}{T_1 + T_2} = -\frac{\gamma P x(0)}{\gamma h} \tag{5.181}$$

From (5.180) and (5.181) it is easy to obtain the expression

$$\frac{x(T_1 + T_2) - x(0)}{T_1 + T_2} = P x(0) - \frac{1}{\gamma h} h \gamma P x(0) \tag{5.182}$$

The matrix expression (5.182) establishes in first approximation a relation between any two successive points lying on the upper face of the box when, in the course of the motion, the successive switchings are given by (5.171) and (5.173). If the hysteresis parameter σ_1^* tends to zero, then T_1 and T_2 will also tend to zero in accordance with (5.179). In this case the left-hand side of

(5.182) will tend towards the derivative $\dot{x}(0)$. Thus, in the limit when $\sigma_1^* = 0$, (5.182) becomes a linear differential equation with constant coefficients of the form

$$\dot{x} = Px - \frac{1}{\gamma h} h\gamma Px \tag{5.183}$$

This equation is identical to (5.115) which describes the sliding motion for an ideal relay characteristic. It is quite natural to regard the ideal relay characteristic as the limiting case of a hysteresis characteristic of the backlash type obtained by letting the parameter σ_1^* tend to zero. This then, is the fundamental explanation for the equation of the sliding motion (5.115). Since, in the limit when $\sigma_1^* = 0$, we also have, according to (5.179) $T_1 = T_2 = 0$, the frequency of switching of the relay control element in the sliding mode becomes infinite.

5.8.1 Relay characteristics with dead zone

When it is necessary to make allowance for a dead zone, expressions can be derived in closed form which determine the periodic free oscillation modes and the sliding motions of the system. A relay characteristic with a dead zone is illustrated in the fourth graph of Table 1.1; the fifth graph represents a characteristic with backlash and a dead zone.

For characteristics with a dead zone there exist two parallel switching planes $\gamma x = \sigma_2^*$ and $\gamma x = -\sigma_2^*$ which divide the phase space into three parts: $\gamma x < -\sigma_2^*$, $-\sigma_2^* < \gamma x < \sigma_2^*$ and $\gamma x > \sigma_2^*$. The control function in these parts assumes the values -1, 0 and 1. The three parallel planes $\gamma Px - \gamma h = 0$, $\gamma Px = 0$ and $\gamma Px + \gamma h = 0$ divide the parts of state space in the sequence indicated into regions in which the projection $\gamma \dot{x}$ of the velocity vector $\dot{x}$ to the vector γ has a fixed sign. A sliding zone can exist on each of these switching planes. The boundaries of these zones are given by parallel straight lines which are the lines of intersection of the plane $\gamma x = -\sigma_2^*$ with the planes $\gamma Px - \gamma h = 0$ and $\gamma Px = 0$, and of the plane $\gamma x = \sigma_2^*$ with the planes $\gamma Px = 0$ and $\gamma Px + \gamma h = 0$. On those portions of the planes $\gamma x = -\sigma_2^*$ and $\gamma x = \sigma_2^*$ lying outside the sliding zone, the conditions of normal switching are fulfilled. If σ_2^* tends to zero, then in the limit the two poles of the sliding mode of the switching phase abut on to each other along the common boundary, thus forming the sliding zone for an ideal relay characteristic considered earlier.

The equations for the sliding motion (5.183) are valid also for relay systems which have a dead zone. The same is also true of the characteristic polynomial $\tilde{\Delta}_x(\lambda)$ and the adjoint matrix $\tilde{F}(\lambda)$ given by (5.123). However, the form of the solution is somewhat different since the second term on the right-hand side of the first of the formulae in (5.123) does not vanish when the columns $\tilde{F}(\tilde{\lambda}_i)\,x(0)$ are being formed by virtue of the fact that the scalar quantity $\beta K^{-1}x(0) = \gamma x(0)$ is now equal to σ_2^* or $-\sigma_2^*$.

The final results for determining the symmetric periodic oscillations now assume a more cumbersome form since, in the present case there are two normal switchings of the control element during a half-period. In constructing the characteristic equation a new problem now emerges: it is necessary to multiply out the determinant of the special matrix which is equal to the sum of a unit matrix and two degenerate matrices each representing the product of a column and a row. A similar problem arises also when analysing the steady-state motions of a relay system acted upon by a constant external disturbance. This problem will be discussed in Section 6.2.

The computational formulae needed for finding the periodic solutions and for investigating their stability have been given in some earlier publications by the author [30].

5.9 PERIODIC MOTIONS CONTAINING PORTIONS OF THE SLIDING MODE

In relay systems with hard switching, periodic motions containing portions only of the sliding mode can arise. A closed state trajectory corresponding to such a periodic motion contains trajectories located not only in the half-spaces $\sigma > 0$ and $\sigma < 0$ but in the sliding zone also. Symmetric periodic oscillations occur in this class of relay systems also.

Let us now seek in general form the periodic solutions for such symmetric oscillations.

Suppose that at $t = 0$ the image point with coordinate $x(0)$ is in the sliding zone and that after a lapse of time T_1 it encounters the bounding half-line of this zone which it must cross in order to reach the half-space $\sigma > 0$. Furthermore, let us suppose that after an interval of time T_2 the image point has returned once more to the sliding zone and falls on to the point symmetric with respect to the initial point, i. e. on to the point with matrix coordinate $-x(0)$. Under these circumstances, a symmetric periodic mode of oscillations will arise in the system. The sliding mode is given by (5.7). The image point moves out of the sliding zone into the half-space $\sigma > 0$ when the left-hand group of conditions (5.8c) is fulfilled.

Consequently, points with coordinates $x(0)$ and $x(T_1)$ should satisfy, respectively, the conditions

$$\gamma x(0)=0 \quad \gamma Px(0)+\gamma h<0 \quad \gamma Px(0)-\gamma h>0 \tag{5.184}$$

$$\gamma x(T_1)=0 \quad \gamma Px(T_1)+\gamma h=0 \quad \gamma P^2x(T_1)+\gamma Ph>0 \tag{5.185}$$

The column $x(T_1)$ is expressed in terms of the column $x(0)$ by (5.133) in the form

$$x(T_1)=-\sum_{i=1}^{n-1} e^{\tilde{\lambda}_i T_1} \frac{\tilde{F}(\tilde{\lambda}_i)}{\Delta(\tilde{\lambda}_i)} h\gamma \frac{F(\tilde{\lambda}_i)}{\Delta'_*(\tilde{\lambda}_i)} x(0) \tag{5.186}$$

where $\tilde{\lambda}_i$ are the roots of (5.132). In the interval of time $0<t<T_1$ conditions (5.134) must be fulfilled. The position of the point at time $t=T_1+T_2$ will be given by the expression

$$x(T_1+T_2)=KM(T_2)K^{-1}x(T_1)+KN(T_2)K^{-1}h \tag{5.187}$$

which is obtained from (2.129) by putting $t_0=T_1$ and $\tau=T_2$.

The square matrices on the right-hand side of (5.187) appear in expanded form as:

$$\left.\begin{aligned} KM(T_2)K^{-1}&=\sum_{j=1}^{n-2} e^{\lambda_j T_2}\frac{F(\lambda_j)}{\Delta'(\lambda_j)}+\left[\frac{F(0)}{\Delta_{n-1}(0)}\right]^{(1)}+T_2\frac{F(0)}{\Delta_{n-1}(0)} \\ KN(T_2)K^{-1}&=\sum_{j=1}^{n-2}\frac{e^{\lambda_j T_2}-1}{\lambda_j}\frac{F(\lambda_j)}{\Delta'(\lambda_j)}+T_2\left[\frac{F(0)}{\Delta_{n-1}(0)}\right]^{(1)}+\frac{T_2^2}{2}\frac{F(0)}{\Delta_{n-1}(0)} \end{aligned}\right\} \tag{5.188}$$

Formulae (5.188) are valid when there is only one double root $\lambda_{n-1}=\lambda_n=0$ among the roots λ_j. If all the roots λ_j are simple, then in (5.188) one should retain only the symmetric part, having taken the sum as far as n.

We will substitute in (5.187) the $x(T_1)$ given by (5.186) and put $x(T_1+T_2)=-x(0)=-x^*(0)$. We then obtain an expression determining the initial condition $x^*(0)$ of the periodic solution we are seeking in the form

$$-x^*(0)=-\sum_{i=1}^{n-1} e^{\tilde{\lambda}_i T_1} KM(T_2)K^{-1}\frac{F(\tilde{\lambda}_i)h}{\Delta(\tilde{\lambda}_i)}\frac{\gamma F(\tilde{\lambda}_i)x^*(0)}{\tilde{\Delta}'_*(\tilde{\lambda}_i)}+KN(T_2)K^{-1}h \tag{5.189}$$

When the column $x(T_1)=x^*(T_1)$ corresponds to the initial condition we shall denote it through $x(0)=x^*(0)$. Three equalities appear in (5.184) and (5.185). However, the important ones are just the two scalar equalities

$$\gamma x^*(0)=0 \tag{5.190}$$

$$\gamma Px^*(T_1)+\gamma h=0 \tag{5.191}$$

which we have written down for the initial value $x(0)=x^*(0)$. The first equality in (5.185) is fulfilled automatically from (5.184). Equality (5.191) is identical to the left-hand boundary condition (5.134) with $t=T_1$, so it can be written in the form

$$\sum_{i=1}^{n-1} e^{\tilde{\lambda}_i T_1} \frac{\gamma F(\tilde{\lambda}_i)\, x^*(0)}{\tilde{\Delta}'_*(\tilde{\lambda}_i)} + 1 = 0 \tag{5.192}$$

The matrix equation (5.189) and the scalar equations (5. 190), (5.191) form a system of $(n+2)$ scalar equations for finding the $(n+2)$ unknowns $x_1^*(0)$, $x_2^*(0)$, ..., $x_n^*(0)$, $T_1>0$ and $T_2>0$. When (5.134) and the inequalities appearing in conditions (5.184) are fulfilled, these $(n+2)$ quantities determine the initial condition $x^*(0)$ of the periodic solution and its half-period T_1+T_2. There should, of course, be no additional normal switchings present.

To investigate the stability of the periodic solutions thus found, it is necessary to vary the matrix equalities (5.186), (5.187) and the scalar expressions (5.190) and (5.191). We shall then have

$$\left.\begin{aligned}
\delta x(T_1) = &-\sum_{i=1}^{n-1} e^{\tilde{\lambda}_i T_1} \frac{F(\tilde{\lambda}_i)\, h}{\Delta(\tilde{\lambda}_i)} \frac{\gamma F(\tilde{\lambda}_i)\, \delta(x_0)}{\tilde{\Delta}'_*(\tilde{\lambda}_i)} \\
&-\sum_{i=1}^{n-1} e^{\tilde{\lambda}_i T_1} \frac{\tilde{\lambda}_i F(\tilde{\lambda}_i)\, h}{\Delta(\tilde{\lambda}_i)} \frac{\gamma F(\tilde{\lambda}_i)\, x^*(0)}{\tilde{\Delta}'_*(\tilde{\lambda}_i)} \delta T_1 \\
\sum_{i=1}^{n-1} e^{\tilde{\lambda}_i T_1} &\frac{\gamma F(\tilde{\lambda}_i)\, h}{\Delta(\tilde{\lambda}_i)} \frac{\gamma F(\tilde{\lambda}_i)\, \delta x(0)}{\tilde{\Delta}'_*(\tilde{\lambda}_i)} \\
&+\sum_{i=1}^{n-1} e^{\tilde{\lambda}_i T_1} \frac{\tilde{\lambda}_i \gamma F(\tilde{\lambda}_i)\, h}{\Delta(\tilde{\lambda}_i)} \frac{\gamma F(\tilde{\lambda}_i)\, x^*(0)}{\tilde{\Delta}'_*(\tilde{\lambda}_i)} \delta T_1 = 0 \\
-\delta x(T_1+T_2) = &\, KM(T_2)K^{-1}\delta x(T_1) \\
&+ KM(T_2)\left[JK^{-1}x^*(T_1) + K^{-1}h\right]\delta T_2 \\
\gamma KM(T_2)K^{-1}&\delta x(T_1) \\
&+ \gamma KM(T_2)\left[JK^{-1}x^*(T_1) + K^{-1}h\right]\delta T_2 = 0
\end{aligned}\right\} \tag{5.193}$$

Equations (5.193) establish a relation between $\delta x(T_1+T_2)$ and $\delta x(0)$, i.e. they form a point transformation in the neighbourhood of the invariant point $x^*(0)$ corresponding to the periodic solution. The roots of the characteristic equation of this transformation determine the stability. When solving actual problems it is sometimes useful to write (5.189) in the form

$$\sum_{i=1}^{n-1} e^{\tilde{\lambda}_i T_1} \frac{F(\tilde{\lambda}_i)\, h}{\Delta(\tilde{\lambda}_i)} \frac{\gamma F(\tilde{\lambda}_i)\, x^*(0)}{\Delta'_*(\tilde{\lambda}_i)}$$

$$= KM(-T_2)K^{-1}x^*(0) + KM(-T_2)N(T_2)K^{-1}h \qquad (5.194)$$

This expression can be obtained from (5.189) by multiplying both sides by the matrix $KM(-T_2)K^{-1}$ and making some obvious substitutions. Matrix $KM(-T_2)K^{-1}$ is obtained from the first formula in (5.188) by replacing T_2 by $-T_2$, and the matrix $KM(-T_2)N(T_2)K^{-1}$ for the case of a doble zero root $\lambda_{n-1} = \lambda_n = 0$ has the form

$$KM(-T_2)N(T_2)K^{-1} = \sum_{j=1}^{n-2} \frac{1 - e^{\lambda_j T_2}}{\lambda_j} \frac{F(\lambda_j)}{\Delta'(\lambda_j)}$$

$$+ T_2 \left[\frac{F(0)}{\Delta_{n-1}(0)}\right]^{(1)} - \frac{T_2^2}{2} \frac{F(0)}{\Delta_{n-1}(0)} \qquad (5.195)$$

As an example, consider a system for stabilising an aircraft whose motion is described by (5.141). Sliding motions of the type considered in Section 5.7 are possible in such a system. Let us find the periodic conditions of the first type which can exist in the system when the conditions of normal switching are fulfilled. Using (5.149)-(5.152), we find expressions

$$\left.\begin{aligned} \frac{\gamma F(\lambda_1)\, h}{\Delta'(\lambda_1)} &= \left(-\frac{1}{M^2} + \frac{\alpha_1}{M} - \alpha_2\right) N h_2' \\ \gamma \left[\frac{F(0)}{\Delta_2(0)}\right]^{(1)} h &= \left(\frac{1}{M^2} - \frac{\alpha_1}{M}\right) N h_2' \\ \frac{\gamma F(0)\, h}{\Delta_2(0)} &= -\frac{N h_2'}{M} \end{aligned}\right\} \qquad (5.196)$$

With account of (5.146), (5.150) and (5.196) we write the equation of periods (5.47) and the normal switching inequalities (5.48), respectively, in the form

$$-\frac{1}{M}\left(-\frac{1}{M^2} + \frac{\alpha_1}{M} - \alpha_2\right) \operatorname{th} \frac{MT}{2} - \left(\frac{1}{M^2} - \frac{\alpha_1}{M}\right)\frac{T}{2} = 0 \qquad (5.197)$$

$$\left(-\frac{1}{M^2} + \frac{\alpha_1}{M^2} - \alpha_2\right) \operatorname{th} \frac{MT}{2} + \frac{T}{2M} > \alpha_2 \qquad (5.198)$$

where the common factor $Nh_2' > 0$ has been discarded and we have put $\sigma_1^* = 0$. When the inequality

$$\alpha_1 < \frac{1}{M} \qquad (5.199)$$

is fulfilled, the equation of periods (5.197) has one positive root T.

With the fulfilment of (5.199), the left-hand side of the equation of periods will be positive for small T and negative for sufficiently large T. Graphically, the left-hand side can be represented in the form of the curve given for region IV in Fig 5.2. Such an analogy is not accidental, since the equation of periods (5.197) is limiting for (5.97) with $\Theta=0$, and (5.199) is fulfilled in region IV (see Fig 4.7) with the same condition.

For values of T satisfying the equation of periods (5.197), (5.198) can be reduced to the form

$$\alpha_1 T/2 > \alpha_2 \tag{5.200}$$

By virtue of the remark just made concerning the nature of the variation of the left-hand side of (5.197), (5.200) can hold only when the condition

$$\left(\frac{1}{M^2} - \frac{\alpha_1}{M} + \alpha_2\right) \operatorname{th} \frac{M\alpha_2}{\alpha_1} + \left(\alpha_1 - \frac{1}{M}\right) \frac{\alpha_2}{\alpha_1} > 0 \tag{5.201}$$

is fulfilled. Thus, when (5.199) and (5.201) are fulfilled simultaneously, there exists one periodic solution of the type we are considering.

To analyse the stability, let us investigate the transformed characteristic equation (5.89). In this problem, the last coefficient p_2' of this equation has the form

$$p_2' = -\frac{T}{2} \operatorname{th} \frac{\lambda_i T}{2} \frac{\gamma F(0)\, h}{\Delta_2(0)} = -\frac{N h_2'}{M} \frac{T}{2} \operatorname{th} \frac{MT}{2} \tag{5.202}$$

It is clear that the coefficient $p_2' < 0$ while the leading coefficient p_0' is, on the contrary, greater than zero, since the curve of periods in this case cuts the abscissa axis from above to below (see Section 5.4). The two coefficients of the transformed characteristic equation have opposite signs, so the periodic solution is unstable.

The closed state trajectory corresponding to this periodic solution consists of two symmetric bundles situated entirely within the half-spaces $\sigma > 0$ and $\sigma < 0$. The ends of these trajectories rest on the switching plane at points lying outside the sliding zone.

With the fulfilment of (5.199) and the inverse inequality (5.201), periodic modes containing portions of the sliding motion arise in the case we are considering.

The periodic solutions should be sought with the aid of the general formulae given in this section. However, it is possible to anticipate the result by appealing to the following arguments. The system of equations (5.141) determining the motion of the system is obtained from (5.96) with $\Theta=0$. In Section 5.5 it was shown that, at any point in region IV (see Figs 4.7 and 5.2), an unstable periodic condition exists. This condition exists for any Θ and, in particular, when the value of this quantity is made as small as one pleases.

For $\Theta = 0$, region IV is determined by (5.199). One can therefore expect that, with the fulfilment of (5.199) an unstable periodic mode exists regardless of whether it consists of portions of the sliding motion or not.

A solution of the problem in the case of an automatic pilot with hard negative feedback obtained by Andronov and Bautin [2] in a study of the structure of the division of space into trajectories confirms the conclusion given above. The argument of the control used by these authors had the form

$$\sigma = \varphi + \beta\dot{\varphi} - \frac{1}{a}\eta$$

The results corresponding to the case being considered here can be obtained from Andronov and Bautin's work [2] by setting $\beta = \alpha_1 - M\alpha_2$ and $1/a = N\alpha_2$. It is interesting to note that we have here an example where the final results can be more simply obtained for a system of equations of higher order.

Chapter 6

RELAY SYSTEMS SUBJECT TO EXTERNAL ACTIONS AND DISTURBANCES

6.1 STEADY-STATE MOTIONS SUBJECT TO A CONSTANT ACTION AND A DISTURBANCE

Desired response signals are fed into control systems by changing in an appropriate manner the argument of the control function σ. The argument in this case is given by (1.19) or the equivalent matrix expression (4.155). In the case of a constant action, the scalar quantity γx_3 and the column element x_3 are constants. Later it will be more convenient to solve the problem in a more general form, in which an external disturbance is assumed to be imposed upon the relay system.

The equations of motion of such a system can be written in the form

$$\dot{x} = Px + h\psi(\sigma) + g \quad \sigma = \gamma x \tag{6.1}$$

which is the matrix version of (1.15). In (6.1) the column elements g specifying the external disturbance are all constants. For a constant action it is easy to reduce the equations of motion to the form (6.1) if the column elements x_3 are taken as the origin of the coordinates of the system. In this case, column g is written in the form

$$g = Px_3 \tag{6.2}$$

The problem of investigating the motion of a relay system with an asymmetric characteristic of the kind illustrated in general form by the sixth graph in Table 1.1 can also be reduced to an analysis of (6.1). In this case, g is given by the formula

$$g = \varepsilon h \tag{6.3}$$

where ε is a positive number giving the upwards displacement of the relay characteristic. Let us therefore consider (6.1). Applying the arguments given in Chapter 5, let us assume that, in the normal switching regions, the relay characteristic includes a hysteresis loop of the backlash type; in the zones where these conditions are not fulfilled, let us assume that we have an ideal relay characteristic. With this approach we obtain, in the first case, general results which are equally suitable for either type of relay characteristic. In the second case we will be able to make use of a more convenient mathematical model, namely the equations of the sliding motion. For normal switching, the sudden change in the value of the control function from $\psi(\sigma)=-1$ to $\psi(\sigma)=1$ is realised when the following conditions are fulfilled:

$$\gamma x=\sigma_1^* \quad \gamma Px-\gamma h+\gamma g>0 \quad \gamma Px+\gamma h+\gamma g>0 \tag{6.4}$$

For the conditions

$$\gamma x=-\sigma_1^* \quad \gamma Px+\gamma h+\gamma g<0 \quad \gamma Px-\gamma h+\gamma g<0 \tag{6.5}$$

a transition takes place from the value $\psi(\sigma)=1$ to the value $\psi(\sigma)=-1$. In this case the conditions for the sliding mode are written in the form

$$\gamma x=0 \quad \gamma Px+\gamma h+\gamma g<0 \quad \gamma Px-\gamma h+\gamma g>0 \tag{6.6}$$

For an ideal relay characteristic, when conditions (6.4) and (6.5) apply, one should take $\sigma_1^*=0$. In relay systems with hard switching conditions, (6.6) can be fulfilled only when $\gamma h<0$. The sliding zone determined by (6.6) is not symmetric with respect to the origin of the coordinates. If the scalar quantity γg changes, the sliding zone moves along the switching plane parallel to itself. For $|\gamma g|<|\gamma h|$, the coordinate origin lies inside the sliding zone.

In analysing (6.1) it is necessary to distinguish two cases which depend upon the solubility of algebraic equations

$$Px+g=0 \quad \gamma x=0 \tag{6.7}$$

determining the equilibrium position of the system. For a non-singular matrix P, (6.7) is soluble if the column g satisfies the equality

$$\gamma P^{-1}g=0 \tag{6.8}$$

The system will execute natural motions of the kind discussed in the last chapter about the equilibrium position given by (6.7) and (6.8).

If (6.7) is not soluble, periodic oscillations governed by the constant disturbance can arise in the system. These oscillations will not be symmetric over the half-period, which somewhat complicates the computational formulae.

Let us turn now to the problem of finding the periodic solutions to (6.1) corresponding to such periodic oscillations.

6.2 PERIODIC OSCILLATIONS IN STATIC SYSTEMS

In this case there will be no zero roots among the roots λ_i ; we shall, moreover, assume that the roots are simple.

For $\psi(\sigma)=1$, the first equation of (6.1) assumes the form

$$\dot{x}=Px+h+g \tag{6.9}$$

In accordance with (2.129), the solution $x(t)$ of this equation, assuming the initial value for $t=t_0$ is $x(t_0)$, is given by the expression

$$x(t_0+\tau)=KM(\tau)K^{-1}x(t_0)+KN(\tau)K^{-1}(h+g) \tag{6.10}$$

in which we put $t=t_0+\tau$. For $\psi(\sigma)=-1$ we find the equation of motion in the form

$$\dot{x}=Px-h+g \tag{6.11}$$

and its solution in the form

$$x(t_0+\tau)=KM(\tau)K^{-1}x(t_0)+KN(\tau)K^{-1}(-h+g) \tag{6.12}$$

Suppose that for $t_0=0$ the coordinate $x(t_0)=x(0)$ satisfies the switching condition (6.4). Then the state of the system will be determined by (6.10) up to the instant of time closest to $t=T_1$ at which the next switching takes place, i.e. at which $x(T_1)$ satisfies (6.5). From (6.10), with $t_0=0$ and $\tau=T_1$, we obtain

$$x(T_1)=KM(T_1)K^{-1}x(0)+KN(T_1)(h+g) \tag{6.13}$$

In the next interval of time $T_1\leqslant t\leqslant T_1+T_2$ the state of the system is given by (6.12). For $t_0=T_1$ and $\tau=T_2$, we obtain from (6.12)

$$x(T_1+T_2)=KM(T_2)K^{-1}x(T_1)+KN(T_2)K^{-1}(-h+g) \tag{6.14}$$

where $t=T_1+T_2$ is assumed to be the instant of time closest to $t=T_1$ at which the next switching occurs, i.e. the instant at which $x(T_1+T_2)$ satisfies the switching conditions (6.4). The periodicity condition has the form

$$x(T_1+T_2)=x(0)=x^*(0) \tag{6.15}$$

From (6.13)-(6.15) it is easy to obtain the expression

$$\begin{aligned}K[E-M(T_1+T_2)]K^{-1}x^*(0)&=K[M(T_2)N(T_1)-N(T_2)]K^{-1}h\\&+K[M(T_2)N(T_1)+N(T_2)]K^{-1}g\end{aligned} \tag{6.16}$$

The relations obtained for $x^*(0)$ are quite general: they are valid for any root λ_i. Consider now the case when the roots are simple and include no zero roots. In this case, the canonical matrix $J=\Lambda$ for P and the matrices $M(t)$, $N(t)$ are given, respectively, by (2.44), (2.94) and (2.120). With the assumptions just made, the matrix $E-M(T_1+T_2)$ is non-singular. Moreover, all the matrices mentioned above commute with each other. With the above assumptions, (6.16) can be solved with respect to $x^*(0)$ by multiplying both sides by the matrix $K[E-M(T_1+T_2)]^{-1}K^{-1}$. Completing this operation and using (2.110) to carry out some elementary transformations, we obtain

$$x^*(0)=-2K[E-M(T_1+T_2)]^{-1}N(T_2)K^{-1}h \\ -K\Lambda^{-1}K^{-1}(h+g) \qquad (6.17)$$

By making the substitution $x(0)=x^*(0)$ in (6.13) it can be transformed to

$$x^*(T_1)=2K[E-M(T_1+T_2)]^{-1}N(T_1)K^{-1}h \\ -K\Lambda^{-1}K^{-1}(-h+g) \qquad (6.18)$$

Here, the column $x(T_1)$ corresponding to the initial value $x(0)=x^*(0)$ will be denoted through $x^*(T_1)$.

The columns $x^*(0)$ and $x^*(T_1)$ must satisfy the switching conditions (6.4) and (6.5), respectively. The equalities appearing first in these conditions form the periodic equations (in the given case there will be two of these) which are written in the form

$$\left.\begin{aligned} &-2\gamma K[E-M(T_1+T_2)]^{-1}N(T_2)K^{-1}h \\ &\qquad -\gamma K\Lambda^{-1}K^{-1}(h+g)=\sigma_1^* \\ &2\gamma K[E-M(T_1+T_2)]^{-1}N(T_1)K^{-1}h \\ &\qquad -\gamma K\Lambda^{-1}K^{-1}(-h+g)=-\sigma_1^* \end{aligned}\right\} \qquad (6.19)$$

The remaining inequalities can be reduced to the form

$$\left.\begin{aligned} &-2\gamma K\Lambda[E-M(T_1+T_2)]^{-1}N(T_2)K^{-1}h \\ &\qquad -\gamma h-|\gamma h|>0 \\ &2\gamma K\Lambda[E-M(T_1+T_2)]^{-1}N(T_1)K^{-1}h+\gamma h+|\gamma h|<0 \end{aligned}\right\} \qquad (6.20)$$

where $|\gamma h|$ denotes the absolute magnitude of γh. To obtain (6.20), it is necessary first to substitute $P=K\Lambda K^{-1}$ in (6.4) and (6.5) and then, respectively $x^*(0)$ and $x^*(T_1)$ form (6.17) and (6.18). In the interval $0\leqslant t\leqslant T_1+T_2$ corresponding to the values found for $x^*(0)$ (6.17) and $x^*(T_1)$ (6.18) the solution is obtained from (6.10) and (6.12) in the form

$$\left.\begin{aligned} x(t) &= x(\tau) \\ &= -2KM(\tau)[E - M(T_1+T_2)]^{-1}N(T_2)K^{-1}h \\ &\qquad - K\Lambda^{-1}K^{-1}(h+g), \quad 0 \leqslant \tau \leqslant T_1 \\ x(t) &= x(T_1+\tau) = \\ &= 2KM(\tau)[E - M(T_1+T_2)]^{-1}N(T_1)K^{-1}h \\ &\qquad - K\Lambda^{-1}K^{-1}(-h+g), \quad 0 \leqslant \tau \leqslant T_2 \end{aligned}\right\} \tag{6.21}$$

Expressions (6.21) determine the periodic solution over an interval of time equal to the period, provided that the positive quantities T_1 and T_2 satisfy the equation of periods (6.19) and switching inequalities (6.20), and provided that there are no additional switchings. The last condition can be written in the form of the inequalities

$$\left.\begin{aligned} \sigma(t) &= \gamma x(t) = \gamma x(\tau) > -\sigma_1^*, \qquad 0 \leqslant \tau \leqslant T_1 \\ \sigma(t) &= \gamma x(t) = \gamma x(T_1+\tau) < \sigma_1^*, \qquad 0 \leqslant \tau \leqslant T_2 \end{aligned}\right\} \tag{6.22}$$

in which $x(\tau)$ and $x(T_1+\tau)$ are given by (6.21).

The periodic solutions that are found are not symmetric over the half-period, so that the average value of the coordinates of the system during an interval of time equal to the period will not be equal to zero. The average value of the coordinates is given by the formula

$$x_{av} = \frac{1}{T_1+T_2}\left\{\int_0^{T_1} x(\tau)\,d\tau + \int_0^{T_2} x(T_1+\tau)\,d\tau\right\} \tag{6.23}$$

where $x(\tau)$ and $x(T_1+\tau)$ are given by (6.21). Allowing for the fact that, in accordance with (2.109) and (2.110),

$$\int_0^{T_i} M(\tau)\,d\tau = N(T_i)$$

it is not difficult to establish for x_{av} the expression

$$x_{av} = \frac{T_1 - T_2}{T_1+T_2} K\Lambda^{-1}Kh - K\Lambda^{-1}K^{-1}g \tag{6.24}$$

However, $K\Lambda^{-1}K^{-1} = P^{-1}$, so that (6.24) can also be written in the form

$$Px_{av} - \frac{T_1 - T_2}{T_1+T_2} h + g = 0 \tag{6.25}$$

The factor in front of column h is equal to the average value of the control function $\psi(\sigma)$ over an interval of time equal to one period. Thus, (6.25) is obtained by equating to zero, on average, the right-hand side of the first equation in (6.1).

Formula (6.25) expresses analytically the fact that the compensation of a constant disturbance takes place in the dynamic mode of the periodic oscillations. The column x_{av}, which contains the controlled coordinate, can be regarded as the statistical error of the relay system we are considering.

Once again, the stability of the periodic solution can be examined by constructing difference equations as a first approximation and by investigating the manner in which the roots of their characteristic equation are distributed. Consider a solution of equations (6.1), close to the periodic solution, in the interval of time between two successive switchings.

Suppose that $x^*(0)+\delta x(0)$ is the initial value of this solution satisfying the switching condition (6.4) for numerically small values of the column elements $\delta x(0)$. The last two switchings will occur at time $T_1+\delta T_1$ and $T_2+\delta T_2$ and the matrix coordinates that correspond to them, which we shall denote by $x^*(T_1)+\delta x(T_1)$ and $x^*(0)+\delta x(T_2)$, will satisfy switching conditions (6.5) and (6.4), respectively. Here $x^*(0)$, $x^*(T_1)$, T_1 and T_2 are parameters of the solution that has been found; δT_1, δT_2 and elements of the columns $\delta x(T_1)$ and $\delta x(T_2)$ are numerically small, together with the elements of the column $\delta x(0)$. For the new initial condition, from (6.13) and (6.14), it is easy to establish the relations

$$x^*(T_1)+\delta x(T_1)=KM(T_1+\delta T_1)K^{-1}[x^*(0)+\delta x(0)] + KN(T_1+\delta T_1)K^{-1}(h+g)$$

$$x^*(0)+\delta x(T_2)=KM(T_2+\delta T_2)K^{-1}[x^*(T_1)+\delta x(T_1)] + KN(T_2+\delta T_2)K^{-1}(-h+g)$$

Variation of these relations yields the first approximation equation in the form

$$\left.\begin{aligned}\delta x(T_1)&=KM(T_1)K^{-1}\delta x(0)+K[\dot{M}(T_1)K^{-1}x^*(0)\\&\qquad+\dot{N}(T_1)K^{-1}(h+g)]\delta T_1\\\delta x(T_2)&=KM(T_3)K^{-1}\delta x(T_1)+K[\dot{M}(T_2)K^{-1}x^*(0)\\&\qquad+\dot{N}(T_2)K^{-1}(-h+g)]\delta T_2\end{aligned}\right\} \tag{6.26}$$

Using (4.198), (6.17) and (6.18), expressions (6.26) can be reduced to the form

$$\left.\begin{aligned}\delta x(T_1) &= KM(T_1)K^{-1}\delta x(0)\\ &\quad - 2KM(T_1)\Lambda[E - M(T_1+T_2)]^{-1}N(T_2)K^{-1}h\,\delta T_1\\ \delta x(T_2) &= KM(T_2)K^{-1}\delta x(T_1)\\ &\quad + 2KM(T_2)\Lambda[E - M(T_1+T_2)]^{-1}N(T_1)K^{-1}h\,\delta T_2\end{aligned}\right\} \quad (6.27)$$

The variations $\delta x(0)$ and $\delta x(T_1)$ satisfy the additional conditions

$$\gamma\,\delta x(T_1) = 0 \qquad \gamma\,\delta x(T_2) = 0 \quad (6.28)$$

which are obtained by variation of the first equalities in the switching conditions (6.4) and (6.5).

Let us substitute $\delta x(T_1)$ and $\delta x(T_2)$ from (6.27) into (6.28), solve these expressions with respect to δT_1 and δT_2, and substitute the expressions obtained into (6.27). We then obtain

$$\left.\begin{aligned}\delta x(T_1) &= K\Bigg\{M(T_1)\\ &\quad - \frac{M(T_1)\Lambda[E-M(T_1+T_2)]^{-1}N(T_2)K^{-1}h\gamma KM(T_1)}{\gamma KM(T_1)\Lambda[E-M(T_1+T_2)]^{-1}N(T_2)K^{-1}h}\Bigg\}\\ &\qquad\qquad \times K^{-1}\delta x(0)\\ \delta x(T_2) &= K\Bigg\{M(T_2)\\ &\quad - \frac{M(T_2)\Lambda[E-M(T_1+T_2)]^{-1}N(T_1)K^{-1}h\gamma KM(T_2)}{\gamma KM(T_2)\Lambda[E-M(T_1+T_2)]^{-1}N(T_1)K^{-1}h}\Bigg\}\\ &\qquad\qquad \times K^{-1}\delta x(T_1)\end{aligned}\right\} \quad (6.29)$$

Finally, we substitute $\delta x(T_1)$ from the first equality in (6.29) into the second. We shall then have

$$\delta x(T_2) = K[M(T_1+T_2) - u'\beta' - u''\beta'']K^{-1}\delta x(0) \quad (6.30)$$

where rows β', β'' and columns u', u'' are given by the expressions

$$\left.\begin{aligned}\beta' &= \beta M(T_1+T_2)\\ \beta'' &= \beta M(T_1)\\ u' &= \frac{1}{\alpha_1}M(T_2)\Lambda[E-M(T_1+T_2)]^{-1}N(T_1)u\\ u'' &= \Big\{\frac{1}{\alpha_2}M(T_1+T_2)\Lambda[E-M(T_1+T_2)]^{-1}N(T_2)\\ &\quad - \frac{\alpha_3}{\alpha_1\alpha_2}M(T_2)\Lambda[E-M(T_1+T_2)]^{-1}N(T_1)\Big\}u\end{aligned}\right\} \quad (6.31)$$

in which α_1, α_2 and α_3 are the scalar quantities

$$\left.\begin{aligned}\alpha_1 &= \beta M(T_2)\Lambda[E - M(T_1 - T_2)]^{-1}N(T_1)u \\ \alpha_2 &= \beta M(T_1)\Lambda[E - M(T_1 + T_2)]^{-1}N(T_2)u \\ \alpha_3 &= \beta M(T_1 + T_2)\Lambda[E - M(T_1 + T_2)]^{-1}N(T_2)u\end{aligned}\right\} \qquad (6.32)$$

and row β and column u are, as before, given by (4.21). Relation (6.14) with condition (6.13) provides a connection between the coordinates of two successive points at which the trajectory cuts the switching plane $\gamma x = \sigma_1^*$ or, to use another terminology, it determines the point transformation of the switching plane $\gamma x = \sigma_1^*$ into itself. The column $x^*(0)$ determines the invariant point of the transformation to which the periodic solution corresponds.

In this respect, the linear difference equation (6.30), represented in matrix form, will determine the point transformation of the switching plane $\gamma x = \sigma_1^*$ in the neighbourhood of the invariant point.

The characteristic equation of this transformation, which can be written in the form

$$\det[\lambda E - M(T_1 + T_2) - u'\beta' - u''\beta''] = 0 \qquad (6.33)$$

determines the stability of the invariant point and, consequently, the stability of the limit cycle. As has been mentioned several times before, for stability to be present it is necessary that the moduli of all the roots of (6.33) shall be less than unity. The principal difficulty now is that of multiplying out the determinant on the left-hand side of (6.33). The trouble arises from the fact that the matrix to which it corresponds has a more complex structure compared with that of the analogous expressions which we came across in the preceding sections. However, the specific character of the unit rank matrices $u'\beta'$ and $u''\beta''$ permit us to obtain the final result by the following method.

Introducing the column matrices

$$\left.\begin{aligned}[\lambda E - M(T_1 + T_2)]^{-1}u' &= \tilde{u}' \\ [\lambda E - M(T_1 + T_2)]^{-1}u'' &= \tilde{u}'' \\ [E - \tilde{u}'\beta']^{-1}\tilde{u}'' &= \tilde{\tilde{u}}''\end{aligned}\right\} \qquad (6.34)$$

Then, in the matrix on the left-hand side of (6.33), we take out first the matrix $\lambda E - M(T_1 + T_2)$ in the form of a factor, and then the matrix $E - \tilde{u}'\beta'$. Using the rule for forming the determinant of a product, we obtain

$$\det[\lambda E - M(T_1 + T_2)]\det(E - \tilde{u}'\beta')\det(E - \tilde{\tilde{u}}''\beta'') = 0 \qquad (6.35)$$

From (4.133), (6.35) quickly reduces to the form

$$\{\det[\lambda E - M(T_1+T_2)]\}(1-\beta'\tilde{u}')(1-\beta''\tilde{\tilde{u}}'')=0 \tag{6.36}$$

or

$$\{\det[\lambda E - M(T_1+T_2)]\}\{(1-\beta'\tilde{u}')(1-\beta''\tilde{u}'')-\beta''\tilde{u}'\beta'\tilde{u}''\}=0 \tag{6.37}$$

provided we substitute into (6.36) the $\tilde{\tilde{u}}''$ obtained from the third equality in (6.34) having previously inverted the matrix $E-\tilde{u}'\beta'$ in this equality with the help of (4.137).

A constant external action does not destroy the autonomicity of the relay systems under discussion. As in the previous chapter, therefore, the stability of the periodic solution will depend on the $(n-1)$ roots of the characteristic equation (6.37) since one of the roots will automatically be equal to zero. In order to establish this result it is necessary to substitute successively into (6.37) $\tilde{u}'$ and $\tilde{u}''$ from (6.34) and β', β'', u' and u'' from (6.31) and (6.32), after which the brackets should be multiplied out and cancellations made. When this has been done in all the expressions containing simultaneously the matrices $\lambda E - M(T_1+T_2)$ and $M(T_1+T_2)$ as factors it is necessary to put the second matrix in the form

$$M(T_1+T_2)=\lambda E-[\lambda E-M(T_1+T_2)]$$

and carry out the appropriate decompositions [it is assumed here that all square matrices commute and that $M(T_1+T_2)=M(T_1)M(T_2)$].

Next, the characteristic equation (6.37) is converted to the form

$$\lambda\{\det[\lambda E-M(T_1+T_2)]\}(\lambda d_1 d_2 - d_3 d_4)=0 \tag{6.38}$$

where

$$\left.\begin{aligned}
d_1 &= \beta[\lambda E - M(T_1+T_2)]^{-1}M(T_2)\Lambda \\
&\qquad \times [E-M(T_1+T_2)]^{-1}N(T_1)\,u \\
d_2 &= \beta[\lambda E - M(T_1+T_2)]^{-1}M(T_1)\Lambda \\
&\qquad \times [E-M(T_1+T_2)]^{-1}N(T_2)\,u \\
d_3 &= \beta[\lambda E - M(T_1+T_2)]^{-1}M(T_1+T_2)\Lambda \\
&\qquad \times [E-M(T_1+T_2)]^{-1}N(T_1)\,u \\
d_4 &= \beta[\lambda E - M(T_1+T_2)]^{-1}M(T_1+T_2)\Lambda \\
&\qquad \times [E-M(T_1+T_2)]^{-1}N(T_2)\,u.
\end{aligned}\right\} \tag{6.39}$$

Multiplying out the matrix relations (6.19)-(6.21) and (6.38) according to the rules we have used in similar situations in earlier chapters, we obtain the following computational formulae for finding the periodic solutions and for analysing their stability:

the equations of periods [according to (2.43) and (2.55) for

$B=P$, $J=\Lambda$ and $\lambda=0$ we have $K\Lambda^{-1}K=P^{-1}=-\frac{F(0)}{\Delta(0)}$]

$$\left.\begin{aligned} -2\sum_{i=1}^{n}\frac{e^{\lambda_i T_2}-1}{\lambda_i\left(1-e^{\lambda_i T'}\right)}\frac{\gamma F(\lambda_i)\,h}{\Delta'(\lambda_i)}+\frac{\gamma F(0)(h+g)}{\Delta(0)}=\sigma_1^*\\ 2\sum_{i=1}^{n}\frac{e^{\lambda_i T_1}-1}{\lambda_i\left(1-e^{\lambda_i T'}\right)}\frac{\gamma F(\lambda_i)\,h}{\Delta'(\lambda_i)}+\frac{\gamma F(0)(-h+g)}{\Delta(0)}=-\sigma_1^* \end{aligned}\right\} \quad (6.40)$$

the switching inequalities [here the switching inequalities (6.20) are expanded for the case $\gamma h\leqslant 0$]

$$\left.\begin{aligned} -\sum_{i=1}^{n}\frac{e^{\lambda_i T_2}-1}{1-e^{\lambda_i T'}}\frac{\gamma F(\lambda_i)\,h}{\Delta'(\lambda_i)}>0\\ \sum_{i=1}^{n}\frac{e^{\lambda_i T_1}-1}{1-e^{\lambda_i T'}}\frac{\gamma F(\lambda_i)\,h}{\Delta'(\lambda_i)}<0 \end{aligned}\right\} \quad (6.41)$$

the periodic solution in an interval of time equal to one half-period of the oscillation

$$\left.\begin{aligned} x(t)=x(\tau)=-2\sum_{i=1}^{n}\frac{e^{\lambda_i\tau}\left(e^{\lambda_i T_2}-1\right)}{\lambda_i\left(1-e^{\lambda_i T_1}\right)}\frac{\gamma F(\lambda_i)\,h}{\Delta'(\lambda_i)}\\ +\frac{F(0)(h+g)}{\Delta(0)},\quad 0\leqslant\tau\leqslant T_1\\ x(t)=x(T_1+\tau)=2\sum_{i=1}^{n}\frac{e^{\lambda_i\tau}\left(e^{\lambda_i T_1}-1\right)}{\lambda_i\left(1-e^{\lambda_i T_1}\right)}\frac{\gamma F(\lambda_i)\,h}{\Delta'(\lambda_i)}\\ +\frac{F(0)(-h+g)}{\Delta(0)},\quad 0\leqslant\tau\leqslant T_2 \end{aligned}\right\} \quad (6.42)$$

the characteristic equation of the $(n-1)$ th degree [that is we have discarded the root $\lambda_n=0$ which is not important in a stability analysis]

$$\prod_{j=1}^{n}\left(\lambda_i-e^{\lambda_j T'}\right)\left[-\lambda\sum_{i=1}^{n}\frac{e^{\lambda_i T_1}\left(e^{\lambda_i T_2}-1\right)}{\left(1-e^{\lambda_i T'}\right)\left(\lambda-e^{\lambda_2 T'}\right)}\frac{\gamma F(\lambda_i)\,h}{\Delta'(\lambda_i)}\right.$$

$$\times\sum_{k=1}^{n}\frac{e^{\lambda_1 T_2}\left(e^{\lambda_k T_1}-1\right)}{\left(1-e^{\lambda_k T'}\right)\left(\lambda-e^{\lambda_k T'}\right)}\frac{\gamma F(\lambda_k)\,h}{\Delta'(\lambda_k)}+\sum_{i=1}^{n}\frac{e^{\lambda_i T'}\left(e^{\lambda_i T_2}-1\right)}{\left(1-e^{\lambda_i T'}\right)\left(\lambda-e^{\lambda_i T'}\right)}\frac{\gamma F(\lambda_i)\,h}{\Delta'(\lambda_i)}$$

$$\left.\times\sum_{k=1}^{n}\frac{e^{\lambda_k T'}\left(e^{\lambda_k T_1}-1\right)}{\left(1-e^{\lambda_k T'}\right)\left(\lambda-e^{\lambda_k T'}\right)}\frac{\gamma F(\lambda_k)\,h}{\Delta'(\lambda_k)}\right] \quad (6.43)$$

In (6.40)-(6.43) we have, for brevity, introduced the notation $T' = T_1 + T_2$. In order to obtain a characteristic equation to which the Hurwitz conditions are applicable, it is necessary to make use of the substitution (4.53). Taking into account the fact that, with this substitution,

$$\lambda - e^{\lambda_s T'} = \frac{1}{\nu - 1}\left(1 - e^{\lambda_s T'}\right)\left(\nu - \operatorname{cth}\frac{\lambda_s T'}{2}\right) \tag{6.44}$$

we can write the transformed characteristic equation in the form

$$\prod_{j=1}^{n}\left(\nu - \operatorname{cth}\frac{\lambda_j T'}{2}\right)\Bigg[-(\nu+1)\sum_{i=1}^{n}\frac{e^{\lambda_i T_1}\left(e^{\lambda_i T_2}-1\right)}{\left(1-e^{\lambda_i T'}\right)^2\left(\nu - \operatorname{cth}\frac{\lambda_i T'}{2}\right)}$$
$$\times\frac{\gamma F(\lambda_i)\,h}{\Delta'(\lambda_i)}\sum_{k=1}^{n}\frac{e^{\lambda_k T_2}\left(e^{\lambda_k T_1}-1\right)}{\left(1-e^{\lambda_k T'}\right)^2\left(\nu - \operatorname{cth}\frac{\lambda_k T'}{2}\right)}\frac{\gamma F(\lambda_k)\,h}{\Delta'(\lambda_k)}$$
$$+(\nu-1)\sum_{i=1}^{n}\frac{e^{\lambda_i T'}\left(e^{\lambda_i T_2}-1\right)}{\left(1-e^{\lambda_i T'}\right)^2\left(\nu - \operatorname{cth}\frac{\lambda_i T'}{2}\right)}\frac{\gamma F(\lambda_i)\,h}{\Delta'(\lambda_i)}$$
$$\times\sum_{k=1}^{n}\frac{e^{\lambda_k T'}\left(e^{\lambda_k T_1}-1\right)}{\left(1-e^{\lambda_k T'}\right)^2\left(\nu - \operatorname{cth}\frac{\lambda_k T'}{2}\right)}\frac{\gamma F(\lambda_k)\,h}{\Delta'(\lambda_k)}\Bigg] = 0 \tag{6.45}$$

Finally, in practical applications, we are helped by the following. Denoting the left-hand sides of the periodic equation (6.40) by $\sigma_1(\tau_1, \tau_2)$ and $\sigma_2(\tau_1, \tau_2)$ in that order, and treating them as functions of the parameters τ_1 and τ_2, the following equality holds:

$$\frac{1}{2}p_0' = \left[\frac{\partial\sigma_1(\tau_1, \tau_2)}{\partial\tau_1}\frac{\partial\sigma_2(\tau_1, \tau_2)}{\partial\tau_2} - \frac{\partial\sigma_1(\tau_1, \tau_2)}{\partial\tau_2}\frac{\partial\sigma_2(\tau_1, \tau_2)}{\partial\tau_1}\right]_{\substack{\tau_1=T_1\\ \tau_2=T_2}} \tag{6.46}$$

where p_0' is the coefficient of the transformed characteristic equation (6.45) for the highest power of ν. Formula (6.46) has a geometrical interpretation. On the plane τ_1, τ_2 (τ_1 is the abscissa, τ_2 the ordinate) the equations

$$\left.\begin{aligned}\sigma_1(\tau_1, \tau_2) &= \overset{*}{\sigma_1}\\ \sigma_2(\tau_1, \tau_2) &= -\overset{*}{\sigma_1}\end{aligned}\right\} \tag{6.47}$$

determine the two curves of periods in implicit form. The coordinates $\tau_1 = T_1$ and $\tau_2 = T_2$ of their points of intersection give the periodic solutions when certain supplementary conditions are fulfilled.

The angle δ between the gradients of the functions $\sigma_1(\tau_1, \tau_2)$ and $\sigma_2(\tau_1, \tau_2)$ is given by the formula

$$\sin\delta = \frac{\dfrac{\partial\sigma_1(\tau_1,\tau_2)}{\partial\tau_1}\dfrac{\partial\sigma_2(\tau_1,\tau_2)}{\partial\tau_2} - \dfrac{\partial\sigma_1(\tau_1,\tau_2)}{\partial\tau_2}\dfrac{\partial\sigma_2(\tau_1,\tau_2)}{\partial\tau_1}}{\sqrt{\left(\dfrac{\partial\sigma_1(\tau_1,\tau_2)}{\partial\tau_1}\right)^2 + \left(\dfrac{\partial\sigma_1(\tau_1,\tau_2)}{\partial\tau_2}\right)^2}\sqrt{\left(\dfrac{\partial\sigma_2(\tau_1,\tau_2)}{\partial\tau_1}\right)^2 + \left(\dfrac{\partial\sigma_2(\tau_1,\tau_2)}{\partial\tau_2}\right)^2}} \quad (6.48)$$

The angle δ will be regarded as positive if the gradient of the function $\sigma_1(\tau_1, \tau_2)$ rotates anti-clockwise as it moves through the shortest distance to take up the same direction as the gradient of the functions $\sigma_2(\tau_1, \tau_2)$. It will be recalled that the gradient vector of the functions $\sigma_1(\tau_1, \tau_2)$ and $\sigma_2(\tau_1, \tau_2)$ along the curve of periods coincides with the normal and that they indicate the direction of deformation of these curves when the constants on the right-hand sides of the equations in (6.47) increase in the positive direction.

From (6.46) and (6.48) it follows that, with a change in the sign of the angle δ, the sign of the coefficient p_0' also changes, so that the formulation of the Hurwitz conditions also changes [see (4.60) and (4.61)]. This situation recalls the similar situation which we discussed in Section 5.4.

6.3 PERIODIC OSCILLATIONS IN NEUTRAL SYSTEMS

Let us seek the periodic solutions for the case when there is one root $\lambda_n = 0$ among the simple roots λ_i. This category includes, in particular, the important class of neutrally stable control systems in the open state. Such systems are encountered in practice.

The matrix equation (6.16) which determines the initial value $x^*(0)$ of the periodic solution is a general relation valid for any roots λ_i. In the case we are considering this equation cannot be solved directly since, when $\lambda_n = 0$, the matrix $[E - M(T_1 + T_2)]$ becomes irreversible. However, this is a computational difficulty, not a fundamental one.

In fact, the matrix coordinate $x^*(0)$ must satisfy the matrix equation (6.16) and the switching conditions (6.4). In expanded form, these relations form a system of $n+1$ scalar algebraic equations which are linear with respect to n unknown coordinates $x_1^*(0), x_2^*(0), \ldots, x_n^*(0)$, i.e. with respect to the elements of column $x^*(0)$. The unknown coordinates can be uniquely determined from any n equations if their determinant is non-zero. The equation left after the unknowns $x_1^*(0), x_2^*(0), \ldots, x_n^*(0)$ have been eliminated will be one of the periodic equations. The second periodic equation is obtained from the switching conditions (6.5).

We shall adopt the following procedure of calculation. We shall apply the expansion formula (2.24) to the general matrix relations, noting that all the matrices in terms of whose elements this expansion is carried out are diagonal, since $M(t)$ and $N(t)$ are given by (2.94) and (2.120) for simple roots λ_i. Using this procedure we will rewrite (6.13) and (6.14) in the form

$$x(T_1) = \sum_{i=1}^{n} e^{\lambda_i T_1} k_i \varkappa_i x(0) + \sum_{i=1}^{n} \frac{e^{\lambda_i T_1} - 1}{\lambda_i} k_i \varkappa_i (h + g) \tag{6.49}$$

$$x(T_1 + T_2) = \sum_{i=1}^{n} e^{\lambda_i T_2} k_i \varkappa_i x(T_1) + \sum_{i=1}^{n} \frac{e^{\lambda_i T_2} - 1}{\lambda_i} k_i \varkappa_i (-h + g) \tag{6.50}$$

We now introduce new columns made up of the variables z_i and the constants $\tilde{h}_i$ and $\tilde{g}_i$, and also a row with elements equal to unity, i.e.

$$z = \left\| \begin{matrix} z_1 \\ z_2 \\ \cdot \\ \cdot \\ \cdot \\ z_n \end{matrix} \right\| \quad \tilde{g} = \left\| \begin{matrix} \tilde{g}_1 \\ \tilde{g}_2 \\ \cdot \\ \cdot \\ \cdot \\ \tilde{g}_n \end{matrix} \right\|, \quad \tilde{h} = \left\| \begin{matrix} \tilde{h}_1 \\ \tilde{h}_2 \\ \cdot \\ \cdot \\ \cdot \\ \tilde{h}_n \end{matrix} \right\|, \quad \tilde{\gamma} = \| 1 \;\; 1 \; \ldots \; 1 \| \tag{6.51}$$

We shall now determine the elements of these columns with the help of the formulae

$$z_i = \gamma k_i \varkappa_i x, \quad \tilde{h}_i = \gamma k_i \varkappa_i h, \quad \tilde{g}_i = \gamma k_i \varkappa_i g \qquad i = 1, 2, \ldots, n \tag{6.52}$$

These formulae do not depend on the character of the roots λ_i. For the simple roots λ_i considered at this stage, the matrix identities (2.64) yield the expressions

$$z_i = \frac{\gamma F(\lambda_i) x}{\Delta'(\lambda_i)}, \quad \tilde{h}_i = \frac{\gamma F(\lambda_i) h}{\Delta'(\lambda_i)}, \quad \tilde{g}_i = \frac{\gamma F(\lambda_i) g}{\Delta'(\lambda_i)} \qquad i = 1, 2, \ldots, n \tag{6.53}$$

In all cases, the following equalities hold for the elements $\tilde{h}_i$

$$\tilde{h}_i = \beta_i u_i, \quad i = 1, 2, \ldots, n \tag{6.54}$$

where β_i and u_i are the elements of the row β and the column u, determined by (4.22). The variables z_i belong to the type of canonical variable first introduced into control theory by Lur'e [14]. Multiplying both sides of (6.49) and (6.50) from the left by the rows $\gamma k_i \varkappa_i$, and applying rule (2.71), we obtain the expressions

$$\left. \begin{aligned} z_i(T_1) &= e^{\lambda_i T_1} z_i(0) + \frac{e^{\lambda_i T_1} - 1}{\lambda_i} (\tilde{h}_i + \tilde{g}_i) \\ z_n(T_1) &= z_n(0) + T_1 (\tilde{h}_n + \tilde{g}_n) \\ i &= 1, 2, \ldots, n-1 \end{aligned} \right\} \tag{6.55}$$

$$\left.\begin{aligned} z_i(T_1+T_2) &= e^{\lambda_i T_2} z_i(T_1) + \frac{e^{\lambda_i T_2}-1}{\lambda_i}(-\tilde{h}_i+\tilde{g}_i) \\ z_n(T_1+T_2) &= z_n(T_1) + T_2(-\tilde{h}_n+\tilde{g}_n) \\ i &= 1,\ 2,\ \ldots,\ n-1 \end{aligned}\right\} \quad (6.56)$$

Expressions (6.55) and (6.56) determine the point transformation in the new variables. We shall write (6.16) in the form

$$\sum_{i=1}^{n}(1-e^{\lambda_i T'})\,k_i\varkappa_i x^*(0) = \sum_{i=1}^{n}\frac{e^{\lambda_i T_2}(e^{\lambda_i T_1}-1)-(e^{\lambda_i T_2}-1)}{\lambda_i}\,k_i\varkappa_i h$$

$$+\sum_{i=1}^{n}\frac{e^{\lambda_i T_2}(e^{\lambda_i T_1}-1)+(e^{\lambda_i T_2}-1)}{\lambda_i}\,k_i\varkappa_i g \quad (6.57)$$

Here $T'=T_1+T_2$. Whence we obtain

$$(1-e^{\lambda_i T'})\,z_i^*(0) = \frac{e^{\lambda_i T_2}(e^{\lambda_i T_1}-1)-(e^{\lambda_i T_2}-1)}{\lambda_i}\,\tilde{h}_i$$

$$+\frac{e^{\lambda_i T_2}(e^{\lambda_i T_1}-1)+(e^{\lambda_i T_2}-1)}{\lambda_i}\,\tilde{g}_i, \quad i=1,\ 2,\ \ldots,\ n-1 \quad (6.58)$$

$$0=(T_1-T_2)\,\tilde{h}_n+(T_1+T_2)\,\tilde{g}_n \quad (6.59)$$

Expressions (6.58) are reduced to the form

$$z_i^*(0) = -\frac{2(e^{\lambda_i T_2}-1)}{\lambda_i(1-e^{\lambda_i T'})}\,\tilde{h}_i - \frac{1}{\lambda_i}(\tilde{h}_i+\tilde{g}_i) \quad (6.60)$$

$$i=1,\ 2,\ \ldots,\ n-1$$

Expression (6.59) is one of the periodic equations and (6.60) determines $n-1$ initial values of the periodic solution in the new variables.

Next, using (2.25), we will rewrite the equalities appearing in the switching conditions (6.4) and (6.5) in the form

$$\sum_{i=1}^{n}\gamma k_i\varkappa_i x(0) = \sigma_1^* \quad (6.61)$$

$$\sum_{i=1}^{n}\gamma k_i\varkappa_i x(T_1) = -\sigma_1^* \quad (6.62)$$

In the new variables we shall have

$$\sum_{i=1}^{n} z_i(0) = \sigma_1^* \quad (6.63)$$

$$\sum_{i=1}^{n} z_i(T_1) = -\sigma_1^* \quad (6.64)$$

Equality (6.63) can be used to find $z_n^*(0)$. In fact, assuming in (6.63) that $z_i(0)=z_i^*(0)$, and using (6.60), we obtain

$$z_n^*(T_1)=2\sum_{i=1}^{n-1}\frac{e^{\lambda_i T_1}-1}{\lambda_i\left(1-e^{\lambda_i T'}\right)}\tilde{h}_i+\sum_{i=1}^{n-1}\frac{1}{\lambda_i}(\tilde{h}_i+\tilde{g}_i)+\sigma_1^* \tag{6.65}$$

We substitute into (6.65) $z_i(0)=z_i^*(0)$ from (6.60) and (6.65). We obtain for $z(T_1)=z^*(T_1)$ the expressions

$$z_i^*(T_1)=\frac{2\left(e^{\lambda_i T_1}-1\right)}{\lambda_i\left(1-e^{\lambda_i T'}\right)}\tilde{h}_i-\frac{1}{\lambda_i}(-\tilde{h}_i+\tilde{g}_i). \tag{6.66}$$

$$i=1,\ 2,\ \ldots,\ n-1$$

$$z_n^*(T_1)=2\sum_{i=1}^{n-1}\frac{e^{\lambda_i T_2}-1}{\lambda_i\left(1-e^{\lambda_i T'}\right)}\tilde{h}_i+\sum_{i=1}^{n-1}\frac{1}{\lambda_i}(\tilde{h}_i+\tilde{g}_i)+T_1(\tilde{h}_n+\tilde{g}_n)+\sigma_1^* \tag{6.67}$$

From (6.63) and (6.64) we form the expression

$$\sum_{i=1}^{n}[z_i(0)-z_i(T_1)]=2\sigma_1^* \tag{6.68}$$

Substituting in here $z_j=z_j^*(0)$ and $z_j(T_1)=z_j^*(T_1)$ from (6.60) and (6.65)-(6.67), and using (6.59), we obtain the second periodic equation which we can write in the form

$$\sum_{i=1}^{n-1}\frac{\left(e^{\lambda_i T_1}-1\right)\left(e^{\lambda_i T_2}-1\right)}{\lambda_i\left(1-e^{\lambda_i T_1'}\right)}\tilde{h}_i-\frac{T_1T_2}{T_1+T_2}\tilde{h}_n=\sigma_1^* \tag{6.69}$$

For $\gamma h<0$ the two inequalities appearing in the switching conditions (6.4) can be replaced by the single equivalent inequality

$$\gamma Px(0)+\gamma h+\gamma g>0 \tag{6.70}$$

Substituting in here $P=K\Lambda K^{-1}$, using (2.25) and noting that $\lambda_n=0$, we put (6.70) in the form

$$\sum_{i=1}^{n-1}\lambda_i\gamma k_i\varkappa_i x(0)+\sum_{i=1}^{n}\gamma k_i\varkappa_i h+\sum_{i=1}^{n}\gamma k_i\varkappa_i g>0 \tag{6.71}$$

which, in the new variables (6.51), will have the form

$$\sum_{i=1}^{n-1}\lambda_i z_i(0)+\sum_{i=1}^{n}(\tilde{h}_i+\tilde{g}_i)>0 \tag{6.72}$$

We substitute in here $z_i(0)=z_i^*(0)$ from (6.60). Then, using (6.59), we write the switching inequality in the form

$$-\sum_{i=1}^{n-1}\frac{e^{\lambda_i T_2}-1}{1-e^{\lambda_i T'}}\tilde{h}_i+\frac{T_2}{T_1+T_2}\tilde{h}_n>0 \tag{6.73}$$

In a similar way, for $\gamma h<0$, the inequalities appearing in (6.5) can be reduced to the form

$$\sum_{i=1}^{n-1}\frac{e^{\lambda_i T_1}-1}{1-e^{\lambda_i T'}}\tilde{h}_i-\frac{T_1}{T_1+T_2}\tilde{h}_n<0 \tag{6.74}$$

Finally, it is necessary to establish that, for the parameters found for the periodic solution, there are no additional switchings. For this purpose it is necessary to construct an argument of the control $\sigma(t)$ in the interval $0\leqslant t\leqslant T_1+T_2$. The argument $\sigma(t)$ can be readily found from the following considerations. Formulae (6.55) and (6.56) determine $z_j(t)$ in the interval $0\leqslant t\leqslant T_1+T_2$ provided that T_1 and T_2 in these formulae are replaced by τ, and correspondingly τ is varied within the intervals $0\leqslant\tau\leqslant T_1$ and $0\leqslant\tau\leqslant T_2$.

If this exchange is carried out, and if the sum of the coordinates $z_j(t)$ is taken over all the indices, then for initial values corresponding to the periodic solution we obtain expressions for $\sigma(t)$ in the corresponding intervals of time. The expressions for $\sigma(t)$ obtained will be given later in a general summary of the final results. If the inequalities in (6.22) are fulfilled there will be no additional switchings.

Thus, in the new variables, the parameters determining the periodic solution are found in the form of the final expressions. Let us turn now to the problem of analysing the stability. For this purpose we shall vary (6.55) and (6.56) taking as the centre of the expansion the periodic solution we have found. We then obtain

$$\left.\begin{aligned}\delta z_i(T_1)&=e^{\lambda_i T_1}\delta z_i(0)+e^{\lambda_i T_1}[\lambda_i z_i^*(0)+(\tilde{h}_i+\tilde{g}_i)]\delta T_1\\ \delta z_n(T_1)&=\delta z_n(0)+(\tilde{h}_n+\tilde{g}_n)\delta T_1\\ &\qquad i=1,\,2,\,\ldots,\,n-1\end{aligned}\right\} \tag{6.75}$$

$$\left.\begin{aligned}\delta z_i(T_2)&=e^{\lambda_i T_2}\delta z_i(T_1)\\ &\qquad+e^{\lambda_i T_2}[\lambda_i z_i^*(T_1)+(-\tilde{h}_i+\tilde{g}_i)]\delta T_2\\ \delta z_n(T_2)&=\delta z_n(T_1)+(-\tilde{h}_i+\tilde{g}_i)\delta T_2\\ &\qquad i=1,\,2,\,\ldots,\,n-1\end{aligned}\right\} \tag{6.76}$$

Here $z_i^*(0)$, $z_i^*(T)$, T_1 and T_2 are the parameters of the periodic solution. Substituting into (6.75) and (6.76), $z_i^*(0)$ and $z_i^*(T_1)$ from (6.60) and (6.65)-(6.67), we convert them to the form

$$\left.\begin{aligned} \delta z_i(T_1) &= e^{\lambda_i T_1}\delta z_i(0) - \frac{2e^{\lambda_i T_1}(e^{\lambda_i T_2}-1)}{1-e^{\lambda_i T'}}\tilde{h}_i\,\delta T_1 \\ \delta z_n(T_1) &= \delta z_n(0) + \frac{2T_2}{T'}h'_n\,\delta T_1 \\ i &= 1,\ 2,\ \ldots,\ n-1 \end{aligned}\right\} \qquad (6.77)$$

$$\left.\begin{aligned} \delta z_i(T_2) &= e^{\lambda_i T_2}\delta z_i(T_1) + \frac{2e^{\lambda_i T_2}(e^{\lambda_i T_1}-1)}{1-e^{\lambda_i T'}}\tilde{h}_i\,\delta T_2 \\ \delta z_n(T_2) &= \delta z_n(T_1) - \frac{2T_1}{T'}\tilde{h}_n\,\delta T_2 \\ i &= 1,\ 2,\ \ldots,\ n-1 \end{aligned}\right\} \qquad (6.78)$$

It is not difficult to show by a direct verification that the system of scalar equations (6.77) and (6.78) can be contracted and written in the matrix form:

$$\delta z(T_1) = M(T_1)\delta z_0 - 2M(T_1)\Lambda[E - M(T_1+T_2)]^{-1}N(T_2)\tilde{h}\,\delta T_1 \qquad (6.79)$$

$$\delta z(T_2) = M(T_2)\delta z(T_1) + 2M(T_2)\Lambda[E - M(T_1+T_2)]^{-1}N(T_1)\tilde{h}\,\delta T_2 \qquad (6.80)$$

In these equations we have written explicitly $T' = T_1 + T_2$ and, in addition, we have assumed here a limiting transition for $\lambda_n = 0$ since, for example,

$$\lim_{\lambda_n=0}\frac{e^{\lambda_n T_1}-1}{1-e^{\lambda_i(T_1+T_2)}} = -\frac{T_1}{T_1+T_2} \qquad (6.81)$$

If we vary (6.63) and (6.64), we obtain the expressions

$$\sum_{i=1}^{n}\delta z_i(0) = 0 \qquad \sum_{i=1}^{n}\delta z_i(T_1) = 0 \qquad (6.82)$$

which can be written in the matrix form

$$\tilde{\gamma}\,\delta z(0) = 0 \qquad \tilde{\gamma}\,\delta z(T_1) = 0 \qquad (6.83)$$

If now we eliminate from (6.79) and (6.80) the variations δT_1 and δT_2 using (6.83) and form the characteristic equation, we obtain (6.33). The only difference is that in relations (6.31) and (6.32) which determine β', β'', u' and u'' we have $\tilde{\gamma}$ and $\tilde{h}$ respectively, in place of row β and column u. However, the final results will be exactly the same since, because of (6.51)-(6.54), we have

$$\beta_i u_i = \tilde{\gamma}_i \tilde{h}_i = \tilde{h}_i = \frac{\gamma F(\lambda_i) h}{\Delta'(\lambda_i)}, \qquad i = 1, 2, \ldots, n \tag{6.84}$$

Thus, the characteristic equations in the form (6.43) and (6.45) are still valid when $\lambda_n = 0$. The limiting transition in (6.43) and (6.45) when $\lambda_n = 0$ can be carried out quite simply.

Finally, substituting into (6.59), (6.69), (6.73) and (6.74) the $\tilde{h}_i$ and $\tilde{g}_i$ found from (6.53) and taking the limiting transition for $\lambda_n = 0$ in the transformed characteristic equation (6.45), we obtain the following expressions:

the periodic equation

$$\left.\begin{aligned} &\frac{T_1 - T_2}{T_1 + T_2} \frac{\gamma F(0) h}{\Delta'(0)} = -\frac{\gamma F(0) g}{\Delta'(0)} \\ &\sum_{i=1}^{n-1} \frac{\left(e^{\lambda_i T_1} - 1\right)\left(e^{\lambda_i T_2} - 1\right)}{\lambda_i \left(1 - e^{\lambda_i T'}\right)} \frac{\gamma F(\lambda_i) h}{\Delta'(\lambda_i)} - \\ &\qquad - \frac{T_1 T_2}{T_1 + T_2} \frac{\gamma F(0) h}{\Delta'(0)} = \sigma_1^* \end{aligned}\right\} \tag{6.85}$$

the normal switching inequalities

$$\left.\begin{aligned} &-\sum_{i=1}^{n-1} \frac{e^{\lambda_i T_2} - 1}{1 - e^{\lambda_i T'}} \frac{\gamma F(\lambda_i) h}{\Delta'(\lambda_i)} + \frac{T_2}{T_1 + T_2} \frac{\gamma F(0) h}{\Delta'(0)} > 0 \\ &\sum_{i=1}^{n} \frac{e^{\lambda_i T_1} - 1}{1 - e^{\lambda_i T'}} \frac{\gamma F(\lambda_i) h}{\Delta'(\lambda_i)} - \frac{T_1}{T_1 + T_2} \frac{\gamma F(0) h}{\Delta'(0)} < 0 \end{aligned}\right\} \tag{6.86}$$

the transformed characteristic equation

$$\prod_{j=1}^{n-1} \left(\nu - \operatorname{cth} \frac{\lambda_j T'}{2}\right) \left[-(\nu + 1) d_1' d_2' + (\nu - 1) d_3' d_4'\right] \tag{6.87}$$

where

$$\begin{aligned} d_1 &= \sum_{i=1}^{n-1} \frac{e^{\lambda_i T_1}\left(e^{\lambda_i T_2} - 1\right)}{\left(1 - e^{\lambda_i T'}\right)^2 \left(\nu - \operatorname{cth} \frac{\lambda_i T'}{2}\right)} \frac{\gamma F(\lambda_i) h}{\Delta'(\lambda_i)} \\ &\qquad + \frac{T_2}{2T'} \frac{\gamma F(0) h}{\Delta'(0)} \\ d_2 &= \sum_{k=1}^{n-1} \frac{e^{\lambda_k T_2}\left(e^{\lambda_k T_1} - 1\right)}{\left(1 - e^{\lambda_k T'}\right)^2 \left(\nu - \operatorname{cth} \frac{\lambda_k T'}{2}\right)} \frac{\gamma F(\lambda_k) h}{\Delta'(\lambda_k)} \\ &\qquad + \frac{T_1}{2T'} \frac{\gamma F(0) h}{\Delta'(0)} \end{aligned} \tag{6.88}$$

$$\left.\begin{aligned} d_3 &= \sum_{i=1}^{n-1} \frac{e^{\lambda_i T'}\left(e^{\lambda_i T_2}-1\right)}{\left(1-e^{\lambda_i T'}\right)^2\left(\nu-\operatorname{cth}\frac{\lambda_i T'}{2}\right)} \frac{\gamma F(\lambda_i) h}{\Delta'(\lambda_i)} \\ &\quad + \frac{T^2}{2T'} \frac{\gamma F(0) h}{\Delta'(0)} \\ d_4 &= \sum_{k=1}^{n-1} \frac{e^{\lambda_k T'}\left(e^{\lambda_k T_1}-1\right)}{\left(1-e^{\lambda_k T'}\right)^2\left(\nu-\operatorname{cth}\frac{\lambda_k T'}{2}\right)} \frac{\gamma F(\lambda_k) h}{\Delta'(0)} \\ &\quad + \frac{T_1}{2T'} \frac{\gamma F(0) h}{\Delta'(0)} \end{aligned}\right\} \tag{6.88}$$

the argument of the control $\sigma(t)$ and the inequalities which exclude the additional switchings

$$\left.\begin{aligned} \sigma(t) &= -2\sum_{i=1}^{n-1} \frac{\left(e^{\lambda_i \tau}-1\right)\left(e^{\lambda_i T_2}-1\right)}{\lambda_i\left(1-e^{\lambda_i T'}\right)} \frac{\gamma F(\lambda_i) h}{\Delta'(\lambda_i)} \\ &\quad + \frac{2\tau T_2}{T'} \frac{\gamma F(0) h}{\Delta'(0)} + \sigma_1^* > -\sigma_1^* \\ &\quad 0 \leqslant t = \tau \leqslant T_1, \\ \sigma(t) &= \sigma(T_1+\tau) \\ &= -2\sum_{i=1}^{n-1} \frac{\left(e^{\lambda_i T_2}-e^{\lambda_i \tau}\right)\left(e^{\lambda_i T_1}-1\right)}{\lambda_i\left(1-e^{\lambda_i T'}\right)} \frac{\gamma F(\lambda_i) h}{\Delta'(\lambda_i)} \\ &\quad + \frac{2T_1(T_2-\tau)}{T'} \frac{\gamma F(0) h}{\Delta'(0)} + \sigma_1^* < \sigma_1^*, \quad 0 \leqslant \tau \leqslant T_2 \end{aligned}\right\} \tag{6.89}$$

Relations (6.85)-(6.89) determine the characteristics of the periodic oscillations of the system; they are invariant with respect to the linear transformations of the coordinates. To obtain the form of the oscillations it is necessary to return to the original coordinates x_i , i. e. it is necessary to carry out the inverse transformation. The inverse transformation is given by the solution of the system of linear algebraic equations which are represented in matrix form by the equalities appearing first in (6.53).

Thus, by transforming to new coordinates and parameters (6.52) it is possible to obtain a complete solution to the problem for the case when multiple roots λ_i are present. In particular, when $\lambda_{n-1}=\lambda_n=0$, the periodic equation can be written in the form

$$\left.\begin{aligned} &\frac{T_1-T_2}{T_1+T_2} \frac{\gamma F(0) h}{\Delta_{n-1}(0)} = -\frac{\gamma F(0) g}{\Delta_{n-1}(0)} \\ &\sum_{i=1}^{n-2} \frac{\left(e^{\lambda_i T_1}-1\right)\left(e^{\lambda_i T_2}-1\right)}{\lambda_i\left(1-e^{\lambda_i T'}\right)} \frac{\gamma F(\lambda_i) h}{\Delta'(\lambda_i)} \\ &\quad - \frac{T_1 T_2}{T_1+T_2}\gamma\left[\frac{F(0)}{\Delta_{n-1}(0)}\right]^{(1)} h = \sigma_1^* \end{aligned}\right\} \tag{6.90}$$

For the case when the roots λ_i include zero roots, one of the periodic equations has the form

$$\frac{T_1 - T_2}{T_1 + T_2} \gamma F(0) h = -\gamma F(0) g \tag{6.91}$$

This equality is meaningful when condition

$$|\gamma F(0) g| < |\gamma F(0) h| \tag{6.92}$$

is fulfilled, indicating that the effect of the controlling action of a relay controller ought to be larger than the corresponding effect of a constant external disturbance.

If column g is written in the form

$$g = \alpha g_* \tag{6.93}$$

where α is a scalar quantity and g_* is a column of constant elements then, from (6.91), we obtain

$$\alpha = -\frac{\gamma F(0) h}{\gamma F(0) g_*} \frac{T_1 - T_2}{T_1 + T_2} \tag{6.94}$$

Equality (6.94) analytically expresses the property by means of which the pulsed measuring devices function.

In this section we have found the asymmetric periodic solutions and consequently also the corresponding asymmetric periodic oscillations which arise in relay systems subject to a constant disturbance. However, the results obtained are of more general use. Formally, all the arguments we have given remain valid also when $g \equiv 0$, in which case the formulae obtained determine the asymmetric natural periodic oscillations of relay systems and, in particular, their asymmetric self-oscillations.

On the basis of (6.85) and (6.90) it can be deduced that, when the roots λ_i include zero roots, and in particular in relay systems that are neutrally stable in the open state, only symmetric self-oscillations can exist.

6.4 SLIDING MODES IN THE PRESENCE OF A CONSTANT EXTERNAL ACTION

In the presence of a constant external action or disturbance, a sliding mode arises when the conditions in (6.6) are fulfilled. For this case we must redefine the control function using the formula

$$\psi(\sigma) = -\frac{\gamma P x}{\gamma h} - \frac{\gamma g}{\gamma h} \tag{6.95}$$

Expression (6.95) can be obtained from (6.1) if we apply the same procedure to these equations as we did earlier in connection with (5.114). We substitute $\psi(\sigma)$ from (6.95) into the first equation in (6.1). We then obtain a matrix equation for the sliding mode for a system that is subject to a constant disturbance of the form

$$\dot{x} = Px - \frac{1}{\gamma h} h\gamma Px + g - \frac{1}{\gamma h} h\gamma g \tag{6.96}$$

From this equation it follows that $\gamma\dot{x} \equiv 0$. Therefore, because of (6.96), the image point moves in the plane $\gamma x(t) = \gamma x(0)$, where $x(0)$ is the initial value. In particular, the image point will slide along the switching plane $\gamma x(t) = 0$ provided that for $t = 0$ the equality $\gamma x(0) = 0$ holds.

The homogeneous equation obtained from (6.96) for $g = 0$ is the same as (5.115), so that it is only necessary to construct anew the particular solution of (6.96) satisfying the zero initial conditions, since the general solution of the homogeneous equation is given by (5.133). [The solution of the homogeneous equation in the form (5.133) is constructed for the initial value $x(0)$ satisfying the condition $\gamma x(0) = 0$.] The particular solution we are seeking is determined according to the type of the second sum in (2.131). Using the notation introduced with (5.117), and taking into account the form of the constant column in (6.96), we write the required particular solution in the form

$$[x(t)]_{\text{part}} = \sum_{i=1}^{n-1} \frac{e^{\tilde{\lambda}_i t} - 1}{\tilde{\lambda}_i} \frac{\tilde{F}(\tilde{\lambda}_i)}{\tilde{\lambda}_i \tilde{\Delta}'(\tilde{\lambda}_i)} \left(g - \frac{1}{\gamma h} h\gamma g\right) + t\frac{\tilde{F}(0)}{\tilde{\Delta}(0)} \left(g - \frac{1}{\gamma h} h\gamma g\right) \tag{6.97}$$

Using (4.21), we can write

$$g - \frac{1}{\gamma h} h\gamma g = \left(E - \frac{1}{\beta u} Ku\beta K^{-1}\right) g \tag{6.98}$$

Next, by a direct verification we establish that

$$\beta K^{-1} \left(E - \frac{1}{\beta u} Ku\beta K^{-1}\right) g = 0 \tag{6.99}$$

Taking into account the second formula in (5.123) and the fact that $\tilde{\Delta}(\tilde{\lambda}_i) = 0$, we obtain

$$\beta(\tilde{\lambda}_i E - J)^{-1} K^{-1} \left(E - \frac{1}{\beta u} Ku\beta K^{-1}\right) g = \beta(\tilde{\lambda}_i E - J)^{-1} K^{-1} g \tag{6.100}$$

On the basis of (5.123) and the expressions just given, it is easy to establish the relations

$$\tilde{F}(\tilde{\lambda}_i)\left(g - \frac{1}{\gamma h} h\gamma g\right)$$

$$= -\frac{\tilde{\lambda}_i \Delta(\lambda_i)}{\beta u} K(\tilde{\lambda}_i E - J)^{-1} u\beta (\tilde{\lambda}_i E - J)^{-1} K^{-1} g \qquad (6.101)$$

The right-hand side of (6.101) vanishes for $\tilde{\lambda}_n = 0$. The right-hand sides of (5.125) and (6.101) have an identical structure; they differ only in respect of the columns $x(0)$ and g at the extreme right of these expressions. Therefore, on the basis of the form of (4.21), (5.131) and (6.97), we can write down the particular solution we are seeking in the form

$$[x(t)]_{\text{part}} = -\sum_{i=1}^{n-1} \frac{e^{\tilde{\lambda}_i t} - 1}{\tilde{\lambda}_i} \frac{F(\tilde{\lambda}_i)}{\Delta(\tilde{\lambda}_i)} h\gamma \frac{F(\tilde{\lambda}_i)}{\tilde{\Delta}'_*(\tilde{\lambda}_i)} g \qquad (6.102)$$

Combining (5.133) and (6.102), we obtain a general solution of (6.96)

$$x(t) = -\sum_{i=1}^{n-1} e^{\tilde{\lambda}_i t} \frac{F(\tilde{\lambda}_i)}{\Delta(\tilde{\lambda}_i)} h\gamma \frac{F(\tilde{\lambda}_i)}{\tilde{\Delta}'_*(\tilde{\lambda}_i)} x(0)$$

$$-\sum_{i=1}^{n-1} \frac{e^{\tilde{\lambda}_i t} - 1}{\tilde{\lambda}_i} \frac{F(\tilde{\lambda}_i)}{\Delta(\tilde{\lambda}_i)} h\gamma \frac{F(\tilde{\lambda}_i)}{\tilde{\Delta}'_*(\tilde{\lambda}_i)} g \qquad (6.103)$$

This solution is constructed for an initial condition $x(0)$ that satisfies equality $\gamma x(0) = 0$. The solution (6.103) will determine the sliding motion up to the time when the inequalities of (6.6) are fulfilled. By analogy with (5.134), these inequalities may be written in the form

$$-1 < \frac{\gamma g}{\gamma h} + \sum_{i=1}^{n-1} e^{\tilde{\lambda}_i t} \frac{\gamma F(\tilde{\lambda}_i) x(0)}{\tilde{\Delta}'_*(\tilde{\lambda}_i)}$$

$$+ \sum_{i=1}^{n-1} \frac{e^{\tilde{\lambda}_i t} - 1}{\tilde{\lambda}_i} \frac{\gamma F(\tilde{\lambda}_i) g}{\tilde{\Delta}'_*(\tilde{\lambda}_i)} < 1 \qquad (6.104)$$

If the sliding motion is stable and the initial value $x(0)$ is such that (6.104) is fulfilled in the interval $0 \leqslant t < \infty$, then the matrix coordinate of the equilibrium position is obtained from (6.103) for $t \to \infty$ in the form

$$x_{\text{st}} = \sum_{i=1}^{n-1} \frac{F(\tilde{\lambda}_i)}{\tilde{\lambda}_i \Delta(\tilde{\lambda}_i)} h\gamma \frac{F(\tilde{\lambda}_i)}{\tilde{\Delta}'_*(\tilde{\lambda}_i)} g \qquad (6.105)$$

The point with coordinate x_{st} lies in the switching plane i.e. $\gamma x_{\text{st}} = 0$ [see (5.132)].

Static equilibrium in the sliding zone can be regarded as a periodic process with an infinitely large frequency where the oscillations have infinitely small amplitude. In fact, in this case the relay control element continues to operate in the vibration mode of fast oscillations and the system itself suffers no change in its position as these oscillations proceed. It should be noted, for example, that for systems which are neutrally stable in the open state, the mechanism by which the effect of external disturbances is equalised remains as discussed previously. Analytically, this mechanism is described by the first formulae in (6.85) and (6.90) with infinitely small values of T_1 and T_2.

6.5 AUTOMATIC PILOT WITH A RELAY SERVO-MOTOR

An automatic pilot with a relay servo-motor was described in Section 1.2 and a block diagram of this system is given in Fig 1.4. The motion of the corresponding system for stabilising an aircraft is given by (1.8). In this section we shall assume that control is by angle and angular velocity, so that the equations of motion are written in the form

$$\left.\begin{aligned} \ddot{\varphi} + M\dot{\varphi} &= -h_1'\psi(\sigma) + g_1' \\ \sigma &= \varphi + \alpha_1\dot{\varphi} \end{aligned}\right\} \tag{6.106}$$

Assuming

$$x_1 = \varphi \qquad x_2 = \dot{\varphi} \tag{6.107}$$

we can rewrite (6.106) in the normal form

$$\left.\begin{aligned} \dot{x}_1 &= x_2, \\ \dot{x}_2 &= -Mx_2 - h_1'\psi(\sigma) + g_1' \\ \sigma &= x_1 + \alpha_1 x_2 \end{aligned}\right\} \tag{6.108}$$

The matrices corresponding to this system of equations have the form

$$\left.\begin{aligned} &P = \begin{Vmatrix} 0 & 1 \\ 0 & -M \end{Vmatrix} \qquad x = \begin{Vmatrix} x_1 \\ x_2 \end{Vmatrix} \qquad h = \begin{Vmatrix} 0 \\ -h_1' \end{Vmatrix} \\ &g = \begin{Vmatrix} 0 \\ g_1' \end{Vmatrix} \qquad \gamma = \| 1, \alpha_1 \| \end{aligned}\right\} \tag{6.109}$$

The matrix and scalar expressions of the open system are represented in the form:

the characteristic and adjoint matrices

$$\lambda E - P = \begin{Vmatrix} \lambda & -1 \\ 0 & \lambda + M \end{Vmatrix} \qquad F(\lambda) = \begin{Vmatrix} \lambda + M & 1 \\ 0 & \lambda \end{Vmatrix} \tag{6.110}$$

the characteristic polynomial and its roots

$$\Delta(\lambda) = \lambda(\lambda + M) \qquad \lambda_1 = -M, \quad \lambda_2 = 0 \tag{6.111}$$

The matrix and scalar characteristics of the closed system have the form:

hardness of switching index

$$\gamma h = -\alpha_1 h_1' \tag{6.112}$$

scalar coefficients

$$\frac{\gamma F(\lambda_1) h}{\Delta'(\lambda_1)} = \frac{(1 - M\alpha_1) h_1'}{M} \quad \frac{\gamma F(\lambda_2) h}{\Delta'(\lambda_2)} = -\frac{h_1'}{M} \quad \frac{\gamma F(\lambda_2) g}{\Delta'(\lambda_2)} = \frac{g_1'}{M} \tag{6.113}$$

canonical coordinates z_i (6.53)

$$\left.\begin{aligned} z_1 &= \frac{\gamma F(\lambda_1) x}{\Delta'(\lambda_1)} = -\frac{1 - M\alpha_1}{M} x_2 \\ z_2 &= \frac{\gamma F(\lambda_2) x}{\Delta'(\lambda_2)} = \frac{M x_1 + x_2}{M} \end{aligned}\right\} \tag{6.114}$$

columns

$$\frac{F(\lambda_1) h}{\Delta'(\lambda_1)} = \begin{Vmatrix} 1 \\ -M \end{Vmatrix} \frac{h_1'}{M} \qquad \frac{F(\lambda_2) h}{\Delta'(\lambda_2)} = \begin{Vmatrix} -1 \\ 0 \end{Vmatrix} \frac{h_1'}{M} \tag{6.115}$$

The characteristics of the sliding mode have the form:

characteristic polynomial (6.132) and its roots

$$\tilde{\Delta}_*(\lambda) = -(1 + \alpha_1 \lambda) h_1' \quad \tilde{\lambda}_1 = -\frac{1}{\alpha_1} \tag{6.116}$$

the column

$$\frac{F(\tilde{\lambda}_1) h}{\Delta(\tilde{\lambda}_1)} = \begin{Vmatrix} -1 \\ \dfrac{1}{\alpha_1} \end{Vmatrix} \frac{h_1' \alpha_1^2}{1 - M\alpha_1} \qquad \frac{F(\tilde{\lambda}_1) h}{\tilde{\Delta}_*'(\lambda_1)} = -\begin{Vmatrix} -1 \\ \dfrac{1}{\alpha_1} \end{Vmatrix} \frac{1}{\alpha_1} \tag{6.117}$$

scalar expressions

$$\frac{\gamma F(\tilde{\lambda}_1) x(0)}{\Delta_*'(x_1)} = \frac{(1 - M\alpha_1) x_1(0)}{\alpha_1^2 h_1'} \qquad \frac{\gamma F(\tilde{\lambda}_1) g}{\tilde{\Delta}_*'(\tilde{\lambda}_1)} = 0 \tag{6.118}$$

auxiliary quantities used in analysing the structure of the partitioning of state space

$$\gamma P = \| 0 \;\; 1 - M\alpha_1 \| \qquad \gamma P^2 = \| 0 \;\; -M(1 - M\alpha_1) \| \tag{6.119}$$

$$\left.\begin{aligned}\gamma P x &= (1 - M\alpha_1)\, x_2, \quad \gamma P^2 x = -M(1 - M\alpha_1)\, x_2 \\ \gamma P h &= -(1 - M\alpha_1)\, h_1'\end{aligned}\right\} \qquad (6.120)$$

Phase picture

The state of the system is determined by the position of the image point on the phase plane $x_1 x_2$. For an ideal relay characteristic the switching of the control element will take place while the image point is traversing the straight line

$$\gamma x = x_1 + \alpha_1 x_2 = 0 \qquad (6.121)$$

This straight line divides the phase plane into the half-planes $\sigma > 0$ and $\sigma < 0$ as shown in Fig 6.1. According to the general theory (see Section 5.1) the direction of the phase trajectories abutting on the switching lines (6.121) is determined by the sign of the expressions given below:

for the plane $\sigma > 0$

$$\gamma \dot{x} = \gamma P x + \gamma h = (1 - M\alpha_1)\, x_2 - \alpha_1 h_1' \qquad (6.122)$$

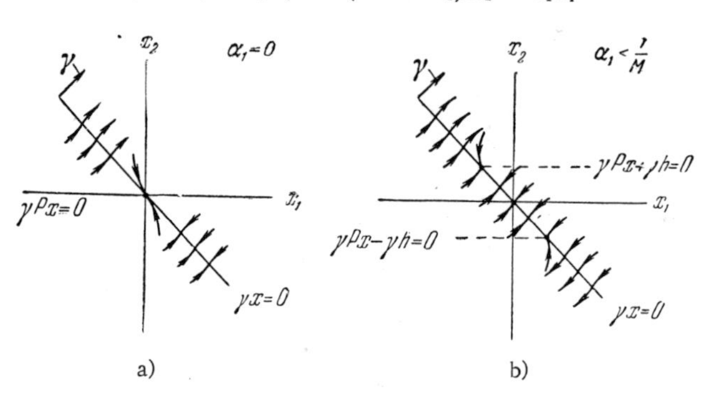

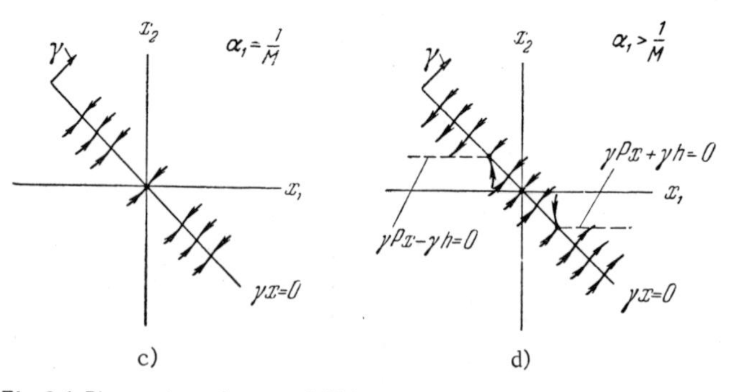

Fig 6.1 Phase plane for a stabilising system with a relay servo-motor.

for the half-plane $\sigma < 0$

$$\gamma x = \gamma P x - \gamma h = (1 - M\alpha_1) x_2 + \alpha_1 h_1' \tag{6.123}$$

The boundary is provided by the straight lines

$$\gamma \dot{x} = \gamma P x + \gamma h = (1 - M\alpha_1) x_2 - \alpha_1 h_1' = 0 \tag{6.124}$$

$$\gamma \dot{x} = \gamma P - \gamma h = (1 - M\alpha_1) x_2 + \alpha_1 h_1' = 0 \tag{6.125}$$

For $\gamma \dot{x} > 0$ and $\gamma \dot{x} < 0$ the velocity vector makes an acute and an obtuse angle with the vector γ, respectively; for $\gamma \dot{x} = 0$ the corresponding piece of the trajectory is tangential to the switching line. The components of the vector γ are the row elements γ; this vector is perpendicular to the switching straight line (6.121) and is directed towards the half-plane $\sigma > 0$ (see Fig 6.1).

The structure of the partitioning into trajectories of that part of the plane adjacent to the switching line is given in Fig 6.1 for various quantitative relations between the system parameters α_1 and M. According to (6.112), for $\alpha_1 = 0$ we have $\gamma h = 0$, and for $\alpha_1 > 0$ and $h_1' > 0$ we have $\gamma h < 0$, so that the case illustrated in Fig 6.1a corresponds to soft switching, while the remaining cases correspond to hard switching. In Figs 6.1b and 6.1d we have finite sliding zones, and in Fig 6.1c the whole switching line is a sliding zone.

In the last case, (6.106) can be written in the form

$$\dot{\sigma} = \frac{1}{M}\left[-h_1'\psi(\sigma) + g_1'\right] \qquad \sigma = \varphi + \frac{1}{M}\dot{\varphi} \tag{6.126}$$

To complete the picture it is necessary to determine the character of the motion of the system at the boundary points of the sliding zone, i. e. to establish whether or not the boundary points belong to the sliding zone itself or to the region of normal switching. This question is decided by the sign of the inequalities in conditions (5.8a) and (5.8b). We now form the expressions

$$\gamma P^2 x + \gamma P h = -M\left[(1 - M\alpha_1) x_2 - \alpha_1 h_1'\right] - h_1' \tag{6.127}$$

$$\gamma P^2 x - \gamma P h = -M\left[(1 - M\alpha_1) x_2 + \alpha_1 h_1'\right] + h_1' \tag{6.128}$$

It is obvious that, when conditions (6.124) and (6.125) are fulfilled, (6.127) and (6.128) will be smaller and larger than zero, respectively. Then, by virtue of (5.8b), the boundary points will belong to the sliding zone. From this we can deduce that the sliding motion takes place in isolation. During the course of this motion the image point cannot leave the sliding zone. For a system with soft switching there is only one point corresponding to the sliding mode. This is the limiting case when $\alpha_1 \to 0$.

The phase picture illustrated in Fig 6.1 can easily be extended to include the case when the relay characteristic contains a hysteresis loop of the back-lash type. We suggest that the reader constructs a diagram for this case, and for this purpose it is recommended that he uses the construction given in Fig 5.5.

Stability of the equilibrium position

The stability of the equilibrium position should be investigated in accordance with the results obtained at the end of Section 5.6. For $\alpha_1 > 0$ the hardness of switching index $\gamma h < 0$, and the only root of the polynomial $\tilde{\Delta}_*(\lambda)$ (6.116) is less than zero; consequently, the equilibrium position possesses asymptotic stability. This result can be verified also for the case of soft switching – i.e. for $\alpha_1 = 0$. Let us now compose the corresponding linear solution, having substituted into (6.106), $\psi(\sigma) = c\sigma$ with $c > 0$.

The characteristic equation of this system

$$\lambda^2 + M\lambda + ch_1' = 0 \tag{6.129}$$

has roots with a negative real part for any coefficient $c > 0$.

Self-oscillations of the system

The system we are discussing is stable in the open state with the result that, in general, only symmetric self-oscillations can exist in it. On the basis of the general formulae (5.35), (5.36), (5.38), (5.88) and the parameters of the system (6.111), (6.113), we obtain:

the periodic equation

$$-\frac{(1 - M\alpha_1)\, h_1'}{M^2} \operatorname{th} \frac{MT}{2} + \frac{h_1'}{M}\frac{T}{2} = \sigma_1^* \tag{6.130}$$

the normal switching inequality

$$\frac{(1 - M\alpha_1)\, h_1'}{M} \operatorname{th} \frac{MT}{2} > \alpha_1 h_1' \tag{6.131}$$

the initial coordinates of a point on the switching line

$$\left.\begin{aligned} x_1^*(0) &= \left(-\frac{1}{M} \operatorname{th} \frac{MT}{2} + \frac{T}{2}\right) \frac{h_1'}{M} \\ x_2^*(0) &= \frac{h_1'}{M} \operatorname{th} \frac{MT}{2} \end{aligned}\right\} \tag{6.132}$$

the transformed characteristic equation

$$\left[\left(1-\operatorname{th}^2\frac{MT}{2}\right)(1-M\alpha_1)-1\right]\nu-\operatorname{th}\frac{MT}{2}=0 \qquad (6.133)$$

The derivative of the left-hand side of (6.130) with respect to T is positive, so that it monotonically increases with $T>0$ from zero to infinity. As a result, the periodic equation (6.130) has only one positive root which vanishes along with the hysteresis parameter σ_1^*. Thus, in a stabilisation system with an ideal relay characteristic, no periodic modes of behaviour can arise and we can therefore expect to find that the equilibrium position will be stable in the large.

The periodic solutions we are seeking exist for $\sigma_1^*>0$ only if (6.131) is fulfilled, for only then will the point with coordinates $x_i^*(0)$ lie outside the sliding zone. For $\alpha_1=0$ this inequality is automatically fulfilled; in fact we could have anticipated this from Fig 6.1a. For $\alpha_1=0$, the characteristic equation (6.133) has a negative root, so that the periodic mode will be stable. The system will execute self-oscillations. This oscillatory process is represented on the phase plane by a closed trajectory which encircles the origin of the coordinates. Such a closed trajectory is often called a limit cycle. The external and internal trajectories wind themselves around the stable limit cycle. The equilibrium position in this case is unstable. However, in the limit when $\sigma_1^*=0$, both $x_1^*(0)$ and $x_2^*(0)$, which are given by (6.132), vanish as the half-period T tends to zero. The limit cycle shrinks into the origin of the coordinates and transmits its stability to the equilibrium position.

When $M\alpha_1>1$, (6.131) can never be fulfilled. In this case the point with coordinates $x_1^*(0)$, $x_2^*(0)$ will lie inside the sliding zone. The bounding case will occur when (6.131) becomes an equality. Eliminating T from this equality and from the periodic equation, we obtain the boundary condition in the form

$$(1-M\alpha_1)\operatorname{th}\left[\frac{M^2\sigma_1^*}{h_1'}+M\alpha_1\right]=M\alpha_1 \qquad (6.134)$$

Figure 6.2 gives the curve (6.134) in terms of the generalised parameters $M\alpha_1$ and $M^2\sigma_1^*/h_1'$; the region of self-oscillations is shaded.

Free sliding motion

In accordance with the general formulae (5.133) and (5.134), and the parameters of the system we are discussing (6.116)-(6.118), we obtain in the sliding motion

$$\begin{Vmatrix} x_1(t) \\ x_2(t) \end{Vmatrix} = -x_1(0)\, e^{-\frac{t}{\alpha_1}} \begin{Vmatrix} -1 \\ \frac{1}{\alpha_1} \end{Vmatrix} \tag{6.135}$$

and the condition for the sliding mode

$$-1 < \frac{(1 - M\alpha_1)\, x_1(0)}{\alpha_1^2 h_1'}\, e^{-\frac{t}{\alpha_1}} < 1 \tag{6.136}$$

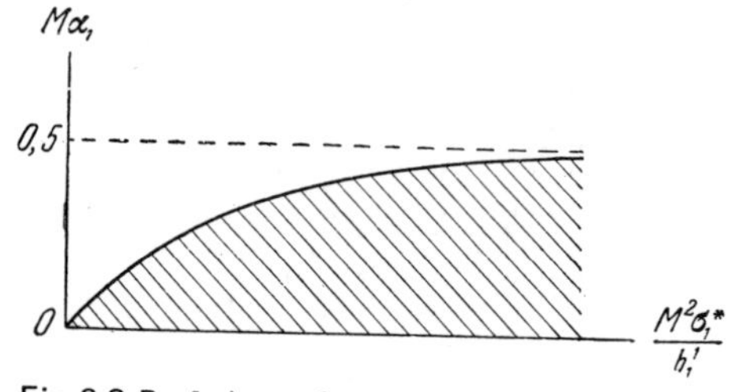

Fig 6.2 Boundary of region of existence of self-oscillations.

Formula (6.135) places a restriction on the choice of the velocity control coefficient, since an excessive increase of α_1 causes the sliding motion to draw in towards the equilibrium position.

Inequalities (6.136) confirm the conclusion that we reached earlier when analysing the structure of the phase plane of the system, since these inequalities will be fulfilled for any $t > 0$ if they are fulfilled for $t = 0$.

Periodic modes in the presence of a constant disturbance

According to the general formulae (6.85)-(6.88), and the system parameters (6.111) and (6.113), we shall have:

periodic equations

$$\left.\begin{aligned} &\frac{T_1 - T_2}{T_1 + T_2} h_1' = g_1', \quad T_1 + T_2 = T' \\ &\frac{(e^{-MT_1} - 1)(e^{-MT_2} - 1)}{M(1 - e^{-MT'})} \frac{(1 - M\alpha_1) h_1'}{M} \\ &\qquad + \frac{T_1 T_2}{T'} \frac{h_1'}{M} = \sigma_1^* \end{aligned}\right\} \tag{6.137}$$

and switching inequalities

$$\left.\begin{aligned} -\frac{e^{-MT_2}-1}{1-e^{-MT'}}\frac{(1-M\alpha_1)h_1'}{M}-\frac{T_2}{T'}\frac{h_1'}{M}>0 \\ \frac{e^{-MT_1}-1}{1-e^{-MT'}}\frac{(1-M\alpha_1)h_1'}{M}-\frac{T_1}{T'}\frac{h_1'}{M}<0 \end{aligned}\right\} \tag{6.138}$$

From the first equation in (6.137), we can express T_2 in terms of T_1:

$$T_2=\frac{h_1'-g_1'}{h_1'+g_1'}T_1, \quad h_1'>g_1'>0. \tag{6.139}$$

We shall, at first, put the second equation in (6.137) into the form

$$-\frac{\operatorname{th}\frac{MT_1}{2}\operatorname{th}\frac{MT_2}{2}}{\operatorname{th}\frac{MT_1}{2}+\operatorname{th}\frac{MT_2}{2}}(1-M\alpha_1)+\frac{\frac{MT_1}{2}\frac{MT_2}{2}}{\frac{MT'}{2}}=\frac{M^2}{2h_1'}\sigma_1^* \tag{6.140}$$

and then reduce it to the form

$$\frac{\frac{MT_1}{2}\operatorname{th}\frac{MT_1}{2}\left(\frac{MT_2}{t_2}-\operatorname{th}\frac{MT_2}{2}\right)+\frac{MT_2}{2}\operatorname{th}\frac{MT_2}{2}\left(\frac{MT_1}{2}-\operatorname{th}\frac{MT_1}{2}\right)}{\frac{MT'}{2}\left(\operatorname{th}\frac{MT_1}{2}+\operatorname{th}\frac{MT_2}{2}\right)}$$

$$+\frac{\operatorname{th}\frac{MT_1}{2}\operatorname{th}\frac{MT_2}{2}}{\operatorname{th}\frac{MT_1}{2}+\operatorname{th}\frac{MT_2}{2}}M\alpha_1=\frac{M^2}{2h_1'}\sigma_1^* \tag{6.141}$$

We shall transform the switching inequalities as below

$$\left.\begin{aligned} \frac{\operatorname{th}\frac{MT_2}{2}\left(1+\operatorname{th}\frac{MT_1}{2}\right)}{\operatorname{th}\frac{MT_1}{2}+\operatorname{th}\frac{MT_2}{2}}(1-M\alpha_1)-\frac{T_2}{T_1+T_2}>0 \\ -\frac{\operatorname{th}\frac{MT_1}{2}\left(1+\operatorname{th}\frac{MT_2}{2}\right)}{\operatorname{th}\frac{MT_1}{2}+\operatorname{th}\frac{MT_2}{2}}(1-M\alpha_1)+\frac{T_1}{T_1+T_2}<0 \end{aligned}\right\} \tag{6.142}$$

If, in the left-hand side of (6.141), we substitute T_2 from (6.139) and note that $y-\operatorname{th} y>0$ when $y>0$, it is not difficult to show that this equation will have a single positive root when $\sigma_1^*>0$ which will vanish when the parameter σ_1^* vanishes. Thus, for $\sigma_1^*=0$, periodic modes do not exist and the effect of the constant disturbance will be compensated for in the sliding mode. We shall discuss this mode of behaviour later. For $\sigma_1^*>0$ and $M\alpha_1>1$ $\alpha_1=0$ (6.142) can never be fulfilled, but these inequalities are satisfied for the value $\alpha_1=0$. The last result can be obtained quite quickly if the hyperbolic tangents are replaced by their arguments. For $\alpha_1 = 0$, the transformed characteristic equation (6.88) has the form

$$p'_0 \nu + p'_1 = 0, \tag{6.143}$$

where

$$p'_0 = \frac{e^{-MT_1}(e^{-MT_2} - 1)^2}{(1 - e^{-MT'})^2} \frac{T_1}{2T'} - \frac{e^{-MT_2}(e^{-MT_1} - 1)}{(1 - e^{-MT'})^2} \frac{T_2}{2T} - \frac{1}{2} \frac{T_1^2 T_2^2}{T'^4}$$

$$p'_1 = -\frac{e^{-MT'}(e^{-MT_1} - 1)(e^{-MT_2} - 1)}{(1 - e^{-MT'})^2}$$

$$- \frac{e^{-MT_1}(1 - e^{-2MT_2})}{(1 - e^{-MT'})^2} \frac{T_1}{2T'} - \frac{e^{-MT_2}(1 - e^{-2MT_1})}{(1 - e^{-MT'})^2} \frac{T_2}{2T}$$

$$- \frac{T_1^2 T_2^2}{2T'^4} \frac{1 + e^{-MT'}}{1 - e^{-MT'}}$$

Since the coefficients p'_0 and p'_1 have the same sign, the corresponding periodic mode will be stable.

Thus, we obtain the same qualitative picture as we did for $g = 0$. For small values of α_1 stable periodic oscillations arise, their limit cycle encompassing a segment of the sliding mode and, for large values of α_1, periodic modes are absent.

Sliding motions in the presence of a constant external disturbance

According to the general formulae (6.103) and (6.104), and the system parameters (6.116)-(6.118), we obtain a formula for the sliding motion which is identical to (6.135). For the inequalities defining the sliding mode, we obtain the expression

$$-1 < -\frac{g'_1}{h'_1} + \frac{(1 - M\alpha_1) x_1(0)}{\alpha_1^2 h'_1} e^{-\frac{t}{\alpha_1}} < 1 \tag{6.144}$$

When $t \to \infty$ the image point asymptotically approaches the origin of the coordinates along a segment of the sliding mode. Inequalities (6.144) are fulfilled during this motion provided they are satisfied for $t = 0$. Thus, the position of static equilibrium in the sliding mode coincides with the controlled position point. The constant disturbance is compensated for in the system without any static deviation from the controlled position.

The following is a general rule which, when fulfilled, ensures that the sliding motion will be independent of the magnitude of the constant external disturbance. Let the columns h and g (determining, respectively, the control vectors and disturbances) differ by a constant scalar factor. Then the roots $\tilde{\lambda}_i$ of the polynomial

$\tilde{\Delta}_*(\lambda) = \gamma F(\lambda) h$ will simultaneously be the roots of the polynomial $\gamma F(\lambda) g$. In this case the second sum in (6.103) vanishes, i. e. the particular solution stipulated by the constant disturbance vanishes.

The above rule is fulfilled in the case of our stabilisation problem, since the columns h and g are given by (6.109).

6.6 FORCED OSCILLATIONS OF RELAY SYSTEMS

Consider the two systems of equations

$$\left.\begin{aligned} \dot{x} &= Px + h\psi(\sigma) \\ \sigma &= \gamma x - a\sin(\omega t + \varepsilon) \end{aligned}\right\} \quad (6.145)$$

and

$$\left.\begin{aligned} \dot{x} &= Px + h\psi(\sigma) + g'\cos(\omega t + \delta) + g''\sin(\omega t + \delta) \\ \sigma &= \gamma x, \end{aligned}\right\} \quad (6.146)$$

in which g' and g'' are columns of constants, and ω, ε and δ are scalar quantities expressing the amplitude, angular frequency and initial phase, respectively, of a harmonic action and harmonic disturbance.

Forced modes of periodic oscillations can arise in a relay system whose motion is described by (6.145) and (6.146). In contrast to the linear case, these oscillations can take place not only at the frequency ω of the external harmonic action or distrubance, but also at frequencies which are whole number fractions of this frequency, i. e. $\omega/2$, .$\omega/3$ etc. Oscillations at fractional frequencies are called subharmonic oscillations of the second kind, third kind etc. Several forms of forced oscillations can exist in a relay system, the particular form that actually appears depending on the previous history of motion of the system or, to be more specific, on the initial conditions. From a mathematical point of view, the problem of determining the forced mode of the oscillations amounts to finding the periodic solutions and analysing their stability.

Of the two cases we have been considering, it turns out that (6.145) yields the simpler form for the periodic solutions. In the discussion that follows, therefore, we shall be assuming that the equations have this form. We should note however that the results obtained are quite general, for it is not difficult to reduce (6.146) to the form (6.145). For this purpose we can make use of the substitution $z = x - x'$, where x' is the particular solution of the linear part of the first equation in (6.146), i. e. the part that determines the forced oscillation mode of the corresponding linear system. Thus, for example, the particular solution $x''(t)$ of the above equation which satisfies the zero initial condition has the form

$$x''(t)=\sum_{k=1}^{n}\frac{(\lambda_k\cos\delta-\omega\sin\delta)e^{\lambda_k t}+\lambda_k\cos(\omega t+\delta)+\omega\sin(\omega t+\delta)}{\lambda_k^2+\omega^2}\frac{F(\lambda_k)g'}{\Delta'(\lambda_k)}$$
$$+\sum_{k=1}^{n}\frac{(\lambda_k\sin\delta+\omega\cos\delta)e^{\lambda_k t}-\lambda_k\sin(\omega t+\delta)-\omega\cos(\omega t+\delta)}{\lambda_k^2+\omega^2}\frac{F(\lambda_k)g''}{\Delta'(\lambda_k)} \quad (6.147)$$

The particular solution x' is obtained from (6.147) by discarding terms containing factors of the exponential function $e^{\lambda_k t}$. This point of view was adopted by Tsipkin in his investigations of forced oscillations in relay systems [25].

Let us therefore consider (6.145). For these equations the normal switching conditions that ensure a transition from $\psi(\sigma)=-1$ to $\psi(\sigma)=1$, and vice versa, are represented, respectively, in the form

$$\left.\begin{aligned}&\gamma x-a\sin(\omega t+\varepsilon)=\sigma_1^*\\&\gamma Px-\gamma h-a\omega\cos(\omega t+\varepsilon)>0\\&\gamma Px+\gamma h-a\omega\cos(\omega t+\varepsilon)>0\end{aligned}\right\} \quad (6.148)$$

and

$$\left.\begin{aligned}&\gamma x-a\sin(\omega t+\varepsilon)=-\sigma_1^*\\&\gamma Px+\gamma h-a\omega\cos(\omega t+\varepsilon)<0\\&\gamma Px-\gamma h-a\omega\cos(\omega t+\varepsilon)<0\end{aligned}\right\} \quad (6.149)$$

The sliding mode is given by the conditions

$$\left.\begin{aligned}&\gamma x-a\sin(\omega t+\varepsilon)=0\\&\gamma Px+\gamma h-a\omega\cos(\omega t+\varepsilon)<0\\&\gamma Px-\gamma h-a\omega\cos(\omega t+\varepsilon)>0\end{aligned}\right\} \quad (6.150)$$

Let $T=\pi/\omega$ be the half-period of the external harmonic action. We shall seek a periodic solution of (6.145) that is symmetric over the half-period T. For $t=t_1$, let the column $x(t_1)$ and the phase $\omega t_1+\varepsilon=\varepsilon_0$ be such that the conditions in (6.148) are satisfied. We shall take this instant as the new time origin. Formally, this does not change the form of (6.145); we have only to replace ε in these equations by ε_0 and to assume that ε_0 is an undetermined quantity. We must seek this quantity in the process of constructing the solution.

The symmetric periodic solution can be found if, for $t=0$, the column $x(0)=x^*(0)$ and the initial phase $\varepsilon_0=\varepsilon_0^*$ satisfy conditions (6.148), and if the next switching takes place at time $t=T$ with $x(T)=-x^*(0)$. It is obvious that the expression for determining the initial condition $x(0)=x^*(0)$, and the form of the symmetric periodic solution $x(t)$ during the interval $0\leqslant T\leqslant T$, will formally be the same as the corresponding expressions obtained in Section 5.2 for symmetric self-oscillations. The expressions for the switching conditions (6.148) for $t=0$ can be obtained if we subtract $a\sin\varepsilon_0^*$ and $a\cos\varepsilon_0^*$ from the left-hand sides of the equation of periods and the switching inequality, respectively.

Thus, for example, for the case of a double zero root $\lambda_{n-1}=\lambda_n=0$, in accordance with (5.47) and (5.48) we obtain for the switching conditions (6.148) expressions in the form

$$-\sum_{i=1}^{n-2}\frac{1}{\lambda_i}\operatorname{th}\frac{\lambda_i T}{2}\frac{\gamma F(\lambda_i)h}{\Delta'(\lambda_i)}-\frac{T}{2}\gamma\left[\frac{F(0)}{\Delta_{n-1}(0)}\right]^{(1)}h-a\sin\varepsilon_0^*=\sigma_1^* \tag{6.151}$$

$$-\sum_{i=1}^{n-2}\operatorname{th}\frac{\lambda_i T}{2}\frac{\gamma F(\lambda_i)h}{\Delta'(\lambda_i)}-\frac{T}{2}\frac{\gamma F(0)h}{\Delta_{n-1}(0)}-\frac{a\pi}{T}\cos\varepsilon_0^*>|\gamma h| \tag{6.152}$$

In these expressions we have put $\omega=\pi/T$. The analogous expressions for the case of simple roots λ_i are obtained from (6.151) and (6.152) if we remove terms corresponding to the double zero root, and the sum is taken from 1 to n. Since, in the determination of the forced oscillations, the half-period T is specified, (6.151) determines the initial phase ε_0^*. Physically, the quantity ε_0^* determines the phase difference between the harmonic action and the forced oscillations of the system.

Let us use $\tilde{x}(t)$ and $\tilde{\sigma}(t)$ to denote the periodic solution that is found. To investigate the stability of our solution we shall apply Lyapunov theory. For this purpose we set up the equations of the perturbed motion as a first approximation. The equations we require are obtained by taking a formal variation of the initial differential equations in the neighbourhood of the functions $\tilde{x}(t)$ and $\tilde{\sigma}(t)$. In matrix form, this equation is obtained from (5.1) in the form

$$\delta\dot{x}=P\,\delta x+h\left[\frac{d\psi(\sigma)}{d\sigma}\right]_{\tilde{\sigma}(t)}\gamma\,\delta x \tag{6.153}$$

where, as before, δx denotes the variation of column x. Next, we write the equality

$$\left[\frac{d\psi(\sigma)}{d\sigma}\right]_{\tilde{\sigma}(t)}=\frac{\dot{\tilde{\psi}}(t)}{\dot{\tilde{\sigma}}(t)} \tag{6.154}$$

where $\tilde{\psi}(t)=\psi[\tilde{\sigma}(t)]$ is a symmetric periodic function which is equal to 1 in the interval $0<t<T$, equal to -1 in the interval $T<t<2T$, and so on. Clearly the function $\tilde{\psi}(t)$ can be represented as the series

$$\dot{\tilde{\psi}}(t)=2\sum_{m=0}^{\infty}(-1)^m\,\delta(t-mT) \tag{6.155}$$

where $\delta(t)$ is the Dirac impulse function.

If we use the filtering property of the function $\delta(t)$, then, on the basis of (6.154) and (6.155), the second term on the right-hand side of (6.153) can be represented in the form

$$2h\gamma\sum_{m=0}^{\infty}\frac{(-1)^m}{2}\left[\frac{\delta x\,(mT-0)}{\dot{\tilde{\sigma}}\,(mT-0)}+\frac{\delta x\,(mT+0)}{\dot{\tilde{\sigma}}\,(mT+0)}\right]\delta(t-mT) \tag{6.156}$$

(Note: in non-integral form, the filtering property of the Dirac function can be written in the form

$$f\,(t)\,\delta\,(t-t_a)=\frac{f\,(t_a-0)+f\,(t_a+0)}{2}\,\delta\,(t-t_a)$$

where $f\,(t)$ may be either a scalar or a matrix function.)

Thus, in (6.153), the second term determines the sequence of impulse actions imposed on the system in discrete, equally spaced instants of time $t=mT$. We encountered a similar problem in Section 4.7. Applying the same arguments that we gave there, we can conclude that between two successive impulse actions, i.e. in the time interval $mT+0\leqslant t\leqslant(m+1)\,T-0$, the change in the variations δx proceeds according to the equation

$$\delta x=P\,\delta x \tag{6.157}$$

Passing on to the next interval we find that the variation jumps instantaneously by the amount

$$\delta x\,(mT+0)-\delta x\,(mT-0)$$

$$=(-1)^m\,h\gamma\left[\frac{\delta x\,(mT-0)}{\dot{\tilde{\sigma}}\,(mT-0)}+\frac{\delta x\,(mT+0)}{\dot{\tilde{\sigma}}\,(mT+0)}\right] \tag{6.158}$$

which is equal to the intensity of the corresponding impulse action. Equality (6.158) is equivalent to each of the inequalities

$$\delta x\,(mT+0)-\delta x\,(mT-0)=\frac{(-1)^m\,2h\gamma\,\delta x\,(mT-0)}{\dot{\tilde{\sigma}}\,(mT-0)} \tag{6.159}$$

and

$$\delta x\,(mT+0)-\delta x\,(mT-0)=\frac{(-1)^m\,2h\gamma\,\delta x\,(mT+0)}{\dot{\tilde{\sigma}}\,(mT+0)} \tag{6.160}$$

Formulae (6.159) and (6.160) are obtained quite easily by successively solving (6.158) first with respect to $\delta x\,(mT+0)$ and then with respect to $\delta x\,(mT-0)$, using the obvious equality

$$\dot{\tilde{\sigma}}(mT+0)-\dot{\tilde{\sigma}}(mT-0)=(-1)^m 2\gamma h \qquad (6.161)$$

The corresponding inverse matrices are found quite simply by direct application of (4.137). The final points of the above interval of time are obtained from the solution of (6.157), which yields the expression

$$\delta x[(m+1)T-0]=KM(T)K^{-1}\delta x(mT+0) \qquad (6.162)$$

With the help of (6.159), we transform this expression to

$$\delta x[(m+1)T-0]=KM(T)K^{-1}\left[E+\frac{(-1)^m 2h\gamma}{\dot{\tilde{\sigma}}(mT-0)}\right]\delta x(mT-0) \qquad (6.163)$$

Since the scalar quantity $(-1)^m\dot{\tilde{\sigma}}(mT-0)$ does not depend on the number m, it can be evaluated, for example, by putting $m=0$. From (6.145) we obtain for the periodic solution the expression

$$\dot{\tilde{\sigma}}(-0)=\gamma KJK^{-1}x_0^*-\gamma h-a\frac{\pi}{T}\cos\varepsilon_0^* \qquad (6.164)$$

where J is the canonical form of matrix P, x_0^* is given by (5.20) and ω is expressed in terms of T. If, for the present, we suppose that J^{-1} exists, then, after representing matrix $N(T)$ in accordance with (2.110) and using (5.20), we can reduce (6.164) to the form

$$\dot{\tilde{\sigma}}(-0)=-2\gamma KM(T)[M(T)+E]^{-1}K^{-1}h-a\frac{\pi}{T}\cos\varepsilon_0^* \qquad (6.165)$$

This formula remains valid for the case when there are zero roots among the roots λ_i. (Note: in earlier chapters we avoided carrying out transformations using matrix J^{-1} because this matrix is non-existent if one of the roots λ_i is equal to zero. Consequently, $N(T)$ was not represented in terms of (2.110) in the expressions obtained. However, in Section 2.5 it was shown that, for zero roots λ_i, the matrix $N(T)$ can be obtained from (2.110) by going to the limit. We made use of this fact when we extended (6.165) to include cases where zero roots λ_i were present. One can, in particular, demonstrate in a similar way that matrix (5.57) can be reduced to the form $S(T)=2M(T)[M(T)+E]^{-1}$.) Thus, (6.163) is a difference equation with a constant matrix operator P^* which, when using (4.21) and (6.165), can be written in the form

$$P^*=K\left\{M(T)-\frac{M(T)u\beta}{\beta M(T)[M(T+E]^{-1}+\frac{a\pi}{T}\cos\varepsilon_0^*}\right\}K^{-1} \qquad (6.166)$$

The periodic solution $\tilde{x}(t)$ will be asymptotically stable if the moduli of all the roots of the characteristic equation of matrix

(6.166) are less than unity, and it will be unstable if the modulus of one root only is greater than unity.

In Section 5.5 we alternated in explicit form the signs of the variations $\delta x(mT)$, an operation that is equivalent to replacing λ by $-\lambda$ in the characteristic equation. To maintain the continuity of the results, we will write the characteristic equation of matrix (6.166) in the form

$$\det\left\{\lambda E+M(T)-\frac{M(T)u\beta}{\beta M(T)[M(T)+E]^{-1}u+\frac{a\pi}{2}\cos\varepsilon_0^*}\right\}=0 \qquad (6.167)$$

Using arguments similar to those we applied when transforming (5.61) to the form (5.65), we shall reduce the characteristic equation (6.167) to the form

$$\{\det[\lambda E+M(T)]\}\left\{1-\frac{\beta[\lambda E+M(T)]^{-1}M(T)u}{\beta M(T)[M(T)+E]^{-1}u+\frac{a\pi}{2T}\cos\varepsilon_0^*}\right\}=0 \qquad (6.168)$$

Using the identity

$$\begin{aligned}\beta M(T)[M(T+E)]^{-1}u-\beta[\lambda E+M(T)]^{-1}M(T)u \\ =\beta\{E-[M(T)+E+\lambda E-\lambda E][\lambda E+M(T)]^{-1}\} \\ \times M(T)[M(T)+E]^{-1}u\end{aligned} \qquad (6.169)$$

in which all the square matrices commute, we can transform the characteristic equation (6.168) to the form

$$\begin{aligned}\{\det[\lambda E+M(T)]\}\{(\lambda-1)\beta[\lambda E+M(T)]^{-1}M(T) \\ \times[M(T)+E]^{-1}u+\frac{a\pi}{2T}\cos\varepsilon_0^*\}=0\end{aligned} \qquad (6.170)$$

The rational function $W^*(\lambda)$ corresponding to (6.18) is represented in the form (see Section 4.3)

$$W^*(\lambda)=-\frac{\beta[\lambda E+M(T)]^{-1}M(T)u}{\beta M(T)[M(T)+E]^{-1}u+\frac{a\pi}{2T}\cos\varepsilon_0^*} \qquad (6.171)$$

At this stage it is important to note the following. It is obvious that, when $a=0$, we obtain from (6.170) the equation

$$\begin{aligned}\{\det[\lambda E+M(T)]\}(\lambda-1)\beta[\lambda E+M(T)]^{-1} \\ \times M(T)[M(T)+E]^{-1}u=0\end{aligned} \qquad (6.172)$$

and the stability of the natural periodic oscillations of the system depends on the distribution of the roots of this equation over the unit circle. One of the roots of (6.172) is automatically equal to unity, which means that according to Lyapunov theory this case is critical; i. e. the equations of the first approximation can tell us nothing about the stability of the motion.

We note that the remaining $n-1$ roots of (6.172) are the same as the roots of (5.68) [see the note in brackets following (6.165)]. When the moduli of all the roots of (5.68) are less then unity, the natural periodic oscillations are stable and the corresponding limit cycle is asymptotically stable. This result was obtained in Section 5.3 from our analysis of the point transformation of the switching plane in the neighbourhood of the invariant point.

A comparison of these two different approaches to the analysis of the stability of self-oscillations leads to the following conclusion. The stability of the natural periodic oscillations of a relay system does not depend on the one root of (6.172) which we know in advance must be equal to unity; the system will be stable if the absolute magnitudes of the other $n-1$ roots of this equation are less than unity.

This fact is quite general to all problems pertaining to the analysis of the stability of the periodic solutions of autonomous systems. The general theorem was proved by Andronov and Vittom as long ago as 1933 and it was extended to a system with discontinuous characteristics in the general case by Aizerman and Gantmakher. The equations of the perturbed motion for relay systems of the type whose theoretical description involves the Dirac impulse function were first obtained by Tsipkin.

Let us reduce (6.170) and (6.171) to a form more convenient for practical application. For this purpose we shall adopt the same procedure of calculation as that used in Section 5.3 to transform (5.68) and (5.69b) to the forms (5.89) and (5.87), respectively.

We obtain

$$\nu^2 \prod_{j=1}^{n-2}\left(\nu - \operatorname{th}\frac{\lambda_j T}{2}\right)\left\{\sum_{i=1}^{n-2}\frac{1-\operatorname{th}^2\dfrac{\lambda_i T}{2}}{\nu-\operatorname{th}\dfrac{\lambda_i T}{2}}\,\frac{\gamma F(\lambda_i)\,h}{\Delta'(\lambda_i)}\right.$$

$$\left.+\frac{1}{\nu}\gamma\left[\frac{F(0)}{\Delta_{n-1}(0)}\right]^{(1)} h+\frac{T}{2\nu^2}\,\frac{\gamma F(0)\,h}{\Delta_{n-1}(0)}+a\,\frac{2\pi}{T}\cos\varepsilon_0^*\right\}=0 \tag{6.173}$$

and

$$W^*(\lambda)=-\frac{1}{\alpha}\left\{\sum_{i=1}^{n-2}\frac{e^{\lambda_i T}}{\lambda+e^{\lambda_i T}}\,\frac{\gamma F(\lambda_i)\,h}{\Delta'(\lambda_i)}+\frac{1}{\lambda+1}\gamma\left[\frac{F(0)}{\Delta_{n-1}(0)}\right]^{(1)} h\right.$$

$$\left.+\frac{T\lambda}{(\lambda-1)^2}\,\frac{\gamma F(0)\,h}{\Delta_{n-1}(0)}\right\} \tag{6.174}$$

$$\alpha = \frac{1}{2}\left\{ \sum_{i=1}^{n-2} \left(1 + \operatorname{th}\frac{\lambda_i T}{2}\right) \frac{\gamma F(\lambda_i)\, h}{\Delta'(\lambda_i)} + \gamma \left[\frac{F(0)}{\Delta_{n-1}(0)}\right]^{(1)} h \right.$$

$$\left. + \frac{T}{2} \frac{\gamma F(0)\, h}{\Delta_{n-1}(0)} + a \frac{\pi}{T} \cos \varepsilon_0^* \right\} \qquad (6.175)$$

Expressions (6.173)-(6.175), as written, refer to the case when $\lambda_{n-1} = \lambda_n = 0$. Thus, the procedure for calculating the symmetric periodic solution with half-period T will be as follows. We specify an amplitude a and a frequency $\omega = \pi/T$ for the external harmonic action. Using (6.151), we find the initial phases ε_0^*, discarding those that satisfy (6.152). After this we construct the argument of the control $\sigma(t)$ and check that $\sigma(t) > -\sigma_1^*$ is fulfilled in the interval $0 \leqslant t \leqslant T$, which ensures against additional switchings. The expression for $\sigma(t)$ is obtained from the second equation in (6.145) if we substitute $\varepsilon = \varepsilon_0^*$ and $x(t)$ from (5.49). Stability is given by the Hurwitz conditions for the transformed characteristic equation (6.173), or in accordance with the argument principle for (6.174). The determination and discarding of the initial phases can be done graphically.

Let us rewrite (6.151) and (6.152) in the form

$$-\sum_{i=1}^{n-2} \frac{1}{\lambda_i} \operatorname{th}\frac{\lambda_i T}{2} \frac{\gamma F(\lambda_i)\, h}{\Delta'(\lambda_i)} - \frac{T}{2}\gamma \left[\frac{F(0)}{\Delta_{n-1}(0)}\right]^{(1)} h = \sigma_1^* + a \sin \varepsilon_0^* \qquad (6.176)$$

and

$$\frac{T}{\pi}\left[-\sum_{i=1}^{n-2} \operatorname{th}\frac{\lambda_i T}{2} \frac{\gamma F(\lambda_i)\, h}{\Delta'(\lambda_i)} - \frac{T}{2} \frac{\gamma F(0)\, h}{\Delta_{n-1}(0)} - |\gamma h| \right] > a \cos \varepsilon_0^* \qquad (6.177)$$

In these expressions we replace T by τ and ε_0^* by ε_0, regarding these quantities as variables which vary within the intervals $0 \leqslant \tau < \infty$ and $0 \leqslant \varepsilon_0 \leqslant 2\pi$. For each of the expressions (6.176) and (6.177) we construct on a single graph curves representing the functions on each side of these expressions. In particular, the left-hand side of (6.176) yields a curve to which, in Section 5.4, we gave the name the curve of periods, denoting it by $\sigma^*(\tau)$. By analogy we shall denote the left-hand side of (6.177) by $\sigma^{**}(\tau)$.

The way in which these graphs are used will be made clear by an actual example.

Forced oscillations in a system for stabilising the course of an aircraft

Consider a relay stabilising system which takes the inertia of the servo-motor into account. The self-oscillations of such a system were investigated in Section 5.5. The curves of periods are given in Fig 5.2 for the different regions of variation of its parameters.

In each diagram we superimpose a plot of the displaced sinusoidal curve $\sigma_1^* + a \sin \varepsilon_0$ on the curve of periods, thus obtaining the graphs shown in Fig 6.3. The method of determining the initial phase of the periodic solution ε_0^* is indicated by the broken lines with arrows in example 6.3a. Having specified the half-period $\tau = T$ we find the ordinate $\sigma^*(T)$ using curve 2 and then the initial phases using curve 1. In this case there will be two such phases corresponding to the ordinate $\sigma^*(T)$; i.e. ε_{10}^* and ε_{20}^*.

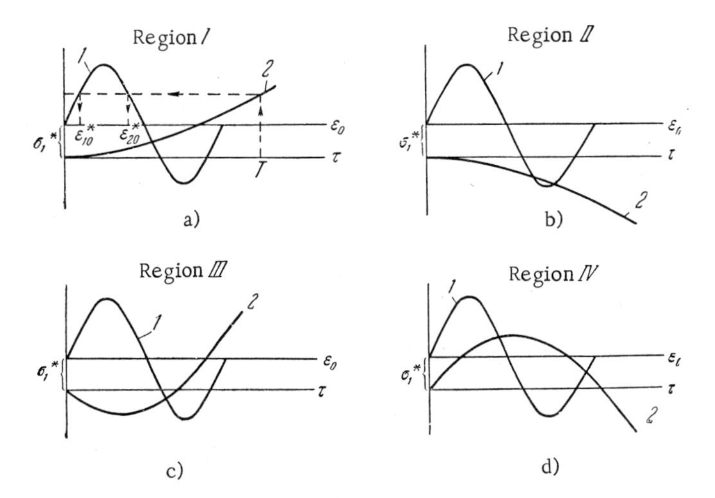

Fig 6.3 Graphical determination of the initial phase of the forced periodic mode. 1 – curve $\sigma_1^* + a \sin \varepsilon_0$; 2 – curve of periods $\sigma^*(T)$.

The function $\sigma^{**}(\tau)$ is obtained from the left-hand side of (5.102) by substituting T for τ and multiplying by τ/π. From the results of the analysis of (5.102) carried out in Section 5.5, we can conclude that, in all four regions, $\sigma^{**}(\tau)$ will be a monotonically increasing function of τ.

Figure 6.4 represents the curve $\sigma^{**}(\tau)$ and also one wave of the cosine curve $a \cos \varepsilon_0$. The initial phase ε_0^* will determine the periodic solution if the corresponding ordinate of curve 1 lies below curve 2. In particular, in the case illustrated in Fig 6.4, (6.158) is satisfied for the initial phase ε_{20}^* only.

From the graphs given in Figs 6.3 and 6.4, it is easy to recognise the various periodic solutions that are possible. Here we will make just one observation. For $\sigma_1^* = 0$, and for any non-resonant frequency of the harmonic action, a value of a_{crit} always exists such that, for $a < a_{crit}$, no periodic solutions can exist. In the resonant case, when the frequencies of the natural periodic

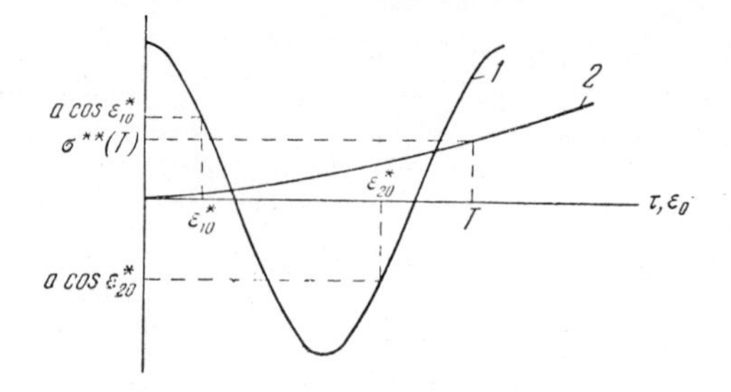

Fig 6.4 Graphical determination of the switching direction. 1 — curve $a\cos\varepsilon_0$; 2 — curve $\sigma^{**}(T)$.

oscillations and harmonic action are equal or differ only slightly from each other, periodic solutions exist for small values of the amplitudes a. When the conditions for stability are fulfilled, the periodic solutions correspond to the forced periodic oscillations of the system.

The characteristic equation can be obtained very simply from (6.173) by making use of the coefficients (5.104) of the transformed characteristic equation (5.103) which we obtained earlier for the purpose of investigating the stability of the natural oscillations of the stabilising system we are considering; we must also take into consideration the roots λ_i given by (4.226). The required characteristic equation can be written in the form

$$p_0\nu^4 + p_1\nu^3 + p_2\nu^2 + p_3\nu + p_4 = 0 \tag{6.178}$$

where

$$\left.\begin{aligned}
p_0 &= a\frac{\pi}{T}\cos\varepsilon_0^* \\
p_1 &= p_0' + a\frac{\pi}{T}\cos\varepsilon_0^*\left(\text{th}\frac{MT}{2} + \text{th}\frac{T}{2\theta}\right) \\
p_2 &= p_1' + a\frac{\pi}{T}\cos\varepsilon_0^*\,\text{th}\frac{MT}{2}\text{th}\frac{T}{2\theta} \\
p_3 &= p_2',\quad p_4 = p_3'
\end{aligned}\right\} \tag{6.179}$$

and the coefficients p_i' in these formulae are given by (5.104). In Section 5.5 it was shown that, in region IV, for $\sigma_1^* = 0$, i. e. for an ideal relay characteristic, an unstable periodic solution of (5.96) exists. We showed there that, in this case, the two coefficients p_0' and p_3' of the characteristic equation (5.103) have opposite signs: $p_0' > 0$, $p_3' < 0$. We note that, for $\sigma_1^* = 0$, the period of the natural oscillations of the system is finite and, according to (5.98), it is determined by the parameters of the system; this period does not depend on the maximum speed of the servo-motor h_2'. Consequently, from (6.179), for the resonant case, we can conclude that, for sufficiently small amplitudes a, the signs of the coefficients

p_1 and p'_0 are the same and the equality $p_4 = p'_3$ always holds. Thus, we have $p_1 > 0$ and $p_4 < 0$, showing that the forced periodic oscillations are unstable.

For these same conditions, self-oscillations exist in region III. In this case (5.106) holds. All coefficients $p'_i < 0$ and $\Delta'_2 = p'_1 p'_2 - p'_0 p'_3 > 0$. For small a the coefficients p_i are negative together with p'_{i-1}, for $i = 1, 2, 3, 4$. The forced oscillations can then be stable only when $p_0 < 0$, and this is possible only for those phase shifts ε_0^* for which $\cos\varepsilon_0^* < 0$. For $\cos\varepsilon_0^* < 0$, the stability condition reduces to the requirement that one Hurwitz inequality holds:

$$\Delta_3 = p_1 p_2 p_3 - p_0 p_3^2 - p_1^2 p_4 < 0 \tag{6.180}$$

which establishes the negativeness of the penultimate Hurwitz determinant for (5.178).

In accordance with (6.180), Δ_3 can be represented in the form

$$\Delta_3 = p'_0(p'_1 p'_2 - p'_0 p'_3) + af_1 \cos\varepsilon_0^* + a^2 f_2 \cos^2\varepsilon_0^* \tag{6.181}$$

where f_1 and f_2 are certain final expressions for the parameters of the system. From (6.181) we can immediately establish that, for small values of a, (6.180) holds. Thus, in the given case the forced oscillations will be stable.

Consequently, on resonance, the signs of the stability of the forced and natural oscillations are the same when a low-power harmonic signal is introduced into the system. It is possible on this basis to stabilise the period of the output oscillations of a self-oscillating system, since such oscillations can be synchronised by an external signal having low power but high frequency stability.

6.7 SLIDING MODE IN THE PRESENCE OF A HARMONIC ACTION

The sliding mode arises when the conditions in (6.150) are fulfilled. The inequalities featuring in these conditions can be rewritten in the form

$$\gamma h < \gamma P x - a\omega \cos(\omega t + \varepsilon) < -\gamma h \tag{6.182}$$

Let us redefine the control function in the sliding zone in terms of the formula

$$\psi(\sigma) = -\frac{1}{\gamma h} \gamma P x + \frac{a\omega}{\gamma h} \cos(\omega t + \varepsilon) \tag{6.183}$$

Then the equation of the sliding motion can be written in the form

$$x = Px - \frac{1}{\gamma h} h\gamma Px + \frac{a\omega}{\gamma h} h \cos(\omega t + \varepsilon) \tag{6.184}$$

Let the origin of time coincide with the beginning of the sliding motion. Then, according to the first equality in (6.150), we have

$$\gamma x(0) = a \sin \varepsilon \tag{6.185}$$

where ε is the initial phase of the harmonic action which corresponds once again to the chosen origin of time. Strictly speaking, we should write ε_0 here, since ε determines the initial phase with respect to the base from which time is measured. The quantity ε_0 can be found by means of a simple transition to the new origin of time, i. e. from the equation

$$\omega t_{\text{new}} + \varepsilon_0 = \omega t_{\text{base}} + \varepsilon$$

The particular solution of (6.184) satisfying the zero initial condition can, with account of (5.115)-(5.117), (6.145)-(6.147) and (6.185), be represented in the form

$$[x(t)]_{\text{part}} = \frac{a\omega}{\gamma h}\left[\sum_{i=1}^{n-1} \frac{(\tilde\lambda_k \cos\varepsilon - \omega \sin\varepsilon)\, e^{\tilde\lambda_k t} - \tilde\lambda_k \cos(\omega t + \varepsilon) + \omega \sin(\omega t + \varepsilon)}{\tilde\lambda_k (\tilde\lambda_k^2 + \omega^2)} \times \frac{\tilde F(\tilde\lambda_i)\, h}{\tilde\Delta'(\tilde\lambda_i)} + \frac{-\sin\varepsilon + \sin(\omega t + \varepsilon)}{\omega} \frac{\tilde F(0)\, h}{\tilde\Delta(0)}\right] \tag{6.186}$$

On the basis of (5.123) and (6.185), and by analogy with the arguments given in Sections 5.6 and 6.4, it is not difficult to establish the relations

$$\left.\begin{aligned} \tilde F(\tilde\lambda_i)\, x(0) &= \frac{F(\tilde\lambda_i)\, h}{\gamma h} a \sin\varepsilon - \frac{\tilde\lambda_i F(\tilde\lambda_i)\, h\gamma F(\tilde\lambda_i)}{\gamma h\, \Delta(\tilde\lambda_i)} x(0) \\ \tilde F(\tilde\lambda_i)\, h &= F(\tilde\lambda_i)\, h, \qquad i = 1, 2, \ldots, n \end{aligned}\right\} \tag{6.187}$$

Let us now combine the general solution of the homogeneous equation (6.184) given by (5.133) with the particular solution (6.186). Then, with account of (6.187) and making some obvious cancellations, we obtain for the general solution (6.184) an expression in the form

$$x(t) = -\sum_{i=1}^{n-1} e^{\lambda_i t} \frac{F(\tilde\lambda_i)}{\Delta(\tilde\lambda_i)} h\gamma \frac{F(\tilde\lambda_i)}{\tilde\Delta'_*(\tilde\lambda_i)} x(0) + \sum_{i=1}^{n-1} e^{\tilde\lambda_i t} \frac{F(\tilde\lambda_i)\, h}{\tilde\lambda_i \tilde\Delta'_*(\tilde\lambda_i)} a \sin\varepsilon$$
$$+ a\omega \sum_{i=1}^{n-1} \frac{(\tilde\lambda_i \cos\varepsilon - \omega \sin\varepsilon)\, e^{\tilde\lambda_i t} - \tilde\lambda_i \cos(\omega t + \varepsilon) + \omega \sin(\omega t + \varepsilon)}{\tilde\lambda_i (\tilde\lambda_i^2 + \omega^2)}$$

$$\times \frac{F(\tilde{\lambda}_i)\,h}{\tilde{\Delta}'_*(\lambda_i)} + \frac{F(0)\,h}{\tilde{\Delta}_*(0)}\, a \sin(\omega t + \varepsilon) \tag{6.188}$$

If we multiply both sides of (6.188) from the left by the row γ and assume that $\tilde{\lambda}_i$ are the roots of the polynomial $\tilde{\Delta}_*(\lambda)$ (5.132), we obtain

$$\gamma x(t) = a \sin(\omega t + \varepsilon)\ ^*) \tag{6.189}$$

[that is, the first equality in (6.150) is fulfilled identically]. Conditions (6.182), which are the conditions under which the sliding mode will exist, can be reduced to the form

$$-1 < \sum_{i=1}^{n-1} e^{\tilde{\lambda}_i t} \frac{\gamma F(\tilde{\lambda}_i)\, x(0)}{\tilde{\Delta}'_*(\tilde{\lambda}_i)} - \sum_{i=1}^{n-1} e^{\tilde{\lambda}_i t} \frac{\Delta(\tilde{\lambda}_i)}{\tilde{\lambda}_i \tilde{\Delta}'_*(\tilde{\lambda}_i)}\, a \sin \varepsilon$$
$$- a\omega \sum_{i=1}^{n-1} \frac{(\tilde{\lambda}_i \cos\varepsilon - \omega \sin\varepsilon)\, e^{\tilde{\lambda}_i t} - \tilde{\lambda}_i \cos(\omega t + \varepsilon) + \omega \sin(\omega t + \varepsilon)}{\tilde{\lambda}_i (\tilde{\lambda}_i^2 + \omega^2)}$$
$$\times \frac{\Delta(\tilde{\lambda}_i)}{\tilde{\Delta}'_*(\tilde{\lambda}_i)} - \frac{\Delta(0)}{\tilde{\Delta}_*(0)}\, a \sin(\omega t + \varepsilon) - \frac{a\omega}{\gamma h} \cos(\omega t + \varepsilon) < 1 \tag{6.190}$$

Sliding mode in a relay servo-motor stabilising system

The equations of motion of this system, and all the matrix and scalar expressions required in the calculation, are given by (6.106)-(6.118). Taking these relations into account, we obtain from (6.188) the result that the over-all motion of the system under sliding conditions proceeds in accordance with the formula

$$\begin{Vmatrix} x_1(t) \\ x_2(t) \end{Vmatrix} = - x_1(0)\, e^{t/\alpha_1} \begin{Vmatrix} -1 \\ \frac{1}{\alpha_1} \end{Vmatrix} - a e^{t/\alpha_1} \sin\varepsilon \begin{Vmatrix} -1 \\ \frac{1}{\alpha_1} \end{Vmatrix}$$
$$- \frac{a\omega \left[\left(-\frac{1}{\alpha_1} \cos\varepsilon - \omega \sin\varepsilon \right) e^{-t/\alpha_1} + \frac{1}{\alpha_1} \cos(\omega t + \varepsilon) + \omega \sin(\omega t + \varepsilon) \right]}{\left(\frac{1}{\alpha_1^2} + \omega^2 \right)}$$
$$\times \begin{Vmatrix} -1 \\ \frac{1}{\alpha_1} \end{Vmatrix} + a \sin(\omega t + \varepsilon) \begin{Vmatrix} 1 \\ 0 \end{Vmatrix}. \tag{6.191}$$

In a similar way we can obtain from (6.190) conditions for the existence of a sliding mode for the given case in the form

$$-1 < -\frac{x_1(0)(1-M\alpha_1)}{\alpha_1^2 h_1'} e^{-t/\alpha_1} - \frac{a(1-M\alpha_1)}{\alpha_1^2 h_1'} e^{-t/\alpha} \sin\varepsilon$$

$$-\frac{a\omega(1-M\alpha_1)\left[\left(-\frac{1}{\alpha_1}\cos\varepsilon - \omega\sin\varepsilon\right)e^{-t/\alpha_1} + \frac{1}{\alpha_1}\cos(\omega t+\varepsilon) + \omega\sin(\omega t+\varepsilon)\right]}{\alpha_1^2 h_1'\left(\frac{1}{\alpha_1^2}+\omega^2\right)}$$

$$+\frac{a\omega}{\alpha_1 h_1'}\cos(\omega t+\varepsilon) < 1 \tag{6.192}$$

In the steady-state mode, we have

$$\left\|\begin{matrix} x_1(t) \\ x_2(t) \end{matrix}\right\| = -\frac{a\omega\left[\frac{1}{\alpha_1}\cos(\omega t+\varepsilon) + \omega\sin(\omega t+\varepsilon)\right]}{\frac{1}{\alpha_1^2}+\omega^2}\left\|\begin{matrix} -1 \\ \frac{1}{\alpha_1} \end{matrix}\right\| + a\sin(\omega t+\varepsilon)\left\|\begin{matrix} 1 \\ 0 \end{matrix}\right\| \tag{6.193}$$

and

$$-1 < -\frac{a\omega}{\alpha_1^2 h_1'\left(\frac{1}{\alpha_1^2}+\omega^2\right)}\left[\frac{1}{\alpha_1}\cos(\omega t+\varepsilon) + \omega\sin(\omega t+\varepsilon)\right]$$

$$+\frac{a\omega}{\alpha_1 h_1'}\cos(\omega t+\varepsilon) < 1 \tag{6.194}$$

In particular, the conditions for the sliding mode (6.194) will be automatically fulfilled if the coefficient h_1' satisfies the inequality

$$h_1' > \frac{a\omega}{\alpha_1}\sqrt{\left(1 - \frac{1-M\alpha_1}{\alpha_1^2\left(\frac{1}{\alpha_1^2}+\omega^2\right)}\right)^2 + \frac{\omega^2(1-M\alpha_1)^2}{\alpha_1^2\left(\frac{1}{\alpha_1^2}+\omega^2\right)^2}} \tag{6.195}$$

Chapter 7

CONTROL SYSTEMS DETERMINED BY SECOND ORDER EQUATIONS

7.1 STATEMENT OF THE PROBLEM

So far it has been assumed that the perturbed motion of a control system about a stationary state is determined by first order differential equations in the normal form (1.15). This form of equations is very convenient and considerably simplifies the calculations. However, in the majority of cases, the initial equations of motion of the class of systems we are considering have the form (1.12). In matrix form they are written

$$\left.\begin{aligned} &(AD^2 + BD + C)\,y = h'\psi(\sigma) + g' \\ &\sigma = (\gamma' + D\gamma'')\,y \end{aligned}\right\} \tag{7.1}$$

where D denotes the differential operator d/dt, A, B, C are constant square matrices.

$$A = \begin{Vmatrix} a_{11} & a_{12} & \dots & a_{1m} \\ a_{21} & a_{22} & \dots & a_{2m} \\ a_{m1} & a_{m2} & \dots & a_{mm} \end{Vmatrix} \qquad B = \begin{Vmatrix} b_{11} & b_{12} & \dots & b_{1m} \\ b_{21} & b_{22} & \dots & b_{2m} \\ b_{m1} & b_{m2} & \dots & b_{mm} \end{Vmatrix}$$

$$C = \begin{Vmatrix} c_{11} & c_{12} & \dots & c_{1m} \\ c_{21} & c_{22} & \dots & c_{2m} \\ c_{m1} & c_{m2} & \dots & c_{mm} \end{Vmatrix} \tag{7.2}$$

y, h', g' are columns and γ', γ'' are rows given by the formulae

$$y = \left\| \begin{matrix} y_1 \\ y_2 \\ \cdot \\ \cdot \\ \cdot \\ y_m \end{matrix} \right\| \quad h' = \left\| \begin{matrix} h'_1 \\ h'_2 \\ \cdot \\ \cdot \\ \cdot \\ h'_m \end{matrix} \right\| \quad g' = \left\| \begin{matrix} g'_1 \\ g'_2 \\ \cdot \\ \cdot \\ \cdot \\ g'_m \end{matrix} \right\|$$

and

$$\gamma' = \| \gamma'_1 \gamma'_2 \ldots \gamma'_m \| \quad \gamma'' = \| \gamma''_1 \gamma''_2 \ldots \gamma''_m \| \tag{7.3}$$

The quantities h'_i, g'_i, γ'_i and γ''_i are constants and the variables y_i determine the coordinates of the system. The matrix

$$\varphi(\lambda) = \lambda^2 A + \lambda B + C \tag{7.4}$$

is characteristic for the linear part of (7.1) and its determinant

$$_{\varphi}\Delta(\lambda) = \det(\lambda^2 A + \lambda B + C) \tag{7.5}$$

is the characteristic polynomial of the linear part of (7.1) or of the open system for $g' = 0$. The roots $_{\varphi}\Delta(\lambda)$ are denoted through λ_i (these roots do not depend on the form in which the equations are represented; also, the characteristic polynomials $\Delta(\lambda)$ and the $_{\varphi}\Delta(\lambda)$ may differ by a constant factor). The adjoint matrix $\Phi(\lambda)$ for the characteristic matrix (7.4) clearly satisfies the equations

$$\varphi^{-1}(\lambda)\Phi(\lambda) = \Phi(\lambda)\varphi^{-1}(\lambda) = {_{\varphi}\Delta}(\lambda) E_m \tag{7.6a}$$

Matrices $\varphi(\lambda)$ and $\Phi(\lambda)$ are square matrices of mth order, and the subscript of the unit matrix indicates its order. We shall suppose that the matrix (7.4) is not singular, i. e. its determinant (7.5) is not identically equal to zero. In addition, we shall assume that the determinant of the coefficients of the higher derivatives of the variables y_i is non-zero. The last condition ensures that it will be possible to reduce the initial system of equations to normal form.

The procedure for reducing these equations to the normal form was described in Section 1.4. Before we can make use of the results obtained in earlier sections it will be necessary first to reduce (7.1) to the normal form (4.2) and then to determine the matrix and scalar quantities appearing in the computational formulae.

We could, however, invert the problem; that is, we could represent the computational formulae in terms of the parameters of (7.1), i. e. in terms of column h' and rows γ' and γ'', the adjoint matrix $\Phi(\lambda)$ and, of course, the roots λ_i which, as we pointed out above, are invariant to the form in which the equations are written. It is possible to reduce the number of intermediate steps when solving this problem, since there is no need to solve (7.1) directly with respect to the higher derivatives. Moreover, the number of cofactors which have to be determined when finding the adjoint matrix is reduced.

We should remember that the orders of the adjoint matrices $\Phi(\lambda)$ and $F(\lambda)$ are m and n, respectively; also, when second derivatives of all the variables y_i are present in (7.1), we have $n=2m$ [see (1.14)].

7.2 CONNECTION BETWEEN THE ADJOINT MATRICES $F(\lambda)$ AND $\Phi(\lambda)$

Consider the case when the matrix A is non-singular, i.e. its determinant $\det A \neq 0$. This means that $\ddot{y}_i$ appears in the initial equations, with $i=1, 2, \ldots, m$, and that the equation can be solved with respect to the higher derivatives.

This case includes, among others, the pulse type and relay stabilising systems considered in Sections 4.10 and 5.5. For example, if we find $\ddot{\varphi}$ from the first equation of (5.96), substitute it into the expression for σ and introduce the notation $\varphi=y_1$ and $\eta=y_2$, then (5.96) can be written in the form

$$\left.\begin{aligned} &\ddot{y}_1+M\dot{y}_1+Ny_2=0\\ &\Theta\ddot{y}_2+\dot{y}_2=h_2'\psi(\sigma)\\ &\sigma=y_1-N\alpha_2y_2+(\alpha_1-M\alpha_2)\dot{y}_1 \end{aligned}\right\} \tag{7.6b}$$

For this system, matrices (7.2) and (7.3) have the form

$$\left.\begin{aligned} &A=\begin{Vmatrix}1 & 0\\ 0 & \Theta\end{Vmatrix} \quad B=\begin{Vmatrix}M & 0\\ 0 & 1\end{Vmatrix} \quad C=\begin{Vmatrix}0 & N\\ 0 & 0\end{Vmatrix}\\ &y=\begin{Vmatrix}y_1\\ y_2\end{Vmatrix} \quad h'=\begin{Vmatrix}0\\ h_2'\end{Vmatrix} \quad \gamma'=\|1, -N\alpha_2\|\\ &\gamma''=\|\alpha_1-M\alpha_2, 0\| \end{aligned}\right\} \tag{7.7}$$

Matrix A in (7.7) is non-singular since $\det A=\Theta\neq 0$. With the assumptions we have made, we can reduce (7.1) to the normal form (4.2). Using the general procedure, we begin by solving the first equation with respect to the higher derivatives. We obtain

$$\ddot{y}=-A^{-1}B\dot{y}-A^{-1}Cy+A^{-1}h'\psi(\sigma)+A^{-1}g' \tag{7.8}$$

We introduce the notation

$$x=\begin{Vmatrix}x'\\ x''\end{Vmatrix} \quad x'=y \quad x''=\dot{y} \tag{7.9}$$

Then, owing to (7.8), we shall have

$$\left.\begin{aligned} \dot{x}' &= x'' \\ \dot{x}'' &= -A^{-1}Cx' - A^{-1}Bx'' + A^{-1}h'\psi(\sigma) + A^{-1}g' \end{aligned}\right\} \qquad (7.10)$$

whence we obtain

$$\left\| \begin{matrix} \dot{x}' \\ \dot{x}'' \end{matrix} \right\| = \left\| \begin{matrix} 0 & E_m \\ -A^{-1}C & -A^{-1}B \end{matrix} \right\| \left\| \begin{matrix} x' \\ x'' \end{matrix} \right\|$$

$$+ \left\| \begin{matrix} 0 \\ A^{-1}h' \end{matrix} \right\| \psi(\sigma) + \left\| \begin{matrix} 0 \\ A^{-1}g' \end{matrix} \right\| \qquad (7.11)$$

The second equation in (7.1) can be written in the form

$$\sigma = \| \gamma'\gamma'' \| \left\| \begin{matrix} x' \\ x'' \end{matrix} \right\| \qquad (7.12)$$

From (7.9), (7.11) and (7.12) the matrices appearing in (4.2) can be written in the form of the block matrices

$$P = \left\| \begin{matrix} 0 & E_m \\ -A^{-1}C & -A^{-1}B \end{matrix} \right\| \qquad (7.13)$$

and

$$h = \left\| \begin{matrix} 0 \\ A^{-1}h' \end{matrix} \right\| \quad g = \left\| \begin{matrix} 0 \\ A^{-1}g' \end{matrix} \right\| \quad \gamma = \| \gamma'\gamma'' \| \qquad (7.14)$$

The square matrices, and the rows and columns appearing in the form of blocks in (7.13) and (7.14) have dimensions $m \times m$, $m \times 1$ and $1 \times m$.

The characteristic matrix for matrix (7.13) has the form

$$\lambda E - P = \left\| \begin{matrix} \lambda E_m & -E_m \\ A^{-1}C & \lambda E_m + A^{-1}B \end{matrix} \right\| \qquad (7.15)$$

Here, E without a subscript denotes a unit matrix of nth order, i.e. a matrix with dimensions $n \times n$. We will convert the matrix (7.15) to triangular form. We will then have

$$(\lambda E - P) \left\| \begin{matrix} E_m & 0 \\ 0 & \lambda E_m \end{matrix} \right\| \left\| \begin{matrix} E_m & E_m \\ 0 & E_m \end{matrix} \right\|$$

$$= \begin{Vmatrix} \lambda E_m & 0 \\ A^{-1}C & \lambda^2 E_m + \lambda A^{-1}B + A^{-1}C \end{Vmatrix} \tag{7.16}$$

The matrix on the right-hand side of (7.16) can be written in the form

$$\begin{Vmatrix} \lambda E_m & 0 \\ A^{-1}C & \lambda^2 E_m + \lambda A^{-1}B + A^{-1}C \end{Vmatrix} = \begin{Vmatrix} \lambda E_m & 0 \\ 0 & \lambda^2 E_m + \lambda A^{-1}B + A^{-1}C \end{Vmatrix}$$
$$\times \begin{Vmatrix} E_m & 0 \\ (\lambda^2 E + \lambda A^{-1}B + A^{-1}C)^{-1} A^{-1}C & E_m \end{Vmatrix} \tag{7.17}$$

The inverse matrices for the quasidiagonal matrices and triangular matrices with identical block elements on the main diagonal can be found by elementary techniques. Applying the rule for inverting the product of matrices to (7.16) and (7.17), and making the appropriate substitution, we obtain

$$\begin{Vmatrix} E_m & -E_m \\ 0 & E_m \end{Vmatrix} \begin{Vmatrix} E_m & 0 \\ 0 & \lambda^{-1}E_m \end{Vmatrix} (\lambda E - P)^{-1}$$
$$= \begin{Vmatrix} E_m & 0 \\ -(\lambda^2 E_m + \lambda A^{-1}B + A^{-1}C)^{-1} A^{-1}C & E_m \end{Vmatrix}$$
$$\times \begin{Vmatrix} \lambda^{-1}E_m & 0 \\ 0 & (\lambda^2 E_m + \lambda A^{-1}B + A^{-1}C)^{-1} \end{Vmatrix} \tag{7.18}$$

Whence we have

$$(\lambda E - P)^{-1} = \begin{Vmatrix} E_m & 0 \\ 0 & \lambda E_m \end{Vmatrix} \begin{Vmatrix} E_m & E_m \\ 0 & E_m \end{Vmatrix}$$
$$\times \begin{Vmatrix} E_m & 0 \\ -(\lambda^2 E_m + \lambda A^{-1}B + A^{-1}C)^{-1} A^{-1}C & E_m \end{Vmatrix}$$
$$\times \begin{Vmatrix} \lambda^{-1}E_m & 0 \\ 0 & (\lambda^2 E_m + \lambda A^{-1}B + A^{-1}C)^{-1} \end{Vmatrix} \tag{7.19}$$

Multiplying out, we obtain

$$(\lambda E - P)^{-1} = \begin{Vmatrix} \lambda^{-1}E_m - \lambda^{-1}(\lambda^2 E_m + \lambda A^{-1}B + A^{-1}C)^{-1} A^{-1}C & (\lambda^2 E_m + \lambda A^{-1}B + A^{-1}C)^{-1} \\ -(\lambda^2 E_m + \lambda A^{-1}B + A^{-1}C)^{-1} A^{-1}C & \lambda(\lambda^2 E_m + \lambda A^{-1}B + A^{-1}C)^{-1} \end{Vmatrix} \tag{7.20}$$

Next, by virtue of the identity

$$\left.\begin{aligned}(\lambda^2 E_m + \lambda A^{-1}B + A^{-1}C)^{-1} = (\lambda^2 A + \lambda B + C)^{-1} A \\ C = (\lambda^2 A + \lambda B + C) - (\lambda^2 A + \lambda B)\end{aligned}\right\}. \tag{7.21}$$

we obtain (7.20) in the form

$$(\lambda E - P)^{-1} = \begin{Vmatrix} (\lambda^2 A + \lambda B + C)^{-1}(\lambda A + B) & (\lambda^2 A + \lambda B + C)^{-1} A \\ -(\lambda^2 A + \lambda B + C)^{-1} C & \lambda(\lambda^2 A + \lambda B + C)^{-1} A \end{Vmatrix} \tag{7.22}$$

From an inspection of the determinants of both sides of (7.16) it is not difficult to establish that

$$\Delta(\lambda) = \det(\lambda E - P) = {}_{\varphi}\Delta(\lambda) \det A^{-1} \tag{7.23}$$

The following two expressions hold:

$$\left.\begin{aligned}(\lambda E - P)^{-1} = \frac{F(\lambda)}{\Delta(\lambda)} \\ (\lambda^2 A + \lambda B + C)^{-1} = \frac{\Phi(\lambda)}{{}_{\varphi}\Delta(\lambda)}\end{aligned}\right\} \tag{7.24}$$

Substituting into (7.22) the appropriate matrices from (7.24) and cancelling the scalar quantity $\Delta(\lambda)$, we obtain the expression

$$F(\lambda) = \begin{Vmatrix} \Phi(\lambda)(\lambda A + B) & \Phi(\lambda) A \\ -\Phi(\lambda) C & \lambda\Phi(\lambda) A \end{Vmatrix} \det A^{-1} \tag{7.25}$$

which establishes a connection between the adjoint matrices $F(\lambda)$ and $\Phi(\lambda)$. In certain cases it may prove more convenient to write this expression in the form

$$F(\lambda) = \begin{Vmatrix} \Phi(\lambda)(\lambda A + B) & \Phi(\lambda) A \\ -{}_{\varphi}\Delta(\lambda) E_m + \lambda\Phi(\lambda)(\lambda A + B) & \lambda\Phi(\lambda) A \end{Vmatrix} \det A^{-1} \tag{7.26}$$

which can be obtained by expressing C in (7.25) by the second formula in (7.21) and by taking (7.4) and (7.6) into account.

Formulae (7.25) and (7.26) establish a direct connection between the adjoint matrices $F(\lambda)$ and $\Phi(\lambda)$ and they permit us to solve the problem which we formulated above.

7.3 FINAL RESULTS EXPRESSED IN TERMS OF THE INITIAL PARAMETERS

The computational formulae which we derived in the preceding sections contain various types of columns and scalar expressions. We shall have solved our problem when we have represented these expressions in terms of the matrix and scalar quantities (7.2)-(7.6).

Thus, according to (7.14), (7.25) and (7.26), we form the columns

$$F(\lambda)h = \left\| \begin{matrix} \Phi(\lambda)h' \\ \lambda\Phi(\lambda)h' \end{matrix} \right\| \det A^{-1} \tag{7.27}$$

$$\frac{d}{d\lambda}\frac{F(\lambda)h}{\Delta_\sigma(\lambda)} = \left\| \begin{matrix} \dfrac{d}{d\lambda}\dfrac{\Phi(\lambda)h'}{{}_\varphi\Delta_\sigma(\lambda)} \\ \lambda\dfrac{d}{d\lambda}\dfrac{\Phi(\lambda)h'}{{}_\varphi\Delta_\sigma(\lambda)} + \dfrac{\Phi(\lambda)h'}{{}_\varphi\Delta_\sigma(\lambda)} \end{matrix} \right\| \tag{7.28}$$

and the scalar expressions

$$\gamma F(\lambda)h = (\gamma' + \lambda\gamma'')\Phi(\lambda)h' \det A^{-1} \tag{7.29}$$

$$\gamma\left[\frac{d}{d\lambda}\frac{F(\lambda)}{\Delta_\sigma(\lambda)}\right]h = (\gamma' + \lambda\gamma'')\left[\frac{d}{d\lambda}\frac{\Phi(\lambda)}{{}_\varphi\Delta_\sigma(\lambda)}\right]h' + \frac{\gamma''\Phi(\lambda)h'}{{}_\varphi\Delta_\sigma(\lambda)} \tag{7.30}$$

From (7.9) and (7.26) we obtain expressions for the column

$$F(\lambda)x(0) = \left\| \begin{matrix} \Phi(\lambda)[(\lambda A + B)y(0) + A\dot{y}(0)] \\ -{}_\varphi\Delta(\lambda)y(0) + \lambda\Phi(\lambda)[(\lambda A + B)y(0) + A\dot{y}(0)] \end{matrix} \right\| \det A^{-1} \tag{7.31}$$

from which we form the expression [see (2.58)]

$$\frac{d}{d\lambda}\frac{F(\lambda)x(0)}{\Delta_\sigma(\lambda)}$$

$$= \left\| \begin{matrix} \dfrac{d}{d\lambda}\dfrac{\Phi(\lambda)[(\lambda A + B)y(0) + A\dot{y}(0)]}{{}_\varphi\Delta_\sigma(\lambda)} \\ -q_\sigma(\lambda - \lambda_\sigma)^{q_\sigma - 1}y(0) + \dfrac{d}{d\lambda}\dfrac{\lambda\Phi(\lambda)[(\lambda A + B)y(0) + A\dot{y}(0)]}{{}_\varphi\Delta_\sigma(\lambda)} \end{matrix} \right\| \tag{7.32}$$

The scalar expression

$$\gamma F(\lambda)x(0) = \det A^{-1}\{(\gamma' + \lambda\gamma'')\Phi(\lambda)$$

$$\times [(\lambda A + B)y(0) + A\dot{y}(0)] - {}_\varphi\Delta(\lambda)\gamma''y(0)\} \tag{7.33}$$

is found from (7.14) and (7.31).

Finally, from (7.14), we find an expression for the hardness of switching index in the form

$$\gamma h = \gamma'' A^{-1} h' \tag{7.34}$$

The scalar computation formulae are expressed in terms of the initial parameters of (7.1) by direct substitution into these equations of the scalar quantities (7.29), (7.30), (7.33) and (7.34) for the appropriate values of the parameters λ.

To find the column y it is necessary to separate out the first m rows in the formulae that determine column x. This operation amounts to the fact that, in the final formulae, only the upper block elements appearing on the right-hand sides of (7.27), (7.28), (7.31) and (7.33) remain.

Let us consider two examples. Formulae (5.47)-(5.49) and (5.89) determine the self-oscillatory mode of a relay system when there is one double zero root among the roots λ_i. In terms of the initial parameters, these formulae are expressed in the form:

the periodic equation

$$-\sum_{i=1}^{n-2}\frac{1}{\lambda_i}\operatorname{th}\frac{\lambda_i T}{2}\frac{(\gamma'+\lambda_i\gamma'')\Phi(\lambda_i)h'}{\varphi\Delta'(\lambda_i)}$$

$$-\frac{T}{2}\gamma\left[\frac{\Phi(0)}{\varphi\Delta_{n-1}(0)}\right]^{(1)}h'-\frac{T}{2}\frac{\gamma''\Phi(0)h'}{\varphi\Delta_{n-1}(0)}=\sigma_1^* \tag{7.35}$$

the normal switching inequalities

$$-\sum_{i=1}^{n-2}\operatorname{th}\frac{\lambda_i T}{2}\frac{(\gamma'+\lambda_i\gamma'')\Phi(\lambda_i)h'}{\varphi\Delta'(\lambda_i)}-\frac{T}{2}\frac{\gamma'\Phi(0)''h'}{\varphi\Delta_{n-1}(0)}>|\gamma''A^{-1}h'| \tag{7.36}$$

the periodic oscillations during a half-period

$$y(t)=\sum_{i=1}^{n-2}\frac{e^{\lambda_i t}}{\lambda_i}\operatorname{th}\frac{\lambda_i T}{2}\frac{\Phi(\lambda_i)h'}{\varphi\Delta'(\lambda_i)}-\frac{T}{2}\left[\frac{\Phi(0)}{\varphi\Delta_{n-1}(0)}\right]^{(1)}h'$$

$$-\frac{tT}{2}\frac{\Phi(0)h'}{\varphi\Delta_{n-1}(0)}+\sum_{i=1}^{n-2}\frac{e^{\lambda_i t}-1}{\lambda_i}\frac{\Phi(\lambda_i)h'}{\varphi\Delta'(\lambda_i)}$$

$$+t\left[\frac{\Phi(0)}{\varphi\Delta_{n-1}(0)}\right]^{(1)}h'+\frac{t^2}{2}\frac{\Phi(0)h'}{\varphi\Delta_{n-1}(0)}\qquad 0\leqslant t\leqslant T \tag{7.37}$$

the transformed characteristic equation

$$\nu^2\prod_{j=1}^{n-1}\left(\nu-\operatorname{th}\frac{\lambda_j T}{2}\right)\left\{\sum_{i=1}^{n-2}\frac{1-\operatorname{th}^2\frac{\lambda_i T}{2}}{\nu-\operatorname{th}\frac{\lambda_i T}{2}}\frac{(\gamma'+\lambda_i\gamma'')\Phi(\lambda_i)h'}{\varphi\Delta'(\lambda_i)}\right.$$

$$\left.+\frac{1}{\nu}\gamma'\left[\frac{\Phi(0)}{\varphi\Delta_{n-1}(0)}\right]^{(1)}h'+\frac{1}{\nu}\frac{\gamma''\Phi(0)h'}{\varphi\Delta_{n-1}(0)}+\frac{T}{2\nu^2}\frac{\gamma'\Phi(0)h'}{\varphi\Delta_{n-1}(0)}\right\}=0 \tag{7.38}$$

Here and below, in accordance with (2.58), we have used the notation

$$\left[\frac{\Phi(\lambda)}{\varphi\Delta_\sigma(\lambda_\sigma)}\right]^{(1)}=\left[\frac{d}{d\lambda}\frac{\Phi(\lambda)}{\varphi\Delta_\sigma(\lambda)}\right]_{\lambda=\lambda_\sigma}\qquad\text{for}\quad\sigma=n-1\text{ and }\lambda_\sigma=0$$

Formulae (5.132)-(5.134) determine the natural sliding motion of a relay system. In terms of the initial parameters these formulae are expressed as follows:

the characteristic polynomial

$$_{\varphi}\tilde{\Delta}_{*}(\lambda)=(\gamma'+\lambda\gamma'')\,\Phi(\lambda)\,h' \tag{7.39}$$

the sliding motion

$$\begin{aligned} y(t) = -\sum_{i=1}^{n-1} e^{\tilde{\lambda}_i t}\frac{\Phi(\tilde{\lambda}_i)}{_{\varphi}\Delta(\tilde{\lambda}_i)}\,h'(\gamma'+\tilde{\lambda}_i\gamma'')\frac{\Phi(\tilde{\lambda}_i)}{_{\varphi}\tilde{\Delta}'_{*}(\tilde{\lambda}_i)} \\ \times[(\tilde{\lambda}_i A+B)\,y(0)+A\dot{y}(0)]-\sum_{i=1}^{n-1} e^{\tilde{\lambda}_i t}\frac{\Phi(\tilde{\lambda}_i)}{_{\varphi}\tilde{\Delta}'_{*}(\tilde{\lambda}_i)}\,h'\gamma''y(0) \end{aligned} \tag{7.40}$$

the conditions for the sliding mode

$$-1<\sum_{i=1}^{n-1} e^{\tilde{\lambda}_i t}\frac{(\gamma'+\tilde{\lambda}_i\gamma'')\,\Phi(\tilde{\lambda}_i)\,[(\tilde{\lambda}_i A+B)\,y(0)+A\dot{y}(0)-{_{\varphi}\Delta}(\tilde{\lambda}_i)\,\gamma''y(0)]}{_{\varphi}\tilde{\Delta}'_{*}(\tilde{\lambda}_i)}<1 \tag{7.41}$$

Matrix and scalar expressions for the values $\lambda=\lambda_i$ occur in the formulae that determine the self-oscillation modes. With account of (4.22), (4.51), (7.23), (7.29) and (7.30) we obtain, for simple roots λ_i, scalar identities in the form

$$\beta_i u_i=\gamma k_i\varkappa_i h=\frac{\gamma F(\lambda_i)\,h}{\Delta'(\lambda_i)}=\frac{(\gamma'+\lambda_i\gamma'')\,\Phi(\lambda_i)\,h'}{_{\varphi}\Delta'(\lambda_i)} \tag{7.42}$$

and, for the double zero root $\lambda_{n-1}=\lambda_n=0$, we obtain an expression in the form

$$\left.\begin{aligned} \beta_{n-1}u_{n-1}+\beta_n u_n=\gamma(k_{n-1}\varkappa_{n-1}+k_n\varkappa_n)\,h \\ =\gamma\left[\frac{F(0)}{\Delta_{n-1}(0)}\right]^{(1)}h=\gamma'\left[\frac{\Phi(0)}{_{\varphi}\Delta_{n-1}(0)}\right]^{(1)}h'+\frac{\gamma''\Phi(0)\,h'}{_{\varphi}\Delta_{n-1}(0)} \\ \beta_n u_{n-1}=\gamma k_n\varkappa_{n-1}h=\frac{\gamma F(0)h}{\Delta_{n-1}(0)}=\frac{\gamma'\Phi(0)\,h'}{_{\varphi}\Delta_{n-1}(0)} \end{aligned}\right\} \tag{7.43}$$

If next we compose the columns $k_i^{(m)}$ with dimensions $m\times 1$ from the first m elements of the columns k_i, and the rectangular matrices $F^{(m)}(\lambda_i)$ of dimensions $m\times n$ from the first m rows of the matrices $F(\lambda_i)$ then, on the basis of (2.64), (2.65b), (7.27) and (7.28), it is not difficult to establish the following identities:

for simple roots λ_i

$$k_i^{(m)}\varkappa_i h=\frac{F^{(m)}(\lambda_i)\,h}{\Delta'(\lambda_i)}=\frac{\Phi(\lambda_i)\,h'}{_{\varphi}\Delta'(\lambda_i)} \tag{7.44}$$

for the double zero root $\lambda_{n-1}=\lambda_n=0$

$$\left.\begin{aligned}(k_{n-1}^{(m)}\varkappa_{n-1}+k_n^{(m)}\varkappa_n)\,h\\ =\left[\frac{F^{(m)}(0)}{\Delta_{n-1}(0)}\right]^{(1)}h=\left[\frac{\Phi(0)}{_{\varphi}\Delta_{n-1}(0)}\right]^{(1)}h'\\ k_n^{(m)}\varkappa_{n-1}h=\frac{F^{(m)}(0)\,h}{\Delta_{n-1}(0)}=\frac{\Phi(0)\,h'}{_{\varphi}\Delta_{n-1}(0)}\end{aligned}\right\} \tag{7.45}$$

Finally, from (4.21) and (7.34), we establish the identity

$$\beta u=\gamma h=\gamma'' A^{-1}h' \tag{7.46}$$

It will be recalled that in our calculations we began by representing many of the relations in terms of the elements of the special matrices β, u, $\varkappa_i$ and k_i. Formulae (7.42)-(7.46) enable us to express these relations in terms of the initial parameters at any stage of their formulation.

Finally, we must ask ourselves whether it is possible to remove the restrictions imposed on the use of the relations we have obtained. These restrictions arise from the fact that the matrix A was assumed to be non-singular. It can be shown that (7.42)-(7.45) retain their significance when more general assumptions are made. It is sufficient to suppose that the determinants made up of the coefficients of the higher derivatives of (7.1) are non-zero.

This can be proved by comparing the solutions of the same differential equations represented in different forms, provided these solutions are obtained independently of each other. Here, of course, we take account of the fact of their uniqueness. Thus, for example, a solution of (7.1) for $\psi(\sigma)=1$, and for the given initial conditions $y(0)$ and $\dot{y}(0)$, can be obtained by the methods of operational calculus constructed on the basis of the Laplace transform. In the general case, i. e. when $\det A=0$, the identity (7.46) loses its meaning.

If we make use of the last identity in (2.73), take account of the note about multiple roots that follows it, and also (7.42) and (7.43), we can obtain relations in the form

$$\beta u=\gamma h=\sum_{i=1}^{n}\frac{\gamma F(\lambda_i)\,h}{\Delta'(\lambda_i)}=\sum_{i=1}^{n}\frac{(\gamma'+\lambda_i\gamma'')\,\Phi(\lambda_i)\,h'}{_{\varphi}\Delta'(\lambda_i)} \tag{7.47}$$

or, in the case of a double zero root $\lambda_{n-1}=\lambda_n=0$, relations in the form

$$\begin{aligned}\beta u=\gamma h=\sum_{i=1}^{n-2}\frac{\gamma F(\lambda_i)\,h}{\Delta'(\lambda_i)}+\gamma\left[\frac{F(0)}{\Delta_{n-1}(0)}\right]^{(1)}h\\ =\sum_{i=1}^{n-2}\frac{(\gamma'+\lambda_i\gamma'')\,\Phi(\lambda_i)\,h'}{_{\varphi}\Delta'(\lambda_i)}+\gamma'\left[\frac{\Phi(0)}{_{\varphi}\Delta_{n-1}(0)}\right]^{(1)}h'+\frac{\gamma''\Phi(0)\,h'}{_{\varphi}\Delta_{n-1}(0)}\end{aligned} \tag{7.48}$$

These formulae yield expressions for the index of the hardness of switching and the scalar quantity $\beta u = \gamma h$ regardless of whether or not the matrix A is singular.

REFERENCES

1. AIZERMAN, M. A., and GANTMAKER, F. R., 'Stability in linear approximation of the periodic solution of systems of differential equations with discontinuous right-hand sides', Prikl. Mat. Mekh., V 21(5) (1957).
2. ANDRONOV, A. A., and BAUTIN, N. N., 'Theory of the stabilisation of the course of an aircraft by an automatic pilot with a fixed-speed servo-motor', Collected papers, Akad. Nauk USSR (1956).
3. ANDRONOV, A. A., VITT, A. A., and KHAIKIN, S. E., 'Theory of oscillations', Fizmatgiz (1959).
4. BELYA, K. K., 'Nonlinear oscillations in systems of automatic regulation and control', Mashgiz (1962).
5. BULGAKOV, B. V., 'Oscillations', Gostekhizdat (1954).
6. GANTMAKER, F. R., 'Theory of matrices', Chelsea, New York (1959).
7. JURY, E., 'Simple data control systems', John Wiley (1959).
8. DOLGOLENKO, Yu. V., 'Sliding modes in relay systems of indirect control', Proceedings of the Second USSR Conference on Automatic Control, V 1, Akad. Nauk USSR (1955).
9. ZHUKOVSKII, N. E., 'Theory of machine operation control', Moscow Technical High School (1909).
10. KUZIN, L. T., 'Calculation and design of discrete control systems', Mashgiz (1962).
11. COURANT, R., and HILBERT, D., 'Methods of mathematical physics', Interscience (1953).
12. LETOV, A. M., 'Stability in nonlinear control systems', Princeton University Press (1961).

13. LETOV, A. M., 'Status of the problem of stability in automatic control theory', Proceedings of the Second USSR Conference on Automatic Control, V 1, Akad. Nauk USSR (1955).
14. LURE, A. I., 'On some nonlinear problems in automatic control theory', HMSO, London (1957).
15. LURE, A. I., 'Operational calculus and its application to problems in mechanics'.
16. LYAPUNOV, A. M., 'Stability of motion', Academic Press (1966).
17. MALKIN, I. G., 'The theory of stability of motion', U.S. Atomic Energy Commission, Washington DC (1952).
18. MARKUSHEVICH, A.I., 'Theory of analytic functions', Gostekhizdat (1950).
19. NEIMARK, Yu. I., 'On the periodic motions of relay systems', Review commemorating A. A. Andronov.
20. OLDENBOURG, R., and SARTORIUS, G., 'Dynamics of automatic control', Am. Soc. Mech. Engrs (1948).
21. POSPELOV, G. S., 'Vibrational linearisation of relay automatic control systems', Proceedings of the Second USSR Conference on Automatic Control, V 1, Akad. Nauk USSR (1955).
22. ROITENBERG, Ya. N., 'Some problems in controlling a motion', Fizmatgiz (1963).
23. SOLODOVNIKOV, V. V., 'Statistical dynamics of linear automatic control systems', Dover Publications (1960).
24. FRAZER, R. A., DUNCAN, W. J., and COLLAR, A. R., 'Elementary matrices and some applications to dynamics and differential equations', Cambridge University Press (1950).
25. TSIPKIN, Ya. Z., 'Theory of relay automatic control systems', Gostekhizdat (1955).
26. TSIPKIN, Ya. Z. 'Theory of pulsed systems', Fizmatgiz (1958).
27. TSIPKIN, Ya. Z., and BROMBERG, P. V., 'On the degree of stability of linear systems', Otdelenie tekhnicheskikh nauk, Akad. Nauk USSR, No 12 (1945).
28. CHETAYEV, N. G., 'Stability of motion', Pergamon Press (1961).
29. CHEZARO, E., 'Elementary textbook of algebraic analysis and the infinitesimal calculus', translated from the German, ONTI (1939).
30. BROMBERG, P. V., 'Stability and self-oscillations of pulsed control systems', Oborongiz (1953).
31. BROMBERG, P. V., 'Control systems with discontinuous characteristics in the presence of a constant external action', Oborongiz (1954).

32. BILHARZ, Z., 'Rollstabilität eines um seine Langsache freien Flugzeugs bei automatisch gesteuerten intermittierenden Konstanten Querrudermomenten', Luftfahrtforschung, V 18(9), pp 317-326 (1941).
33. COTTON, E., 'Sur la notion de nombre characteristique de Liapunoff', Annls Éc. Norm. sup., V 36, p 217 (1919).
34. OPPELT, W., 'Die Flugzeugsteuerung im Geradeausflug', Luftfahrtforschung, V 14(4-5), (1937).

INDEX